Acolytes of Nature

Acolytes of Nature

Defining Natural Science in Germany, 1770–1850

DENISE PHILLIPS

The University of Chicago Press   Chicago and London

Publication of this book has been aided by a grant from the Bevington Fund.

DENISE PHILLIPS is assistant professor of history at the University of Tennessee.

The University of Chicago Press, Chicago 60637
The University of Chicago Press, Ltd., London
© 2012 by The University of Chicago
All rights reserved. Published 2012.
Printed in the United States of America

21 20 19 18 17 16 15 14 13 12   1 2 3 4 5

ISBN-13: 978-0-226-66737-9 (cloth)
ISBN-10: 0-226-66737-5 (cloth)

Library of Congress Cataloging-in-Publication Data

Phillips, Denise, 1974–
  Acolytes of nature : defining natural science in Germany, 1770–1850 / Denise Phillips.
    pages ; cm
  Includes bibliographical references and index.
  ISBN 978-0-226-66737-9 (cloth : alkaline paper) —
ISBN 0-226-66737-5 (cloth : alkaline paper) 1. Science—Germany—History—18th century. 2. Science—Germany—History—19th century. I. Title.
  Q127.G3P45 2012
  500.20943'09034—dc23
                                                                2011050361

♾This paper meets the requirements of ANSI/NISO Z39.48-1992 (Permanence of Paper).

Contents

Acknowledgments vii
Introduction 1

1   Natural Knowledge and the Learned Public in the Enlightenment   27
2   The Expanding Ranks of Nature's Friends   60
3   Defending Learned Dignity   86
4   Nature in a Local Microcosm   115
5   Wooing the Polite Public   149
6   The Nature of the Fatherland   177
7   The Wellspring of Modernity   202
8   The Particularity of Natural Science   228
    Conclusion   254

   Notes 261   Selected Bibliography 307   Index 349

Acknowledgments

In writing this book, I have encountered many long-dead German researchers who were, among other things, highly insightful analysts of intellectual sociability. Reading their reflections on the pleasures and perils of a scholarly life always made me appreciate all the more the friends and advisers who have helped shape my own work. Everett Mendelsohn was always available with guidance and unflagging good humor early on in this project, and his own work on the professionalization of German science helped open up many of the questions with which this study still wrestles. I owe similar intellectual debts to Peter Galison and David Blackbourn, who were both sources of excellent advice as this project was taking shape. Many other people have offered suggestions, references, and helpful criticisms over the years, among them Deborah Coen, Michael Gordin, Theresa Levitt, Jason Glenn, Matt Stanley, Sharrona Pearl, Debbie Weinstein, Alix Cooper, Matt Jones, Liz Lee, Katja Zelljadt, Caitlin Murdoch, Brandon Hunziker, Alon Confino, Glenn Penny, Myles Jackson, Andreas Daum, Harriet Ritvo, Uljana Feest, Ann Johnson, Brian Vick, Kathryn Olesko, Nicolaas Rupke, and Lynn Nyhart. I am particularly grateful to Thomas Broman, whose thoughtful criticisms of an earlier draft of the full manuscript helped me enormously in the final stages of writing.

The University of Tennessee has been a great place to be a German historian, and I feel lucky to be part of such a vibrant local community of scholars. Vejas Liulevicius, Maria Stehle, Gilya Schmidt, Dan Magilow, and the other members of the Faculty Seminar on Modern Germany have provided an invaluable sounding board for ideas in the later stages of this project, and my UT colleagues in science studies, Ted Richards, Stephen Blackwell, Millie Gimmel, and Heather Douglas, have done the same.

I am grateful to the Deutscher Akademischer Austausch Dienst, Harvard's Center for European Studies, the Dibner Institute at MIT, the National Science Foundation, and the University of Tennessee for providing institutional and financial support for my research. My thanks also go to Lorraine Daston for furnishing me with office space at the Max Planck Institute for the History of Science during my research stay in Berlin, and to the many archivists and librarians in Berlin, Dresden, Halle, and Bamberg who have hunted down files and references for me.

Some parts of chapter 8 were previously published in "Epistemological Distinctions and Cultural Politics: Educational Reform and the *Naturwissenschaft/Geisteswissenschaft* Distinction in Nineteenth-Century Germany," in *Historical Perspectives on "Erklären" and "Verstehen,"* ed. Uljana Feest (Berlin: Springer, 2010), 15–35, and are reprinted here with kind permission from Springer Science+Business Media B.V.

Thanks to Catherine, Abigail, Mom, and Dad, for all those years of great family conversations around the dinner table, and for everything since. Most important, thanks to my husband, Walter, for his love, patience, and support; and to my daughter Jordan, for putting everything in its proper perspective.

INTRODUCTION

Science was responsible for the dramatic technological advances that transformed European society in the nineteenth century—the steamship, the railroad, the telegraph, the camera. This, at least, was what the industrialist and physicist Werner von Siemens claimed in his famous 1886 speech "The Scientific Age." Most historians of technology no longer agree with this claim, at least not when stated so baldly, but for Siemens and many of his contemporaries, this boast on science's behalf had the ring of truth.[1] According to Siemens, "the enlivening breath of science" had transformed almost every area of European life. Science was a force like no other in human history, and humanity, he predicted, would be happy under its harmonious rule.[2]

Despite the intellectual and technical gulf that separated the late nineteenth century from the late eighteenth century, there was much in Siemens's speech that would have sounded familiar to eighteenth-century natural philosophers, ignorant though they were of photographs and steamships. Siemens's conviction that knowledge about nature should be *public* knowledge was one that they would have shared. They would have applauded the industrialist's opening salvo, which celebrated natural knowledge's escape from the "closed-off circles" of the traditional learned estate and its transplantation into "public life," though their specific social referents for these terms would have been somewhat different. In fact, the Bavarian official Mathias Flurl had offered a keynote address to the Bavarian Academy of Sciences in 1799 that began on exactly this note. Just like Siemens, Flurl believed himself to be living in an age when the beneficent effects of an improved understanding of nature were everywhere visible.[3]

Much had changed in the study of nature in the century that separated these two speeches, but Flurl would also have been familiar with the basic forms of knowledge

1

that Siemens championed. Like Siemens, an eighteenth-century natural philosopher would have considered the quantitative study of physical mechanics a valuable kind of knowledge. Eighteenth-century researchers also thought it profitable "to ask Nature herself through properly carried-out experiments," to use Siemens's words, though their experiments would not have measured up to nineteenth-century standards of precision.[4] Flurl would likely have been surprised that the physicist Siemens did not have more to say in praise of natural history, but others in the late nineteenth century still shared the passion Flurl espoused for collecting and describing plants, animals, and minerals.

Alongside these congruencies lies an equally important disjuncture. Despite their familiarity with experiment, careful observation, and quantification, eighteenth-century listeners would not have known what to make of Siemens when he spoke of a world imbued with "the spirit of modern science," and the title of the industrialist's speech, "The Scientific Age [*Das naturwissenschaftliche Zeitalter*]," would have been a mystery to them. It would not have been their inexperience with telegraph wires and railroad tracks that would have made this phrase incomprehensible. It would have been the concept at the heart of Siemens's speech, science.

Modern science has roots twisting into many corners of the past. The category "science" itself, however, is a relative historical novelty. In Europe's various languages, the word took on its modern meaning only over the course of the nineteenth century.[5] Before 1800, there was no standard collective term for the sciences of nature taken together. Enlightenment philosophers loved inventing new classificatory schemes, but a unified science of nature rarely appeared as a landmark on their maps of knowledge. The German philosopher Christian Wolff, for example, scattered different facets of the study of nature among three more general categories. The description of specific natural objects belonged to history, and general causal explanations about nature to philosophy. Fields like mechanics that described nature quantitatively were part of mathematics.[6] Other eighteenth-century thinkers offered similar arrangements, dividing the study of nature among the three areas of natural philosophy, natural history, and applied mathematics, sometimes tacking on chemistry as an independent category as well.[7]

In this earlier period, the German word that would later mean "science," "*Naturwissenschaft*," sometimes meant "natural history," and sometimes "natural philosophy"; it was occasionally used to refer to a loose combination of all kinds of knowledge about physical objects and natural processes, from the study of electricity to the invention of new chimney designs. In its modest and imprecise eighteenth-century guise, "*Naturwissenschaft*" was not the sort of word one used to define an age; it was not even a term that necessarily got included in dictionaries. Eighteenth-century writers thought of themselves as living in an Age of Philosophy or Reason, and when they championed Newtonian physics or Linnaean taxonomy as models for other areas of knowledge, they hoped to make the rest of human knowledge similarly rational, not similarly "scientific [*naturwissenschaftlich*]."[8] Siemens and his contemporaries, in contrast, saw science as a powerful and coherent cultural force with a deep history; this category formed an indispensible reference point in their understanding of the world.

This book traces the emergence of this important new category within German-speaking Europe between 1770 and 1850; it follows the evolution of "*Naturwissenschaft*" from an eighteenth-century neologism to a nineteenth-century rallying cry. This was a development that took place over the time period that Reinhart Koselleck called the *Sattelzeit*, the watershed years that transformed the Old Regime society of the mid-eighteenth century into the nascent modernity of the mid-nineteenth. In these decades, Koselleck argued, the key political and social concepts of modern German culture first took shape. The multivolume *Geschichtliche Grundbegriffe* that he authored with a number of colleagues provides an initial road map to the conceptual developments of these years.[9] "*Naturwissenschaft*" did not receive an entry in the *Geschichtliche Grundbegriffe*, but the category belongs very much within this wider history. Science emerged as a classificatory label as the Enlightenment public was evolving into its much larger, more complex, and multilayered nineteenth-century successor, and this temporal parallel is no coincidence. Before developing this argument further, however, I would like to dwell a bit longer on the shifting historical meaning of words, in order to illustrate just how distinctive the German category "*Naturwissenschaft*" was in the moment of its emergence.

Science, *Naturwissenschaft*, and Nineteenth-Century German Peculiarities

English-speaking historians of German intellectual life are used to translation problems. In contemporary scholarship, the term "*Wissenschaft*" is perhaps the most obvious stumbling block. This category in-

INTRODUCTION

cludes all of academic knowledge, both science and scholarship combined, and present-day English has no similar word. At a loss for a suitable equivalent, historians often leave the term in German.

For most of the nineteenth century, Anglophone translators faced a different problem. In English, "science" did not take on its narrower modern meaning until the second half of the nineteenth century, and through at least the 1870s, English dictionaries still defined "science" with the same broad boundaries that their German counterparts used for *Wissenschaft*.[10] As a result, H. E. Lloyd's 1836 German-English dictionary needed only one word to capture *Wissenschaft*: it meant "science." Lloyd used three terms, however, to approximate *Naturwissenschaft*: "science of nature, natural philosophy, physicks."[11] Several decades later, when H. W. Eve wanted to render Hermann von Helmholtz's *Über das Verhältnis der Naturwissenschaften zur Gesammtheit der Wissenschaft* into English, "*Wissenschaft*" posed no problem at all. That was "science." The word that left him stumbling was "*Naturwissenschaft*." "The German word *Naturwissenschaft* has no exact equivalent in modern English, including as it does both the Physical and the Natural Sciences," Eve wrote in an explanatory footnote. By "the natural sciences," Eve meant fields like anatomy, physiology, and the natural historical disciplines (present-day British English still employs the terms "the natural sciences" and "the physical sciences" in approximately the same way). Eve went on to note that the Germans used "*Naturwissenschaft*" to cover roughly the same ground that the seventeenth-century founders of the Royal Society had intended when they talked about "natural knowledge." No phrase in modern English usage, however, seemed a perfect fit, so Eve made do with "natural science" as a literal but unsatisfying translation.[12] Rendering the term into French does not appear to have been any more straightforward. The French-German dictionaries of the mid-nineteenth century chose various equivalents for it. An 1868 edition of Mozin's dictionary chose "*physique*," which was decidedly too narrow to capture the standard German usage, as this term referred only to physical science. An earlier, 1862 edition of the same dictionary chose "*philosophie*." Both editions agreed, however, that "*la science*" meant "*Wissenschaft*."[13]

Another related German word caused similar difficulties for dictionary makers and translators. By the early nineteenth century, German speakers possessed a general label for an investigator of nature, be his field physics, chemistry, or some branch of natural history. This type of person was called a *Naturforscher*. The early nineteenth-century *Naturforscher* was not exactly the same as his later nineteenth-century descendant, the fully professionalized scientist, but the outlines of this later figure were perceptible in his basic features.[14] For most of the nineteenth century, the British and the French struggled to find easy equivalents for this term. To say *Naturforscher* in nineteenth-century French, you had to use two words instead of one. French-German dictionaries defined a *"Naturforscher"* as both a *"physicien"* and a *"naturaliste,"* and these words were not synonyms. The former was a student of *les sciences mathematiques et physiques* (the physical and mathematical sciences), while the latter studied *les sciences naturelles* (anatomy, physiology, and the natural historical disciplines).[15]

In English, "natural philosopher" was probably the best translation of *"Naturforscher,"* but Anglophones at the time realized that this was an inexact rendering of what the Germans meant by their word. The British natural philosophers who admired Germany's national scientific association, the Gesellschaft Deutscher Naturforscher und Ärzte (GDNA), found it difficult to know what to call this group in their native tongue. In 1829 David Brewster called the meeting a "Congress of Philosophers."[16] James Johnston, in his report on the 1830 meeting of the German group, called it various things, all within one single article—"The Society of German Scientific Men," or "The Society of German Naturalists"—and finally explained in a footnote that English really did not have a word that adequately captured the German *"Naturforscher."*[17] A few years later, William Whewell would coin the term "scientist" in response to this dilemma, but coining the term was about all he did. Even he himself did not actually put the word into use, and it remained a controversial neologism through the end of the century.[18]

Of course, one should not overdraw these linguistic differences. After all, when the French talked about *"les sciences"* in the plural (as in the Academie des sciences), they often, though not always, meant to refer collectively to all the disciplines that studied nature. In 1831, the British founded a society in imitation of the Gesellschaft Deutscher Naturforscher und Ärzte, one that brought together roughly the same group of disciplines as the German group, and they managed to find a name for it. They called it the "British Association for the Advancement of Science" (BAAS). But their peers in other learned disciplines also chastised them for their atypically restrictive use of the word "science," pointing out that theology, the study of languages, and the study of history were sciences, too.[19] The BAAS's narrower usage took many decades to assert itself in the language as a whole. *Naturwissenschaft*, in contrast, was already a widely used term in German by the 1830s.

INTRODUCTION

Historians have a long tradition of treating *"Wissenschaft"* as the primary conceptual peculiarity that marked off German-speaking intellectual life from other language traditions throughout the nineteenth century. German definitions of *"Wissenschaft"* certainly had distinctive features that set them apart from discussions of science elsewhere; both speculative philosophy and philology had a larger hand in shaping this category than in many other European language traditions.[20] But at least for the first two-thirds of the nineteenth century, other Europeans did not find this German concept strange enough to render its translation problematic. The category was not peculiar enough, in any case, to require apologetic translators' footnotes, or to make the compilers of dictionaries become prolix looking for multiple equivalents. When early nineteenth-century British and French intellectuals looked at German for an equivalent of their term "science," the word they picked was *"Wissenschaft."* It was the vocabulary Germans used to talk about the study of nature—words like *"Naturwissenschaft"* and *"Naturforscher"*—that left translators scratching their heads.

Several recent studies have argued that the modern notion of the unity of science was a mid-nineteenth-century German invention. The linguistic evidence suggests that there is good reason to take this claim seriously. This invention, however, has so far been assigned to a fairly small number of individuals, all of whom rose to prominence within German science only after 1850. Supposedly, a new, unified ideal of science first crystallized in the neo-Newtonian programs of figures like Emil Du Bois-Reymond and Hermann von Helmholtz.[21] "In 1830 there could as yet be no thought of a unified natural science," Herbert Schnädelbach claimed. Only when a new, more ambitious program of reductive mechanism emerged as a plausible unifying framework did this concept become meaningful. With the discovery of the first law of thermodynamics and the triumph of a kinetic theory of heat, it became possible to imagine that all fields of natural knowledge might be reduced to a set of interlocking mathematical laws. With this promise on the horizon, German thinkers supposedly began to conceive of the natural sciences as a unified thing.[22] The German emphasis on scientific unity, in other words, has been painted as an innovation that was the product of a few leading researchers, made possible through the growing midcentury appeal of mechanistic, physicalist explanations of natural phenomena.

Only a brief survey of how and where this category first emerged in German culture suggests that the story is much larger than that. The popular Brockhaus lexicon already defined *"Naturwissenschaft"* as a unified science of nature in 1824, when Du Bois-Reymond and Helmholtz were still young boys. The first learned society in Germany to call itself a "Natural Scientific Society" was not a club of eminent professors in Berlin but a civic association of obscure provincial naturalists founded in the small town of Blankenburg in 1831. Indeed, Helmholtz and Du Bois-Reymond both advanced a definition of science that was considerably narrower than the standard usage of the term. They thought that only those fields that had been able to formulate their knowledge as mathematical laws truly belonged to science, a definition that left fields like anatomy, botany, zoology, or morphology out in the cold. In reality, the German concept of a unified science seemed distinctive to other Europeans precisely because it *combined* the physical sciences and the natural historical disciplines. That was why Eve found the concept so hard to render into English, and why the French needed two words, *"physicien"* and *"naturaliste,"* to say *"Naturforscher."* Clearly, we need a history of this category that goes further back in time and that includes a broader range of disciplines, both the physical and the life sciences. We also need to look at its history through a wider historical lens.

If we want to understand the emergence of this distinctive new category, civic associations like Blankenburg's Natural Scientific Society offer one promising place to start. In fact, if one searches German intellectual life for social locations in which all of the sciences of nature came together within a single forum, private learned societies offer the most widespread and numerous examples. Like the new concept "science," many local associations joined the different branches of natural knowledge together under one roof. As a result, these societies offer an excellent place to examine the concrete communities that formed under the banner of Nature from the Enlightenment through the revolutions of 1848–49. They were one key location where a new conceptual unity, "the natural sciences," could be related to a demarcated domain of collective social practice. They offer perhaps the best place from which to watch how this concept slowly gained salience within German public culture, and as a result they stand at the center of this book.

In fact, like the concept *Naturwissenschaft*, the general natural scientific society was in many ways a distinctly German phenomenon in the first half of the nineteenth century. At first glance, this may seem like an odd assertion, given the large body of work that exists on the scientific associational life of other countries in this period. Private societies whose activities included the study of nature were no rarity in France and Britain. But it was fairly unusual in these contexts for a society to define its *sole* purpose as the study of nature, nothing more and nothing less. In Britain, the British Association for the Advancement of Science covered ground roughly similar to the German GDNA, on which it had been modeled. Otherwise, the British learned societies founded in this time period did not describe their intellectual goals in terms similar to a typical German natural scientific society. Natural philosophy was of central importance in the Literary and Philosophical Societies so popular in provincial centers from the 1780s forward, but as their names suggest, these groups defined their mission more broadly. They were founded to support philosophy, literature, and the arts, not just the study of nature.[23] Even societies devoted to the study of natural history (*all* of natural history, that is, rather than just one branch of it) were rare creatures in Britain before 1850.[24] A handful of such groups, such as London's Linnaean Society, had appeared in the late eighteenth century. Between 1800 and 1850, however, most provincial British learned societies with natural historical interests also studied local antiquities. Most German cities had two separate groups for these twin pursuits. Like their British counterparts, France's provincial bourgeoisie showed a strong preference for generalist societies that combined science, literature, and the useful arts. For the period between 1800 and 1850, only 3 French provincial learned societies were devoted to science in general, while many dozens of groups had either broader or narrower horizons. And France's less prestigious equivalent of the GDNA and the BAAS, the Congrès scientifiques de France, included historians and antiquarians as well as students of the natural sciences.[25]

In contrast, by 1850 German-speaking Europe had about 45 local societies that made the general study of nature their collective aim. As subsequent chapters will explore, there was a fair amount of variation among these different groups. Nonetheless, natural scientific societies were a clearly discernible group within German associational life by the mid-nineteenth century. When Johannes Müller composed a bibliographic guide to German-speaking Europe's scientific societies in the 1880s, he considered "natural scientific societies" a well-defined type of association. The British Council's contemporaneous publication, *Yearbook of Scientific and Learned Societies of Great Britain*, had no similar category. At first glance, the societies devoted to "science generally" would seem a good match, but on closer inspection, any apparent similarity between the German and the British taxonomies dissolves. The category of "general science" included groups as disparate as London's Society for the Encouragement of Arts, Manufactures, and Commerce; the Royal Asiatic Society; and the Balloon Society of Great Britain. In Germany, these three groups would not have belonged together under any single heading, most certainly not under the heading "*Naturwissenschaft.*"[26]

As we know from past work in science studies, the label "science" refers to a complex set of activities that do not have any obvious or essential unity. Despite the fact that this study charts the history of a unifying category, the internal diversity of the scientific enterprise remains readily apparent at every turn. By the mid-nineteenth century, German researchers shared a collective concept of "the natural sciences" as a meaningful cultural unity; some researchers also thought these sciences shared their own unique natural scientific method. Beyond this very general accord, German *Naturforscher* continued to disagree about many things—the form that general conclusions ought to take, the role of mathematics in the study of nature, the precise kinds of observations and experiments that were most useful, the best way to characterize the scientific method, and the kinds of explanations that counted as legitimate in any given field. If asked to define science, a provincial botanist would answer differently than a university-based morphologist; the morphologist's answer, in turn, would likely differ from the answer of a chemist. A broad cultural investment in the category "science" emerged alongside continued disagreement about the specifics of its content.[27]

Indeed, the power of the term came in part from its continued ambiguity, the fact that it designated a loosely organized cause that many different kinds of people might join. As a result, this book is not about the development of a clearly defined philosophical category; it tracks the cruder and more general history of collective linguistic usage. It focuses on the common reference points of large groups, not the carefully delineated ideas of individual thinkers. Like other recent studies in historical epistemology, it charts the gradual process through which the fundamental collective categories used to define authoritative knowledge develop over time.[28]

To explain how "science" went from descriptive label to rallying cry, this study looks at the reciprocal relationships among a developing concept (the unity of the natural sciences), a word ("*Naturwissenschaft*"), and sites of collective social action. This is a different task than following the "institutionalization of science" along the lines pursued in an older historiography.[29] Rather than assuming we know what science is, and searching through history to find places where people are practicing it, we will need to pay attention to how people themselves described the cause that captured their allegiance. In this study, the connections between concepts and communities will not be as tight as the one imagined by historians who wanted to trace out the emergence of "the scientific community." Examining the emergence of *"Naturwissenschaft"* as a category is not the same thing as documenting the creation of a scientific community in the sense in which sociologists have used the term.[30] As *Naturwissenschaft* came into general use, it did not bring with it a set of Mertonian norms of science that governed behavior across all scientific fields.

The story of this concept's emergence also does not fit into older narratives that cast science as a force of secularization. In the second half of the eighteenth century, the study of nature was frequently praised as a path to piety and a way to contemplate the handiwork of the divine. This basic argument was still in wide circulation a hundred years later, though perhaps not expounded at such great length or with the same frequency.[31] Science's critics sometimes claimed that it fostered a materialist worldview that threatened to destroy all higher spiritual values. The study of nature may have carried the scent of French materialism and atheism to some; this scent could seem exhilarating or noxious, depending on a person's perspective. But before 1850, most of Nature's acolytes were eager to distance themselves from such associations. In the decades under consideration here, the natural sciences were generally seen as working in harmony with other forms of spiritual understanding. We need to know much more about the different theological and philosophical inflections of this basic position, but it is worth recognizing the existence of a broad consensus. Indeed, its lack of specificity may be significant. The chemist and physicist J. S. C. Schweigger, originally a theology student, wrote of his relief at being able to leave behind the quarrels of theologians to contemplate God through mathematics.[32] In a period that had Hegelianism and Strauss's *Life of Jesus* to chew over, the study of nature could often seem blessedly uncontroversial, theologically speaking.[33]

By the 1840s, the word "science" had shed its earlier status as a vague neologism and had become an indispensable intellectual reference point for educated German speakers. The cast of characters that forged science into a powerful new category was broad, and the word's history is much more than a rarified intellectual tale. Its emergence was intertwined with the major political, social, and cultural developments of the eighteenth and nineteenth centuries. Much like other central concepts within nineteenth-century public life, terms such as "nation" and "citizen," "science" gained its meaning against the backdrop of a socially diverse and densely populated public culture.[34]

Science between the Learned and the Educated Public

Calling German science "socially diverse" strikes a strange chord, at least for the first half of the nineteenth century. Most historians of science have agreed that the interesting development in this period was the institutionalization of a new scientific elite within the state-funded universities. As David Cahan put it, German science was "a state affair."[35] For the decades *after* 1850, this generalization no longer holds; we have a body of excellent work that looks beyond the universities to German museums, civic associations, and the activities of scientific popularizers.[36] But for the first half of the century, Cahan's succinct formulation still captures the dominant view. The university has been the main site for the history of German science, the professoriat its primary actor.[37]

The universities, and the university educated, played an important role in the emergence of *"Naturwissenschaft"* as a public category, but sketching this word's history requires bringing a wider intellectual world into view. To provide a first glimpse into this wider world, I would like to enlist the services of a somewhat obscure tour guide—not Goethe, Humboldt, or Helmholtz, but a minor German naturalist named Jacob Sturm. The span of Sturm's long, happy, and productive life coincides almost exactly with the period covered in this book. He entered the world in 1771, the son of a Nuremberg engraver, and followed his father into the engraving trade when he came of age. In turn, Jacob's own two sons would labor alongside their father in the family workshop. Sturm died, still a resident of the city of his birth, in 1849.

At Sturm's graveside, the pastor Johann Wolfgang Hilpert described the engraver as a man "whose name resounds honorably wherever the natural sciences are practiced." Hilbert boasted that Sturm's name was known around the globe, and this was not simply provincial puffery. Over the course of his life, this modest Nuremberg citizen had indeed built up a network of scientific connections that reached across Europe and even across the Atlantic. As the author of the well-regarded *Deutschlands Flora in Abbildungen*, Sturm was a member of over twenty different scientific societies, groups based in foreign cities such as Moscow, Stockholm, and Philadelphia, as well as associations in German academic centers like Berlin, Jena, and Halle. Late in his life, Sturm received an honorary doctorate of philosophy from the University of Breslau.[38]

Sturm's life cuts an odd path through the historiographical consensus outlined above. The primary institutions in his scientific career were not universities, though the faculty of Breslau did eventually recognize his achievements with an honorary degree. They were learned societies, groups like his hometown Natural History Society or Berlin's colorfully named Society of Nature-Researching Friends. These kinds of organizations, central to the story at hand, currently occupy an ambiguous place in our picture of nineteenth-century German science.

While learned societies have been seen as central sites for the study of nature in the seventeenth and eighteenth centuries, their importance supposedly diminished rapidly (at least in the German context) after 1800, when the universities took over as the premier institutions of scientific life. After this sea change, learned societies and academies have been assigned several distinct functions. If they were devoted to a specific scientific discipline, they have been cast as the organs of the period's emerging disciplinary communities. The national GDNA (hereafter Society of German Natural Researchers and Doctors), which divided itself into separate disciplinary sections shortly after its founding, has generally also been slotted into this broader history of an emerging specialized elite. Historians have also noted the continued honorific functions of state academies like the Prussian Academy of Sciences. Such organizations gave prizes and preserved a symbolic relationship among different disciplines, but according to most assessments, the real action was elsewhere.[39]

For the most part, however, nineteenth-century scientific associations have been placed within quite a different history. They have been cast as the vehicles of a middle-class culture of amateur science, a pursuit that emerged in tension with, or even in reaction to, the consolidation of a university-based scientific elite. Scientific voluntary associations, in Andrew Zimmerman's words, "challenged the hegemony of official academic institutions."[40] For Andreas Daum, the growth of scientific associational life represented bourgeois society's appropriation of natural knowledge for the satisfaction of its own needs. Through associational life, the German middle classes sated a desire for scientific information that Germany's academic scientific establishment had left unmet.[41]

In making these claims, both Zimmerman and Daum were describing the second half of the nineteenth century, the period when, as Daum has convincingly shown, pursuits like "amateur science" and "popularization" first came to be clearly defined as distinguishable fields of cultural production. To talk of popularization in the modern sense, Daum points out, requires a corresponding notion of a closed community of professional scientists, a group whose results are then communicated in altered form to a wider lay audience. This dialectic of professionalization and popularization only became fully stabilized after 1848.

But what of the period before 1848, when the dialectical pair of the professional and the popular was still half formed and inconsistently applied? In this earlier period, private learned societies did not belong to a clearly delineated secondary realm of scientific production. Though we can see a more organized "low" scientific culture emerging by the 1830s and 1840s, for most of the period under consideration here, there are significant analytical benefits to seeing all of Germany's natural researchers, nascent university professionals and small-town naturalists alike, as belonging to a single intellectual world, albeit one that had both more and less eminent members. Here again, Jacob Sturm's biography offers a useful introduction to the diverse kinds of people who participated in German scientific networks in this period. From his base in Nuremberg, Sturm corresponded with a number of men who had similar intellectual interests but varied occupational profiles. Sturm's first patron was the Erlangen professor and botanist J. C. D. Schreber; Sturm also knew Schreber's successor in Erlangen, Wilhelm Daniel Joseph Koch. He traded letters with the Regensburg apothecary David Heinrich Hoppe, and with the Bohemian aristocrat and agricultural reformer Count Kaspar von Sternberg. Ludwig Reichenbach, Saxon court councillor and the director of the Dresden Botanical Garden, was another of his correspondents.

Taken collectively, such exchange networks were part of a broader cosmopolitan community that Hilpert, Sturm's eulogizer, called the learned world, also sometimes known as the republic of letters.[42] Based on our current understanding of the history of nineteenth-century science, one might think that the provincial pastor Hilpert was being quaint in evoking this concept. There are differing views on when exactly the republic of letters ceased to be a meaningful reference point for European intellectuals, but there is general agreement that it was long gone, or at least transformed beyond all recognition, by 1850.[43] Anne Goldgar set its demise around 1750, and Anthony La Vopa posited a similar expiration date, arguing that the older, Latinate ideal of the *res publica literaria* dissolved in the second half of the eighteenth century into a broader vernacular public sphere.[44] For Dena Goodman, the republic of letters remained a viable category after 1750, but according to her, the republic now had its center in the fashionable and enlightened world of the Parisian salons, and took a form quite different from its pedantic early modern predecessor. In contrast, L. W. B. Brockliss has shown that the less fashionable networks that had characterized the older republic were alive and well through the end of the century, and even expanding. The *philosophes*, he argued, were a particular camp within this broader republic, not a novel and separate formation.[45] Lorraine Daston also described the republic of letters as a continually evolving entity through the entirety of the eighteenth century.[46] For both Brockliss and Daston, however, the revolutionary and Napoleonic era marked an important moment of rupture, when the nation-state edged out the cosmopolitan republic as the primary organizing force of intellectual life.

To understand the social structure of German science through the middle of the nineteenth century, it is essential to recognize that "the learned world" remained a viable organizational idea well after 1800. This is not a hard claim to support, given that German researchers, even elite ones, continued to talk quite openly about belonging to the learned world far into the nineteenth century. In 1817, Lorenz Oken described his new review journal, the *Isis*, as an organ of "the learned world."[47] About fifty years later, when Rudolf Virchow wanted to make the case that science ought to serve national interests, the first thing he knew he had to do was argue against the countervailing ideal of a cosmopolitan republic of letters.[48] The Nuremberg pastor Hilpert was not being old-fashioned in talking about the world of "learned men"; he was talking about intellectual life in a way that was still common among educated Germans in the middle of the nineteenth century.

Furthermore, it was not just the *concept* of the learned world that survived; many older patterns of learned sociability persisted, too. The early modern republic of letters organized itself through published exchange, but also through letter writing and learned societies. Jacob Sturm built his intellectual reputation using these same materials. He published scientific work, but he also cultivated a diverse network of correspondents. Building from this base, he acquired further intellectual connections in the form of memberships in learned societies. Recognition from a university came only late in his life, but his honorary degree from the University of Breslau suggests that a man could still build a meaningful intellectual reputation in the first half of the nineteenth century using these kinds of resources. Hilpert described Sturm as a serious researcher, and not just an amateur; he belonged to the world of science, not to the world of self-improving hobbyists, and the professors in Breslau apparently agreed.

Of course, the republic of letters had not crossed the threshold of 1800 completely unchanged, and tracing these changes will be a major task of this book. But despite many significant transformations, nineteenth-century learned sociability still had a number of features that would have been familiar to earlier generations, and these continuities are crucial to understanding the particular tensions that characterized German intellectual life in this period.

The nineteenth-century relevance of "the learned world," as both a concept and a set of social norms and practices, has been hard to see for a number of reasons, and these stem in part from the standard explanations offered for the rise of German university science. R. Steven Turner provided one of the most influential interpretations of the changes of this period, and for him, the German research university emerged in close conjunction with the transformations that turned the early modern learned estate into the modern educated middle class. What Turner called the "Great Transition" occurred in precisely the same period covered by this book, and a key feature of this Great Transition, in Turner's version of the story, was the demise of traditional ideals of learnedness.[49]

The early modern learned estate was composed of the tiny group of university-educated men who belonged to one of the traditional learned professions; these men, trained in law, medicine, or theology, enjoyed certain legal privileges, and they shared a common identity grounded in eloquence, Latinity, and a polymathic mastery of a common learned heritage. With the rise of a vernacular reading public in the second half of the eighteenth century, Turner argued, the values of the traditional learned man came to seem increasingly dated, unworldly, and impractical. By 1800, witty enlightened satirists had more or less mocked the learned man, the *Gelehrte*, out of existence. The figure that replaced him was the modern *Bildungsbürger*, the educated middle-class man. This figure, like his early modern counterpart, was a man trained at a university, and there was a fundamental social continuity between the members of these two successive groups. But the new ideal of *Bildung*, or self-cultivation, became the defining cultural property of this newer group, which was now made up of modern professional men. As *Bildung* replaced learnedness, ideals of taste and functional expertise replaced older shared values of erudition.[50]

Turner's account, though it captured many important changes, mischaracterized both the status and the fate of the "learned man" in two ways. First of all, Turner conflated the learned estate and the republic of letters into one single entity; every university-educated man supposedly thought of himself as a member of the republic of letters. That was clearly not true. Belonging to the early modern republic of letters required more than just a university degree. It required the active cultivation of intellectual interests and participation in networks of learned exchange. The learned world, in short, was not the same thing as the learned estate. The latter was understood as a group within the corporate social order; the former was an intellectual community whose correspondence networks might well cut across standard divisions within Old Regime society.[51]

Second, it is misleading to link the fate of the learned man so inextricably with the decline of early modern ideals of polymathic erudition. By the late eighteenth century, the erudite and old-fashioned pedant was definitely a favorite object of parody, but Germans continued to speak in (mostly) respectful tones of "the learned public," and the learned man had by no means disappeared as an authoritative figure. Through the middle of the nineteenth century, educated Germans often distinguished a narrower learned public from a broader educated one. This distinction, in fact, was extremely important in structuring German cultural debates from the late eighteenth century forward. Nineteenth-century ideals of education and refinement (*Bildung*) developed alongside a more narrowly defined conception of scientific expertise, and the language of learnedness remained central to how this expertise was described. In the 1880s, Werner Siemens would speak of the "love of *Wissenschaft* for its own sake" that characterized "the German learned man," and this captures the relationship between these two terms well.[52] In common parlance, a man of *Wissenschaft*, or science, was also a man of learning, a *Gelehrte*. For the Nuremberg pastor Hilpert, the learned world *was* the world of *Wissenschaft*, not its long dead early modern antecedent, and here again, the provincial pastor's usage was entirely typical.[53]

Once we recognize this fact, we can better account for something that has seemed paradoxical about the relationship between elite science and the public sphere in this period. Thomas Broman has argued convincingly that scientific knowledge gained cultural authority in the eighteenth century by embedding itself in the broader discourse of public opinion and the public sphere.[54] Scientific knowledge, like other forms of public knowledge, was seen as trustworthy because it could be subject to rational criticism and debate, and because it was made available to the eyes of all. Broman also points out, however, that science was "universally accessible in principle but recondite in practice." Actually participating in scientific debate required a high level of expertise, and as a result, he argues, modern disciplinary communities "effectively began to withdraw large regions of scientific knowledge from the public sphere almost as soon as it formed," but without totally abandoning the key ideological benefits the public sphere provided.[55]

One way to resolve this seeming paradox is to recognize that when eighteenth- and early nineteenth-century authors wrote about "the public," they could have two rather different kinds of communities in mind. As Heinrich Bosse has shown, Germany's seventeenth- and early eighteenth-century republic of letters possessed many of the same features that Jürgen Habermas ascribed to the public sphere. Its members also saw themselves as forming a public whose opinions were secured through reasoned debate and whose collective judgments ought to command authority.[56] Yet the republic of letters differed from Habermas's version of the public sphere in significant ways, too. The public sphere, in T. C. W. Blanning's description, was conceived of as something "anonymous and unhierarchical," and one gained access to it "solely by the capacity to pay for the cultural commodities it consumes."[57] Similarly, Anthony La Vopa described the German public sphere as consisting primarily of "solitary readers and writers," the purchasers and producers of books and journals.[58]

Membership in the republic of letters required something more. An aspirant for membership in the learned world had to prove that he (and occasionally she) possessed sufficient knowledge and skill to join in the conversation. Unlike the wider public, the learned world was not anonymous or composed of solitary readers; its members were densely interconnected, and they did not talk to one another only through the medium of print. Furthermore, their social interactions were not governed solely by the rules of commodity exchange but also by a reciprocal set of obligations that often had more in common with gift economies than with the market.

In the late eighteenth century, Germans did not always draw a clear distinction between these two kinds of publics. Sometimes they wrote quite openly about the differences between the narrower learned public and a broader educated one; sometimes they conflated these potentially separate groups into a single unified category. Other authors recognized this division but rejected its legitimacy, arguing that the opinion of any rational human was as good as any other.[59] Indeed, throughout the entire period covered in this book, it was not an easy matter to draw a clean line between the truly learned and the merely educated public. As Jacob Sturm's case illustrates, formal educational credentials were not all that mattered in the learned world. People could move up from the more general public to the more expert one; purchasing a book or a natural history collection might be the first step in a process of self-education that could eventually lead to recognition from better-informed peers. The sociological boundary one crossed when one joined the learned world, however, had real weight, and if the precise location of this boundary was open to negotiation, it is not hard to see some basic differences between the relationships constituted through the period's learned networks and the relationships produced within the wider cultural marketplace of Germany's emerging consumer society.

In arguing for the existence of multiple publics, I might seem to be falling prey to a common analytical error. Harold Mah has criticized historians for their excessively spatial uses of the term "public sphere." Too often, he argues, historians have talked about the public sphere as if it were simply a neutral stage that people could enter and leave. Since historians have thought of the public sphere as a kind of empty space, they have been quick to posit the existence of as many "public spheres" as they pleased. For Mah, this way of conceptualizing the public sphere misses a key point—the fact that the public claimed power as a unified collective subject. Against the early modern state, "public opinion" demanded a hearing because it supposedly had the right to speak with a single voice; the public was not a space but a collective entity.[60]

Mah's correction is a valuable one, but there are also dangers in seeing "the public" as too singular a thing. Eighteenth- and nineteenth-century writers often modified the noun "public" with different adjectives. They talked about the learned public, the educated public, the botanical public, or the patriotic public. From this perspective, there were in fact many different publics who might pass judgment on matters of general concern. Each of these various publics was conceived of as a collective unity in the way that Mah described, but each of these publics was also associated with a somewhat different social network; each had different entry requirements and different tacit expectations. These various networks overlapped and intertwined in complicated ways, but there is significant analytical value to keeping this diversity in view. Mah is right to chide historians for treating the public sphere simply as a stage on which various historical actors played their part, but there are similar problems with thinking of "the public" as a term with only a single possible referent.

As a result, we should be careful not to let the structures of the *learned* public simply dissolve into a broader marketplace of cultural goods. In his erudite and entertaining history of the German academic persona, William Clark has shown that early modern university professors achieved their special brand of charisma in part through their reputation within a broader learned world.[61] The German officials who oversaw the universities liked nothing better than a professor who made noise in the republic of letters. Clark's tendency to refer to this learned world simply as "the market," however, lumps together too many disparate forms of exchange under a single homogenizing label.[62] It also obscures the fact that state officials were themselves often members of the learned world; when they listened to its applause, it was in part their own hands they heard clapping.

Dubbing the learned world "the market" also blunts our ability to further explore a crucial issue that troubled German researchers a great deal in this period—namely, in a world where a growing public looked to natural history and natural philosophy as a source of entertainment, and where knowledge was something that could be purchased by anyone who could afford a book or an attractive specimen, how did one keep the structures of expertise stable? What, in other words, was the relationship between the learned public, whose members were vetted, socially interconnected, and bound by mutual obligations, and the broader public, in which no such limiting relationships pertained?

"Science" emerged as a category because it was useful for answering precisely these kinds of questions. And these questions were particularly thorny ones in the period under examination here for several reasons. In the decades after 1770, a large number of new aspirants came knocking at the door of the learned world. These new men, and a much smaller number of women, appeared on the scene as a result of several broader developments within German political, cultural, and economic life. Germany's reading revolution, which shifted most of learned discussion into the vernacular, greatly expanded the audience for natural philosophy and natural history.[63] Growing circles of readers also meant more opportunities (though often not terribly lucrative ones) for authors, engravers, and artists. Already in the seventeenth century, connections to the book trade had introduced people like Maria Sibylla Merian to natural history; the engraver Jacob Sturm followed a similar route to participation in the learned world. As print culture expanded, figures like Sturm also multiplied.[64] By the early nineteenth century, one can even find people making a modest living catering to the public's enthusiasm for natural history collecting.

The schooling revolution that began in the late eighteenth century also created a new stratum of people at the lower edge of what the late eighteenth century called "the educated estates." New technical schools, normal schools, and agricultural academies produced new kinds of educated men of ambiguous social status. Such men joined scientific associations in large numbers in the 1830s and 1840s. Forestry and mining officials, the men who staffed the technical branches of the German states' expanding bureaucracies, did the same, as did other men from the lower ranks of state service.[65] Within manufacturing and certain fields of craft production, one can also find a growing number of aspirants for inclusion in the learned world, men like the famous instrument maker and natural philosopher Joseph Fraunhofer.[66] Natural history and natural philosophy attracted attention among the upper strata of the social order, too; German noble landlords, many of whom were interested in agricultural improvement, often also became enthusiastic collectors of *naturalia*. These improving landlords, and the increasingly commercialized agriculture they practiced, set up the conditions for the emergence of yet another new group of men, estate managers and practical agricultural experts whose knowledge of farming might or might not be based on formal training.[67]

In the passage above, I described this group of new men as "large," and that statement requires clarification. In comparison to the scale of most social history, the number of people involved in the changes outlined above was in fact not really all that high. When placed against the institutions of later periods, the new normal schools and academies of the *Sattelzeit* look quite modest. Most of them taught dozens of students in a given year, not thousands.[68] Even the largest scientific societies of this period did not have more than a few hundred local members. Within the learned world, however, the pressures that this expansion created were significant, and the resulting aftershocks reverberated through public culture as a whole.[69] Indeed, the developments outlined above put significant strain on Germans' basic categories of social classification, and in this period of intense flux, men of varied backgrounds had good reason to take advantage of the status that could be gained through participation in the republic of letters.

Just as it makes sense to keep the distinctive features of the learned public in view, it is also important not to let the distinctive practices that characterized learned societies fade too seamlessly into a broader history of civil society and associational life. When viewed in this way, scientific societies become much less interesting, just one lesser example of familiar processes that can be better examined elsewhere. Historians of German political and social history who have studied associational life in this period have generally deemed their subject important for several reasons. Clubs and societies forged new kinds of social linkages, building up a new middle-class culture out of the more localized and fractious world of the Old Regime. They created forums of debate that were free of state oversight, facilitating the growth of a new political public. They have also been seen as central to the history of modern nationalism.[70] In very general terms, learned societies certainly contributed to each of these developments, as previous historians have noted.[71] When placed within these broader trajectories, however, learned societies do not cut a particularly impressive figure. They were much smaller and less numerous than the period's ubiquitous Freemason lodges; they were less effective as cryptopolitical vehicles than the choral or gymnastic associations, and less promising as nascent economic interest groups than were early agricultural, industrial, or workers' organizations.[72]

The distinctive appeal of learned societies lay in their ties to the old and still prestigious traditions of the republic of letters. Through the early nineteenth century, most German intellectuals proudly listed their various society memberships on the title pages of their books. The lure of learned status best explains the proliferation of these groups, not just a need for information nor a yen to claim a generalized bourgeois identity.[73]

Indeed, in this transitional period, we need to be careful about how we use terms like "the *Bildungsbürgertum*," or the educated middle class. Many of the developments outlined in the following chapters were certainly part of the history of this social category, and as a loose descriptive term, "the educated middle classes" has a certain value.[74] But no one in the first half of the nineteenth century called himself a *Bildungsbürger*.[75] This particular social label came into general use only in the twentieth century, and transposing it back into the early nineteenth century, as Turner and others have done, preemptively answers questions that were still very much open, especially before 1850. Through the middle of the nineteenth century, the term that people *did* use, "the educated estates [*die gebildeten Stände*]," stretched beyond the *Bildungsbürgertum* in several ways. This group included well-read and cultivated noblemen and women; it included prosperous merchants and their wives. It also might include a well-read artisan or entrepreneur.

This heterogeneous group "the educated estates" made up the broader public from which the learned wanted to distinguish themselves; this was also the broader public the learned wanted to court. To make things more complicated still, learned natural researchers also wanted to bring *new* groups into the ranks of the educated. They wanted to educate farmers and craftsmen, the diverse members of "civil society [*bürgerliche Gesellschaft*]" in the broader meaning often given that term in this period.[76] In such a dynamic and socially heterogeneous landscape, defining the boundaries of the learned world was a tricky task. Educated Germans worked out their answers to this conundrum through the varying trajectories of their individual lives, the evolving networks of local learned sociability, and the back-and-forth of published debate. Under the cumulative pressure of these negotiations, early modern habits of classifying knowledge (and knowledge makers) strained to a breaking point, and new categories like science appeared in their stead.

Eighteenth- and nineteenth-century German intellectuals participated in a variety of sophisticated, complicated debates about the nature of natural knowledge. To interpret their complex reflections on this topic merely as the product of social anxieties or collective enthusiasms would be facile. What this book attempts to understand, however, is why intellectually sophisticated and highly individualized people aligned themselves with certain broad, general habits of linguistic usage. In explaining *that* kind of behavior, shared status anxieties, along with shared passions and ambitions, offer a good starting place for analysis.

The Germans in Comparative Context

There are many aspects of the developments I have just outlined above that are *not* particular to Germany. The tensions I have described bear a strong family resemblance, in fact, to the conflicts that other historians have examined within French and British science in the same period. In those places as well, the expansion of the public sphere, along with the accompanying specter of natural knowledge as spectacle, amateur dalliance, or utilitarian slave, raised troubling questions about how to stabilize expertise. Here, too, an emerging scientific elite competed with a "low" scientific culture for the right to define what counted as legitimate knowledge.[77] Furthermore, intellectual traffic across language barriers remained intense in the nineteenth century. Cosmopolitan exchange was an integral feature of scientific life in all parts of Europe.[78] In the following chapters, I have tried to indicate at specific moments the reasons that the Germans' classificatory schemes diverged from those used in other places. At the outset, however, it is possible to make a few generalizations about why developments within German-speaking Europe looked different from developments elsewhere.

In Germany, the social networks of natural knowledge cut across the boundaries of a particularly strong learned tradition that had its central base in the universities.[79] "The learned," as a social and educational group, had particular strength and coherence in German-speaking Europe, but in the case of natural knowledge, the republic of letters extended well beyond the boundaries of the learned estate, much further than was the case in other scholarly fields. *Naturforscher* were a much more socially heterogeneous group than philologists or historians. The people who participated in published discussion, who had collections and did experiments, were not all "learned" in the narrower sense, and, as discussed above, these other kinds of researchers were growing more numerous with each passing decade. Champions of science, particularly if they were men with liberal political sentiments, sometimes considered this social diversity a positive thing, but this heterogeneity also created perennial status problems. Adding to this problem was the fact that German rulers and their officials were often quite interested in taking advantage of practical knowledge possessed by men from outside the university-educated elite.[80]

These crosscutting tensions marked out the peculiar features of the German concept of "*Naturwissenschaft*." They set the natural sciences in stronger contradistinction to other scholarly fields like history, and also differentiated them more clearly from emerging practical fields of discussion. This particular combination of tensions helps to explain, in other words, why in Britain in the 1880s a society devoted to ballooning, Eastern antiquities, or manufacturing could still be classified as part of a broader project called "science," while an educated German would have described the situation very differently.

Finally, Germany was not a unified nation but a polycentric cultural space. Historians of France and Britain have often emphasized the importance of the relationship between province and metropolis in shaping the scientific cultures of those nations.[81] Germany, in contrast, had many competing cultural centers, and many smaller, "provincial" places that were still loath to think of themselves as such.[82] Given Germany's political decentralization, one might be tempted to read Germans' interest in the unity of science simply as a projection of nationalist longings.[83] Science and nation building were indeed sometimes linked, but the former category was not the projection of the latter. The ways in which the following story is a "national" one are more complicated than that, and they have to do with a shared set of conditions that pertained within German-speaking Central Europe, not a conscious common effort to forge a unified nation through the

INTRODUCTION

creation of a unified science. It was the complex interplay among cosmopolitan, national, and other regional loyalties that gave the German context its particular cast. Regional, civic, and dynastic allegiances played an important role in shaping the German scientific landscape, and in focusing just on the nation, we would be missing key elements of the story.[84]

Because of the absence of a clear cultural center, some historians have argued that German intellectuals were more exclusively oriented toward printed exchange than their colleagues in France and Britain, where people could gather in the capital and talk face-to-face.[85] There is certainly something to this argument; in terms of raw numbers of specialized journals, German-speaking Europe eclipsed Britain and France.[86] This thriving print culture did not lessen educated Germans' desire for local intellectual sociability, however; if anything, it heightened it. For if specialized scientific accomplishments were to mean anything in a man's daily life, they needed to be visible in a local setting. Jacob Sturm's eulogizer described how difficult it could be to get cosmopolitan learned fame to register in a local context: "Hardly anyone in his native city noticed him," Hilpert wrote. "And yet, while in Nuremberg hardly anyone even knew of the existence of a Jacob Sturm, he stood in friendly contact with the greatest *Naturforscher* of all the nations of the civilized world."[87] For both Germany's academic elite and more modest figures, local scientific societies provided an answer to this dilemma. They allowed learned men to advertise the collective scientific resources of their city within a landscape full of competing centers; at the same time, they made each member's scientific accomplishments more visible within local cultural life. In nineteenth-century Germany, intellectual reputations were secured first and foremost through publication, but the business of building a name for oneself was not just a matter for the printed page. Learned reputation had important local components as well. It was grounded in the urban social worlds where most German *Naturforscher* spent their daily lives.[88]

This study draws on material from across German-speaking Europe, but in order to examine the complex position of the natural sciences in local urban settings, there is one city to which I return throughout the book more than any other, and that is Dresden. Nineteenth-century Dresden is an interesting place to observe scientific life for several reasons. As the capital of Saxony, it was at the heart a densely urban and precociously industrial state, a place where the early tensions of the emerging industrial world were felt particularly keenly.[89] Though less commercially powerful than its neighbor Leipzig, the Saxon capital was also home to a number of thriving economic enterprises. Widely recognized as an important artistic center, the Saxon court city had an active literary and philosophical life, one that included both polite salons run by aristocratic women and less well-heeled circles of radicalizing young liberal men. From the early nineteenth century forward, Dresden also had a well-regarded medical academy and technical school. It did not have a university; rather, it offers us a chance to look at other kinds of professors, ones who were also part of the elite learned world of this period. Most important, the city had an unusually rich scientific associational life. Dresden had more general natural scientific societies than any other German city in this period.

This is a book about a German particularity; it is not a book about a German *Sonderweg* (special path). Germans did not have a different concept of science because they were more "inward looking," less practical, or less empirical than the French or the British. They did not have a different concept of science because the state and the nobility were too strong, and civil society too weak. They did not have a different concept of science because intellectual activity seduced the middle classes away from the political realm. "*Naturwissenschaft*" was another sort of German peculiarity, one that fits quite well with our revised understandings of nineteenth-century German history.[90]

A Note on Terms and Translation

Writing the history of a German word in English comes with inherent difficulties. To avoid making the preceding introduction too cumbersome, I have up until now followed familiar English usage and used "science" for "*Naturwissenschaft*." For the rest of the book, I have translated this term as "natural science." When I use the adjective "scientific" from here on, the German equivalent would be "*wissenschaftlich*." When the source says "*naturwissenschaftlich*," I have translated this as "natural scientific." Historians of science generally make it a point to avoid the term "scientist" before the late nineteenth century, when this professional persona appeared in its mature form (the Germans, too, started using a new label, "*Naturwissenschaftler*"). As a result, I have often used the word that Germans employed in this period, "*Naturforscher*," without translating it, or have rendered it literally into English as the somewhat awkward "natural researcher."

ONE

Natural Knowledge and the Learned Public in the Enlightenment

"*Naturwissenschaft*" first appeared in the early eighteenth century as a synonym for *physica*, or natural philosophy.[1] Three decades after its debut, it was still not common enough to be included in Christoph Ernst Steinbach's 1734 German dictionary.[2] Just a few years later, however, J. H. Zedler's encyclopedia listed it as one of several possible synonyms for "natural philosophy," and by the 1770s, the term was in wide circulation.[3] By the late eighteenth century, it was also used as a synonym for "natural history," and could mean "natural knowledge" in a more general sense, too.[4]

From our perspective, J. S. T. Gehler made things a bit clearer. In 1790, he called natural history, natural philosophy, and applied mathematics "the three main divisions of *Naturwissenschaft* as a whole."[5] With this definition, we might seem to have already arrived at a crucial watershed. At least in broad terms, Gehler employed the word in a way that looks much like later nineteenth-century usages; "natural science" was an umbrella term for all the disciplines that studied nature. Drawing on previous work on nineteenth-century German science, we might imagine a clear way forward from here. We might predict that this new overarching category grew more powerful as the old tripartite division of natural history, natural philosophy, and applied mathematics continued to decay in the face

27

of modern scientific specialization. That would put us in familiar territory. We could follow Rudolf Stichweh in examining the process of disciplinary specialization that took off around 1800 and see how the hierarchically structured fields of early modern learned knowledge became an array of vertically organized modern scientific disciplines, collectively labeled "the natural sciences."[6] We might even see the stabilization of this new terminology as part of the process whereby science became an independent "system" clearly differentiated from the other constituent systems (the state, the economy) that made up society as a whole.[7]

Yet a quick glance ahead at the early nineteenth century reveals that this is not, in fact, to be the way forward. The following juxtaposition illustrates succinctly why not: by 1824, the Brockhaus lexicon, an encyclopedia written for a broad educated audience, had an entry for *"Naturwissenschaft"* that used this word as the label for a unified natural science.[8] The 1833 edition of Gehler's more technical *Physical Dictionary* did not. The editors who revised Gehler's dictionary chose *"Physik,"* or natural philosophy, as their master label for the scientific study of nature, and *"Naturwissenschaft"* still kept its modest eighteenth-century place as one of nine possible synonyms for this term. Georg Wilhelm Muncke, the author of the article on natural philosophy, had indeed abandoned the triad of natural history, natural philosophy, and applied mathematics for a wider field of disciplines (though he kept other well-worn early modern distinctions, such as the one between historical and philosophical kinds of knowledge).[9] The transition to a more complicated disciplinary array seems to have done nothing in and of itself to raise the status of *"Naturwissenschaft"* as an overarching term.

"Natural science" gained prominence through different channels than these, and its presence in the 1824 Brockhaus lexicon gives us a clue to where we might look to better understand the word's history. As we will see in later chapters, this category was the product of the border zone where learned natural researchers met the broader public; it played an important role in attempts to clarify the relationship between learned experts and an expanding lay audience of readers, authors, and collectors. It was also a term that natural researchers used to rally the troops internally, but in ways that often avoided difficult, specific questions about how *exactly* the different fields of natural knowledge might relate to one another. For the editors of the 1833 Gehler dictionary, who took it as their task to describe the contours of learned natural knowledge with some degree of intellectual precision, *"Naturwissenschaft"* was not yet an indispensable reference point.

To be able to follow the trajectory of this word after 1800, however, we need to first be clear on its late eighteenth-century status. In this regard, statements like the one Gehler made in 1790 are deceptive in their apparent familiarity. Gehler used the phrase "natural science as a whole" to mean all of learned natural knowledge in one of the sentences in his article; just a few lines earlier, however, he had listed the word as a synonym for natural philosophy. The eighteenth-century word *"Naturwissenschaft"* still lacked certain key features of its nineteenth-century descendant, even in cases where it was used to refer to natural knowledge in general. It was still just one of several words one might choose, a word with many possible synonyms. It lacked the strong emotional punch that it would carry several decades later, and it lacked a role as an organizing reference point in intellectual exchange.

When Johannes Müller compiled his guide to German scientific societies in the 1880s, he had little trouble pulling together a group called "natural scientific societies." That would have been a fool's errand in the 1780s and 1790s. In the eighteenth century, the support of "natural science" was not yet a distinct social cause around which people organized themselves. But the *study of nature* was. That was already a recognizable collective enterprise by the late eighteenth century, but one with looser boundaries and a wider purview than what the nineteenth century called natural science. Someone who contributed to this collective enterprise already had a name; he or she was called a *Naturforscher*. German-speaking Europe's *Naturforscher* belonged to a loose, decentralized community, a community whose integrative sinews ran through learned societies, personal correspondence networks, and published journals.[10]

Knowledgeable *Naturforscher* formed part of what late eighteenth-century Germans called "the learned public," or, less frequently, "the republic of letters." This latter term can be confusing; some historians have used it to refer primarily to the early modern predecessor of the enlightened public sphere, while others use it as a synonym for the enlightened public as a whole.[11] The late eighteenth-century "learned public" was decidedly not the same thing as the public as a whole; in its practices and in its constitutive ideals, it differed considerably from the public of readers, theatergoers, and coffee drinkers familiar to us from much previous work on the public sphere. It was a narrower and more exclusive public, embedded within the broader public but distinct from it in many ways, too. Like earlier citizens of the learned world, late eighteenth-century *Naturforscher* described themselves as linked to one another through ties of friendship and mutual obligation. In addition to communicating in print, the learned still assiduously wrote letters and exchanged specimens.[12] When *Naturforscher* spoke of the public that had the right to assess their work, they often designated this narrower learned public as their intended judge. The learned public was a public full of *Kenner*, of experts; its members were vetted and interconnected in ways members of the general public were not.[13]

"*Naturwissenschaft*" and the Study of Nature in the Enlightenment

As Quentin Skinner has argued, when writing the history of a word, tracing changes in meaning is not enough. We also need to look for moments when a word's referents changed, or moments when it began to be used in new kinds of speech acts.[14] Between 1770 and 1850, "natural science" became invested with an emotional intensity it utterly lacked in the eighteenth century; it became a word that communicated fervor and inspired loyalty. In the eighteenth century, the word was a descriptive label for a (loosely specified) field of knowledge; in the nineteenth century, it referred to a powerful cultural force. It stopped being merely a descriptive term and became useful for both acts of praise and acts of condemnation. For Werner von Siemens, "the natural scientific age" was an electric phrase spoken with celebratory zeal. The historian Johann Droysen spat similar words out, one imagines when one reads him, with a strong flavor of disgust in his mouth.[15] Eighteenth-century authors chose different objects for their ire, and shouted different slogans.

Eighteenth-century *Naturforscher* did, however, already see the study of nature as a unified cause. They just did not associate this cause with a single epistemic category, and they often shifted happily between ill-defined synonyms when they sang the praises of learned natural knowledge. The word "*Naturwissenschaft*," which later came to mean "science," did not yet serve as an exclusive reservoir for the collective enthusiasms of enlightened natural researchers. This gap between word and cause is the topic of the following section, which also looks at the terms that eighteenth-century authors *did* invest with rhetorical force.

First, however, we will start with a few more examples of how authors in the late eighteenth century used the term "*Naturwissenschaft*."[16] In 1783, Wenceslaus Johann Gustav Karsten published a textbook he described as a guide to "how natural science ought to be practiced in the future." The book offered an introduction to the study of nature in one volume, breaking with established tradition. The three fields Karsten covered—natural history, natural philosophy, and chemistry—were usually treated as separate sciences, each deserving a textbook of its own. These three subjects had generally recognized parameters, which Karsten described briefly in his introduction. Natural history dealt with specific natural bodies, plant, animal, and mineral. Chemistry described the composition of matter and the rules that governed how different fundamental elements combined, while natural philosophy explained the causes at work in the physical world. Karsten argued that these three *Wissenschaften* (sciences) were in reality so closely interconnected that it made sense to join them together into "a single whole."[17]

Chemistry, natural history, and natural philosophy belonged together as a single whole; the role that Karsten assigned mathematics in this mix was more complicated. In another publication, he noted that the successes of applied mathematics (by which he meant things like Newton's mechanics) had led some of his predecessors to treat mathematics and natural philosophy as if they were one and the same thing. Karsten considered this approach outdated. Advances in chemistry and natural history, he thought, made a solely mathematical natural philosophy untenable. Furthermore, applied mathematics, though valuable, was a field that not all students of nature needed to master. Karsten himself was both a mathematician and a natural philosopher, but he argued that not everyone needed to learn mathematics. The reasons he gave for his position were pragmatic. Many learned young men simply had no interest in or talent for the subject. They would end up "yawning and falling asleep" in a mathematics lecture, and since there were also many nonmathematical ways to study nature, there was no reason to bore people unnecessarily.[18] But according to Karsten, "No one [could] lay claim to the title *Naturforscher*" without understanding the interconnections that bound natural history, chemistry, and natural philosophy together.[19]

Karsten's suggestions differed dramatically from another proposal put on the table in the mid-1780s, Immanuel Kant's *Metaphysical Foundations of Natural Science*. Unlike his colleague in Halle, the Königsberg professor placed natural history and chemistry outside of the circle of true *Wissenschaften*. "Proper" natural science needed the apodictic certainty that came with laws that could be formulated a priori, and chemistry and natural history, which in Kant's view were just experiential (albeit still valuable) forms of knowledge, could never aspire to reach these heights. Kant considered mathematics, in contrast, central to *Naturwissenschaft* as he defined it.[20]

With these two books juxtaposed, one might try to claim that in the mid-1780s, German thinkers were already engaged in a full-fledged debate about how to define a novel concept, modern natural science. That would be anachronistic for several reasons. Karsten described himself as laying out a program for "how natural science should be practiced in the future," but he did not put all that much weight on the term itself. His main concern was not defining "*Naturwissenschaft*," but laying out a new program for *Naturlehre*, natural philosophy. Karsten's primary aim was to lobby for natural philosophy to draw more heavily on chemistry and natural history than it had in the past.

Furthermore, neither Karsten's natural philosophy nor Kant's version of "*Naturwissenschaft*" stretched as far as later standard usages of "the natural sciences." Kant, in contrast to Karsten, was indeed interested in defining *Naturwissenschaft*, but he did so in an idiosyncratic way that broke with the evolving colloquial uses of the word. For Kant, *Naturwissenschaft* properly so called was found only where one found mathematics; there were other, experiential kinds of knowledge about nature, but these were not part of *Naturwissenschaft* in the strict sense. A more common move in the late eighteenth century would have been to separate mathematics from other areas of natural knowledge and save words like "*Naturwissenschaft*" for the latter, not the former. For example, when the poet Friedrich Gottlieb Klopstock listed different types of German learned men, he put *Naturforscher* in a separate group from mathematicians. Johann Bernoulli referred to his esteemed predecessor Johann Heinrich Lambert as a "mathematician and *Naturforscher*" as if those were two different things.[21] In 1790, the University of Jena published a list of its courses for the coming semester, and for the first time a subset of these lectures appeared under the heading "the natural sciences." Lectures in mechanics and other areas of applied mathematics were not on this list.[22]

Even Karsten's more generous concept of natural philosophy did not include all the areas that appeared under the Jena faculty's label, which mapped onto more loose and colloquial uses of the term. Only the *general* parts of chemistry and natural history belonged within his new natural philosophy; "special natural history," the study of individual kinds of plants, animals, and minerals, did not. Similarly, though Kant's philosophy certainly played an important role in later discussions of science, the highly technical definition he offered in the *Metaphysical Foundations* serves as a poor starting point for a general history of the concept as a widely shared cultural reference point.[23] This was partly a matter of raw numbers. Just as Karsten suggested in his textbook, nonmathematical kinds of natural knowledge simply had much broader appeal. Though applied mathematics had its share of practitioners, chemistry, natural history, and nonmathematical areas of natural philosophy like the study of electricity attracted much larger followings.[24]

More important, if we started from Karsten and Kant and tried to trace the fate of specific intellectual proposals for how natural knowledge (or parts thereof) might be brought together as a unity, we would end up with many branching threads but nothing like a whole piece of cloth. As we will see in later chapters, such proposals multiplied, but they did not converge. With only these things in hand, we would be at a loss to explain both the nineteenth-century concept's strange new stability and the timing and manner of its emergence. Complicated philosophical debates about the internal structure of natural knowledge are by no means irrelevant to this concept's history, but the glue that held the natural sciences together did not come from any one intellectual synthesis.

Given the evocatively emotive character of the later nineteenth-century term "natural science," its strength seems, rather, to have something to do with collective enthusiasm and emotion. To that end, the most interesting thing to look for in the late eighteenth century is perhaps not a handful of thinkers (however famous) who seem to be inching toward a more unified concept of natural science. The real thing we need to look for are places where *Naturforscher* were rallying together and claiming to be part of some unified cause.

We do not have to look far. Late eighteenth-century *Naturforscher* disagreed about many things, but they were fond of portraying themselves as partners and friends in a collective enterprise. When they described this collective enterprise, they generally focused on several shared values: studying nature was a useful pursuit; it was also a pious and a pleasurable one. An enlightened *Naturforscher* examined nature with an eye to serving the public weal, but his eye was neither cold nor calculating, even when it measured and counted. A taste for nature and a concern for utility were not opposing values in late eighteenth-century Germany; indeed, they were not necessarily formally distinct. The period's concept of "utility" was broad, blending together moral, aesthetic, and economic meaning.[25] When the naturalist Friedrich Martini wrote of a desire to discover nature's "hidden aspects and treasures," the "treasures" of which he spoke might be valued for their narrow practical utility but also for their rarity, beauty, or complexity.[26]

A *Naturkenner* knew more about nature than other people did; he or she also had a special emotional, spiritual, and aesthetic relationship to the natural world. The enlightened natural researcher, in other words, was not "objective" in the modern sense (his successors in the first half of the nineteenth century would not be either). Late nineteenth-century ideals of objectivity tried to eliminate the subjectivity of the knower from the process of making knowledge. As Lorraine Daston and Peter Galison have shown, eighteenth-century *Naturforscher* had no such aspirations.[27] An eighteenth-century *Kenner* was both an expert and a connoisseur; he had a trained eye and a warm heart. In the late nineteenth century, a "friend of nature" was a conservationist; in the eighteenth century, he was a scientist. It was remarkable, wrote Martini, "how much influence the study of Nature has on the hearts and souls of her admirers, and how powerfully this science [*Wissenschaft*] binds together all its acolytes in ties of friendship." According to Martini, throughout the German provinces and throughout Europe, in Sweden, in Russia, in Italy—all of the friends of nature had "the same sensibility, the same way of thinking, the same manner of comportment."[28] Similarly, the subscribers to Lorenz Crell's chemical journal were all "friends" of the study of chemistry; eighteenth-century German chemists thought of themselves as sharing not just a common set of technical abilities but also a common passion.[29]

Although natural theology as a formal enterprise declined in popularity on the Continent in the second half of the eighteenth century, general references to natural research as a spiritual exercise remained common.[30] The botanist Friedrich Ehrhart, for example, informed the reader of his flora that he had spent "the Sundays of the last three summers engaged in the contemplation of God within the plant kingdom."[31] Friedrich Wilhelm von Leysser argued that the Book of Nature offered a way that Christianity could win back those who had become scornful of revealed religion.[32] The Nature-Researching Society of Halle also ascribed religious meaning to its activities, and Berlin's Society of Nature-Researching Friends was devoted to "a closer understanding and contemplation of Nature and her eternal Creator."[33] Martini even mused that the collective efforts of the group were "a preparation to the study that will occupy [them] through all eternity." Surely humans, so sociable in this life, would not be lonely in the next, and Martini wondered if the bonds forged through the Berlin society might stretch into the afterlife, where the local Berlin group would finally have a chance to meet all the society's many foreign members face-to-face. In heaven, they would continue the communal work they had begun on earth.[34]

The collective enterprise of natural research had its most obvious social organs in groups like the aforementioned Nature-Researching Society of Halle and the Society of Nature-Researching Friends of Berlin. These groups were part of a network of learned and patriotic societies that proliferated across Central Europe in the final third of the eighteenth century.[35] In contrast to the nineteenth century, in this earlier period a division between practical and theoretical knowledge did not yet play an important role in the social organization of intellectual exchange. In their interests, memberships, and activities, the various nature-researching societies or physical societies of the late eighteenth century were very similar to societies devoted to improving agriculture and manufacturing, groups generally referred to as "economic-patriotic societies" in the German context. In its narrowest use, the German term *"Oekonomie"* referred to the study of agriculture, but the word was also employed as a general label for all productive endeavors, as well as for the growing body of literature that dealt with these fields.[36] Both nature-researching and economic societies alike were part of a European-wide network that joined together natural philosophy, natural history, and practical economic improvement.[37]

Both socially and intellectually, the period's learned societies were closely connected with the development of cameralism, an eighteenth-century science that tried to increase state revenues through the improvement of agriculture, forestry, mining, and the handicrafts. Cameralism, a discipline peculiar to northern and central Europe, was supposed to provide officials with the knowledge they needed to superintend the healthy functioning of the state, and one of cameralism's primary aims was to create new practical sciences that could guide bureaucrats in their management of the body politic, knowledge that would be systematic and useful at once. Through the cameralists' efforts, practical activities like forestry, agriculture, and mining all became topics for academic teaching, both at universities and at newly founded specialized academies.[38] "There is no higher calling for a mathematician or a *Naturforscher*," wrote the Göttingen professor Johann Beckmann, "than to use their knowledge to improve manufacturing, which also serves the improvement of the state."[39]

Economic improvement was not just a concern for those working inside the state; it was a joint project of state officials and other members of Germany's emerging civil society. Enlightened journals and patriotic-economic societies carried this banner, too. On the educational front, private academies multiplied alongside state-sponsored ones. Dresden had a small private forestry academy around 1800; the German states also housed a number of small private chemical

CHAPTER 1

schools.[40] Already in the late seventeenth century, regional naturalists had begun cataloging the "natural riches" of individual states, and this enterprise also grew in popularity as the eighteenth century progressed.[41] In Switzerland and southern Germany, several short-lived societies were founded for the study of *Vaterlandskunde*, or "regional science," a pursuit that covered both civil history and natural history, and which included a great deal of information of direct economic relevance as well.[42] In addition to uncovering the native "treasures of the Fatherland," German naturalists were also optimistic about possibilities for import replacement through the acclimatization of useful new exotic species.[43]

In formal philosophical terms, the Enlightenment's new practical sciences were distinct from those parts of natural philosophy and natural history that dealt only with the structure and description of objects. In practice, these two kinds of inquiry were conducted in close concert, and this distinction had little importance as an organizing reference point within forums of communication. The Economic Society of Leipzig described its area of concern as "everything having to do with the productive estate in general, in the widest possible extent." This included "urban and rural economy, manufacturing and trade" but also the "advantageous application of mathematics, natural philosophy and chemistry."[44] While patriotic-economic societies had often been directly involved in economic projects at midcentury, after the 1770s, gathering, ordering, and collecting information became their dominant concern. They built up natural historical collections and collections of physical instruments; they also gathered masses of information on weather.[45]

The statutes of "nature-researching" societies, *Naturforschende Gesellschaften*, typically reversed this order of priorities. Berlin's Society of Nature-Researching Friends promised to explore "natural history in all its aspects, with the help of a good natural philosophy" and to put this knowledge to use for the benefit of humanity.[46] Their revised 1784 statutes requested that their members send in observations about animal husbandry, chemical analyses of rare and useful products, and new inventions or improvements to physical instruments or machines.[47] Other nature-researching societies of the period had a similar range of interests. One of the Leipzig Linnaean Society's first publications was a description of nearby Altenburg's pewter industry.[48] Groups like the Physical-Economic Society in Königsberg (founded in 1789) represented an even more obvious intermediary between these two types of associations.

Patriotic-economic societies, state academies, and private learned societies for natural research shared a common rhetoric of public-spirited (*gemeinnützig*) and patriotic service. A "patriot," as eighteenth-century thinkers defined the term, worked to aid humanity in general as well as his more immediate fatherland, and *Naturkenner* believed they could succeed at both of these tasks.[49] The founder of the Society of Nature-Researching Friends in Berlin described it as a group of "noble thinking, patriotic friends" with an interest in serving the public good.[50] Its members burned with "patriotic eagerness" for their cosmopolitan cause, "the cultivation of natural knowledge [*die Beförderung natürlicher Kenntnisse*]."[51] The Nature-Researching Society in Halle wanted only "good, honorable and public-spirited subjects" as members.[52]

In print forums, too, natural history and natural philosophy were often joined together with practical economic or technical topics. A journal like the *Leipziger Magazin zur Naturkunde, Mathematik, und Oekonomie* was one example among many. Bibliographers also often treated the practical sciences together with natural philosophy and natural history.[53] Even bibliographies that did not advertise that they included works of *Oekonomie* often did. A 1782 *Büchersammlung zur Naturgeschichte* (Collection of books on natural history) contained a subsection for works dealing with "Use" under every major heading.[54] Alternately, late eighteenth-century textbooks on individual branches of learned science sometimes stretched to include related areas of practice; an individual *Wissenschaft* itself could be divided into both theoretical and practical components.[55] This scheme of classification, the roots of which could be traced back to antiquity, took on new life in the late eighteenth century. A number of introductory works on botany, for example, covered both "theoretical" botany (for example, taxonomy) and "practical" botany (gardening, agriculture, and viniculture).[56]

In describing the ways that natural philosophy or natural history could be useful to economic practice, eighteenth-century authors proceeded differently than would their nineteenth-century grandchildren. They did not, for example, argue that there was a special natural philosophical method that could be transferred to other fields. "The natural scientific method" held a central place in later nineteenth-century defenses of science; such a concept was foreign to eighteenth-century German intellectual life. "Natural history," wrote the Bavarian official Mathias Flurl, "is, so to speak, the foundation on which all the other physical sciences must be built." For Flurl, the "physical" sciences included all the fields, both practical and theoretical, that studied nature.[57] In explaining natural history's foundational role, Flurl made

CHAPTER 1

no reference to a transfer of methods; he gave concrete examples of specific useful facts that naturalists might discover. Botanists found new plants; mineralogists, new mineral deposits.[58] When Karsten wrote about teaching natural philosophy according to a "scientific method [*wissenschaftliche Methode*]," he meant a method that followed standards shared across the world of learning, not something that was peculiar to the natural sciences.[59] Natural philosophy needed to measure up to these more general standards; it was not itself their source. A body of knowledge qualified as a learned *Wissenschaft* when it was comprehensive and systematic; Karsten thought his proposed version of natural philosophy could be both of those things.[60]

For eighteenth-century researchers, the power of natural philosophy and natural history came from the more general power ascribed to learned knowledge.[61] In describing what learned natural knowledge supposedly added to fields like mining or forestry, eighteenth-century authors wrote that it made these pursuits "*wissenschaftlich*," "theoretical," "learned," or "philosophical." If farmers studied natural history, natural philosophy, and enlightened agricultural science, they became "philosophical" or "learned" farmers, not "natural scientific" ones. For Kant, the knowledge of general principles relevant to a given sphere of practice made a man worthy of the name of "theoretical doctor, agriculturalist, and the like."[62] When Hans Caspar Hirzel publicized the clever agricultural innovations of the autodidact Jacob Guyer, he styled Guyer "a philosophical farmer."[63]

Learned doctors, jurists, and theologians had always been masters of both theory and practice. The composite phrase "the arts and sciences" was the standard description for learned knowledge in this period; academic knowledge included both *Wissenschaften* and *Künste*, sciences and arts. By the late eighteenth century, the *Wissenschaften* were generally distinguished from the *Künste* in philosophical discussion as follows: "A science [*Wissenschaft*], as opposed to an art, presents those truths that involve the higher powers of our intellect. An art contains those truths that are the object of lower faculties."[64] In more casual speech, however, these two words were often used as if they were synonyms, and both words had learned associations and positive rhetorical force.[65] When eighteenth-century academics defended the value of learning, they defended both *Künste* and *Wissenschaften*.[66]

Enlightened natural history and natural philosophy drew prestige from more general ideas about the value of learned knowledge. This is a statement that requires further elaboration. The kinds of developments that I have been discussing above—the spread of voluntary associations and the enlightened celebration of utility and taste—have sometimes been described as eroding learnedness as a source of public authority. The enlightened *Naturforscher* was certainly a different animal from the erudite, Latinate learned man of the late seventeenth century; he still generally thought of himself, however, as "learned." Groups like the Society of Nature-Researching Friends described their members as patriotic, tasteful, *and* learned all at once. If one looks for it, the language of learnedness still permeates late eighteenth-century public discourse; the next section examines how these references to learnedness ought to be understood.[67]

The Learned World in the Late Eighteenth Century

In 1787, the lawyer and historian Johann Georg Schlosser published a pamphlet entitled *On Pedantry and Pedants*. If his readers opened its cover expecting to find a familiar satire of unsociable learned men, they would have been disappointed. Schlosser had a different target, and a different warning to deliver. Learned men, he wrote, had once quaked at the thought of being labeled pedants; they had feared being thought insufficiently worldly and socially polished. They should do so no more. Now, Schlosser thought, learned men faced a different danger. In chasing after publicity, the learned had lost their way in polite, mixed-gender sociability and an excessively exuberant print culture:

One publishes the sciences [*Wissenschaften*] with alphabetical indexes, writes encyclopedias, . . . overviews of physics, philosophy, even jurisprudence for women, holds lectures for courtly cavaliers and ladies and so on. Every morning gentlemen and ladies read a piece from a journal or a learned dictionary; every evening one takes that little gem out into society, and argues and reasons about it with more energy than one ever showed at school.[68]

When the learned embraced *le monde*, according to Schlosser, they had paid a high price. In trying to cut a pleasing figure before the ladies and gentlemen of good society, they had traded the true quest for knowledge for the shallow pleasures of curiosity. They had abandoned systematic thought for entertaining anecdotes. In the process, the ties that once bound together the "truly learned" had decayed. Learned socializing, correspondence, and travel had once smoothed the sharp edges of critical debate. Under the constant glare of worldly publicity, forced to suffer criticism from the ignorant and half informed, Schlosser thought the truly learned were ironically now more isolated than ever. "Nothing can help this situation," Schlosser wrote, "except pulling the learned circle closed once again, and kicking out the public." The *results* of learned deliberations were relevant to the public, but as long as uncertainties remained, the learned ought to keep their debates among themselves.[69]

Schlosser's commentary might seem to fit easily into a familiar account of late eighteenth-century German cultural life. Through the prism of his negative satire, one might argue, we see traditional ideals of learnedness giving way to enlightened, cultivated ideals of taste. Here, we find an old-fashioned *Gelehrte* lamenting the dissolution of older learned networks into a new, vernacular public sphere.[70] In this reading, Schlosser would be the voice of a lost cause, mourning the demise of a fading world. But Schlosser might also be viewed through a different lens. A historian looking for the roots of the modern scientific community in the late eighteenth century might see him as prescient, not old-fashioned. He might be cast as an early proponent of professional scientific closure, a man looking ahead to the elite communities of nineteenth-century *Wissenschaft*.[71]

Both of these explanations fail to capture the way critiques like Schlosser's fit into the late eighteenth-century cultural landscape. Schlosser was not a disgruntled, old-fashioned outsider, and the distinction he proposed—between the truly learned on the one hand and a mixed-gender public of casual readers and ill-informed writers on the other—was one that was quite commonly invoked within German enlightened circles. Despite his derisive jibes at *Aufklärer* ("enlighteners"), Schlosser himself lived a life that was deeply intertwined with the enlightened public sphere. He was a frequent contributor to review journals and the author of popular history books for a wider reading audience. As Alexander Košenina has argued, Schlosser's position on these issues actually bore a strong resemblance to Kant's.[72] In suggesting that the learned remove the general public from their internal debates, Schlosser did not want to simply return to what he described as the lamentable isolation that had characterized preceding generations of learned men. The learned were still to be in the world, just not absorbed by it. They were to decide serious questions among themselves before broadcasting their results to all and sundry.[73]

Seeing Schlosser as an early voice in favor of a modern scientific community also distorts crucial features of his cultural horizons. Reading his complaints as an early effort to "institutionalize science" obscures the specific forms of status and community that characterized the late eighteenth century. Schlosser was unusually draconian to suggest a rejection of publicity *tout court*. A more common late eighteenth-century solution to the dilemma he described was to draw a distinction not between publicity and learnedness, but between two different kinds of publics—one whose deliberations helped *produce* learned knowledge and one that merely received it.[74] The best way to understand the structure of this learned public is not to ask what nineteenth-century developments it anticipates but to trace from whence it came. To do this we have to question one of Schlosser's primary assertions—namely, that early modern patterns of learned sociability had been erased by the emergence of a vernacular public sphere.

"The learned public" had two older synonyms that were also still in frequent use in the late eighteenth century, "the learned world" and "the republic of letters." In the constellation of these three synonyms, we see the new language of publicity paired with older conceptions of learned community. Particularly in the German context, the relationship between publicity and learnedness has been difficult to articulate clearly. An older literature often equated learnedness with polymathic erudition and used the republic of letters as a label for the early modern learned estate as a whole. Against this backdrop, late eighteenth-century developments could seem more novel than they actually were. According to this older literature, the late eighteenth century supposedly saw the replacement of comparatively static ideals of erudition with modern, forward-looking ideals of research.[75] With the benefit of recent work on early modern scholarly practices, this sharp dividing line now seems untenable. Kristine Haugen has pointed out that many of the practices once attributed to the advent of a novel conception of research around 1800 were actually much older; Martin Gierl has argued that the genre of *historia literaria*, once thought to be the labors of the quaintly erudite, were actually the products of an intellectual world already conversant with the ideas and practices of research.[76] Especially in the history of the natural sciences, this interpretive tradition, which posited a "research revolution" around 1800, made it hard to see the thick continuities that linked the early modern learned world with the scientific culture of the late eighteenth century.

In order for the "republic of letters" to be a useful analytical concept, however, it has to be used with greater specificity than is sometimes the case. As Herbert Jaumann has argued, the republic of letters should not be used as a label for all possible aspects of learned culture or as a synonym for the learned estate as a social group. The republic of letters was an idea used to conceptualize the *communicative struc-*

tures that bound the learned together. Its history should be a history of forms of learned communication, not of the learned estate as a social formation.[77] To use the republic of letters as a label for enlightened public culture in general also stretches its boundaries too far in the wrong directions.

To begin to trace out the relationship between the learned public and the general public sphere, we could start with Klopstock's colorful and idiosyncratic *Deutsche Gelehrtenrepublik*, a 1774 work in which the famous poet offered a lengthy and metaphorical new "constitution" to the German learned world. The citizens of Klopstock's imagined republic were a highly organized bunch, much like the citizens of an incorporated German town. They were divided into guilds according to their various pursuits; they elected aldermen to represent them in their provincial diet. Like the guild masters of a hometown, they carefully screened those who wanted to join their ranks, denying entry to bad authors and half-educated pretenders. Such people belonged not in the citizenry but to the lowly rank and file, the *Volk*. The republic of letters tolerated this mob, it informed them of its results, but it did not allow them to participate in its debates.[78]

This basic distinction can be found in less elaborate regalia throughout German print culture in the final third of the eighteenth century. The term "the learned public" was a common turn of phrase, and it was generally used to designate a kind of public that had a special right to act as an informed judge on certain kinds of subjects. For example, a book announcement for the third volume of a work on the natural history of birds advertised to potential buyers that the earlier volumes in this series had been "received with praise by the learned public." The book's publisher, A. G. Schneider, assured his potential customers that his latest release had been crafted with a desire to "continue to earn the honor that experts [*Kenner*]" had previously given his publications.[79] Similarly, Dr. Moritz Balthasar Borckhausen of Darmstadt presented his *Rheinisches Magazin zur Erweiterung der Naturkunde* "to the learned public for their considered judgment."[80]

When scholarly and scientific topics were under consideration, eighteenth-century writers generally did not imagine a public composed simply of human beings exercising their natural reason, the kind of public that Habermas described in outlining the bourgeois public sphere. They addressed knowledgeable readers who had the preexisting expertise to effectively judge their work. Writers sometimes explicitly evoked this "learned public" when they wanted to distinguish truly competent judges from the audience of all possible readers. In a 1777 work on horses, Johann Gottfried Prizelius criticized a rival author for citing selectively from his sources. This practice was "a clever artifice which will convince much of the riding public, few of whom study this subject [i.e., horses] as a science [*Wissenschaft*]; with the learned public, such a move will likely result in something other than praise."[81] In his 1785 report on a fossilized bone, Ildephon Kennedy referred interchangeably to "the learned public" and "the learned world" as his intended audience. He encouraged anyone who found an interesting fossilized bone to "present it, as soon as possible, clearly and truthfully to the learned world, so that other knowledgeable natural researchers will be in a position to examine it." For Kennedy, the judges of his work were other "knowledgeable natural researchers," members of the *learned* public, not the reading public as a whole.[82] The circle of informed readers might be specified even further; authors sometimes submitted their work to the judgment of "the chemical public," "the botanical public," "the mathematical public," or "the natural historical public."[83]

The distinction between a learned public and a wider public of less informed readers and writers shared certain features with later nineteenth-century patterns of professionalization and popularization. But it mapped onto an intellectual world with decidedly different structures, access points, and internal expectations. The central features of this learned public came from the evolving traditions of the early modern republic of letters. By the late eighteenth century, this knowledgeable, learned public was intertwined with the wider reading public but had not dissolved into it entirely.

An anonymous reviewer in the *Journal aller Journale* described this difference clearly in an article about a recent Hamburg production of Schiller's *Don Carlos*. The production had left out several scenes from Schiller's play, and as a result, the reviewer said, playgoers had not been given crucial historical information that they needed to follow the plot. A viewer who already knew something about Spanish history would understand what happened in the play; someone without this background would not. The reviewer complained that it was unreasonable to expect that sort of preexisting knowledge from a theater audience. "Must one have a familiarity with statistics and church history, and have read travel descriptions, in order to be qualified to understand a play?" he asked. The reviewer then imagined a potential objection to his criticism: isn't the playwright writing for "an educated public [*ein gebildetes Publikum*]"? "But are the educated and the learned publics the same?" the reviewer asked rhetorically. No, he concluded, they were not. "Even when a learned man goes to a play, he does not go there to be a learned man. If there is one place in which we are all equal, then it is in the theater. The poet should write for his public: In the theater one cannot pick one's people."[84]

In the learned public, in contrast, one could indeed pick one's people, at least to a certain extent. Like the theater critic above, historians have often emphasized the homogenizing tendencies of the eighteenth-century public sphere, the ways in which new social and communicative practices worked to erode the status markers of the Old Regime social order. Within a newly formed public of readers, writers, theatergoers, and coffee drinkers, everyone, at least rhetorically, was equal in the way that the anonymous theater critic described above; this was a public where access came in the form of a ticket purchased or a journal read.[85]

As the *Journal aller Journale* pointed out, the entry fee to the learned public was much higher. The learned public had a different structure than a theater audience; it unapologetically vetted its members. Learned societies and academies were the learned world's most prominent and formal vetting structures; these were the organizations that an older historiography described as "institutionalizing science" in this period.[86] In the parlance of the late eighteenth century, these were organizations that managed the learned world. The Academy of Sciences in Paris cut a particularly impressive figure in this role, but organizations like the Society of Nature-Researching Friends also understood themselves to be regulating learned conversation. Friedrich Martini, this society's founder, wanted only experts, or *Kenner*, for his group, not just casual collectors. When Martini first proposed the idea of founding a society for natural research in Berlin, he wrote that members in the groups should be "not just true lovers of nature [*wahre Liebhaber der Natur*]," but should also have "considerable knowledge." The existence of such a group, he thought, would help to set an example for other, less serious collectors.[87] In his essay on pedantry, Johann Schlosser also spoke of learned societies and academies as the guardians of the learned world, though he thought they were doing a lousy job.[88]

Martini pointed out another important criterion for acceptance into the learned public: access to the necessary material resources to do serious learned work. The wise men of the eighteenth century, Martini wrote, were men with a "workshop."[89] Expertise could not be separated from the forms of material culture that allowed its production. To speak with authority before the public as a chemist, one needed a laboratory; a naturalist needed a collection, an experimental philosopher the right kinds of instruments. Everyone needed access to the right kinds of books. A *Naturforscher* could not theorize out of thin air; he needed to collect, compare, and experiment.[90] Ursula Klein has shown how apothecary-chemists in the eighteenth century could use their laboratories to build learned reputations for themselves; John Heilbron's history of electrical research in this period shows the importance of physical cabinets and specialized instruments for scholars with ambitions in these fields.[91] Similarly, naturalists with large and well-ordered collections held a commanding place in late eighteenth-century networks of natural knowledge.[92]

The learned world was also held together by forms of exchange that stretched beyond the world of print. Schlosser was wrong to argue that these older forms of learned sociability and correspondence had completely decayed by the 1780s. As L. W. B. Brockliss has shown through his study of the provincial French intellectual Calvert, the republic of letters continued to develop into the second half of the eighteenth century, and self-styled citizens of the learned world still carefully tended correspondence networks, writing copious letters and describing themselves as friends bound by reciprocal obligations.[93]

Such activities were also important in learned societies. Groups like Berlin's Society of Nature-Researching Friends or Jena's Nature-Researching Society published journals, but they also poured enormous energy into writing letters, and they continued to use older tropes of learned friendship to describe the relationship among their correspondents.[94] For Martini, the many correspondents of the society he founded were all "friends." Martini's conception of friendship owed a heavy debt to the mid-eighteenth-century cult of sentimental friendship furthered by writers like Johann Wilhelm Ludwig Gleim, but in its basic form, his references to his correspondents as "friends" also represented an extension of an older early modern tradition.[95]

When the Society of Nature-Researching Friends described how it made reliable knowledge, the *personal* connections among all its members remained central. According to Martini, networks of friends were networks of observers who carried the society's local Berlin core along with them to the farthest corners of the globe. "One could say that one lives, I believe, in every country where one has the pleasure of having a friend."[96] As the society's revised 1784 statutes put it, the group's non-local members were "friends who see, hear, observe, judge, etc. for us in all the places where we cannot go ourselves." These far-flung friends allowed the group's core members in Berlin to enjoy the benefit of living "in more than one part of Creation."[97] Here, common sensibility supposedly created the conditions for transparent, almost perfect communication; reading the report of a friend was as good as having seen the thing oneself.[98]

Networks of learned friends also helped structure the world of printed exchange. Authors often presented their decision to publish their work as something done at the urging of their "friends." Karsten opened his *Anleitung zur gemeinnützlichen Kenntniß der Natur* with extensive quotations from his correspondence with other *Naturforscher*. When he had first mentioned his ideas in an earlier textbook, he wrote, "Several men who are fully conversant with [applied mathematics and natural philosophy], and who had performed proven services for these sciences, assured me that they were in full agreement with my position." Before moving on to describe his own proposal in detail, Karsten quoted excerpts from the letters of these experts to defend his innovation.[99] Similar rhetorical moves appear in more modest works, as well. The youthful Albrecht Wilhelm Roth had decided to write a natural history textbook for schoolteachers, he said, when a friend had asked him to recommend a suitable introductory work and Roth had realized that none existed.[100] In 1799, Mathias Flurl described conversations with learned friends as a preparatory stage that ought to precede the act of publication. A man who had been "initiated into the shrine of the Muses" had an obligation to be an eager and industrious observer. "He must exert himself to share everything with the scientific circle of his friends, and then to share [his results], if they prove to be truly useful, through his friends with the rest of the world."[101]

Flurl described scholarly friends not just as critics of first resort but also as intermediaries through which a man's work could reach the broader world; a learned man shared his findings "through his friends." This is an intriguing turn of phrase, and there is some evidence to suggest that in certain cases it should be taken literally. In the case of works sold by subscription, learned correspondence networks sometimes seem to have doubled as channels of commercial distribution for an author's work. Johann Bernoulli instructed people to contact him if they were interested in his edition of Lambert's papers (with discounts for purchasers who bought in bulk). He also informed potential buyers that his fellow mathematician Karl Friedrich Hindenburg in Leipzig would take orders.[102] J. A. E. Götze provided a thick paragraph of people who could take subscriptions for his book on animal intestines, and only a few of the names listed were booksellers. Most were fellow naturalists and medical practitioners, and seem likely to have been taken from Götze's circle of correspondents. The list included Friedrich Martini in Berlin, Professor Ernst Gottfried Baldinger in Göttingen, an Apotheker

Meyer in Stettin, and the Graf von Borke in Pomerania, along with several other state officials and professors from other parts of central and northern Europe. Götze also mentioned that "any of my friends and colleagues in the Nature-Researching Society in Berlin," would accept orders for the book.[103] Klopstock explicitly referred to all the people collecting subscriptions for his *Deutsche Gelehrtenrepublik* as his "correspondents."[104] Book announcements also sometimes made a more diffuse request to "friends and patrons" to help spread the reputation of a work.[105] Particularly in the case of specialized publications with limited appeal, the book market could be quite chummy.

In order to fully understand how learned authors moved between exchanges of services and gifts in the republic of letters and the monetary economy of the book market, we need to know much more about the financial and commercial side of learned book and journal making.[106] But we should definitely be cautious about portraying late eighteenth-century learned exchange as "commodified" in any strong sense.[107] Even in cases where things were being bought and sold within the learned world, it is not particularly helpful to speak of commodity exchange, at least if we define a "commodity" in a rigorous way. To make this point, William Reddy's analysis of how the introduction of the metric system helped fully commodify the buying and selling of cloth around 1800 can serve as a useful point of comparison. The Old Regime's units of measurement, Reddy argued, took careful account of the kind of labor that went into making an object. Varied, local units of measurement reflected regional differences in land, material, and work habits. In contrast, the new "universal" metric system, based on a single standard established in Paris, erased these local differences; it erased considerations of differential labor and landscape. In the process, the metric system transformed pieces of cloth into true commodities—they became objects shorn of all features but their exchange value, objects whose price no longer included an explicit consideration of the complex labor that went into their production.[108]

By these standards, eighteenth-century learned books were only very incomplete and imperfect kinds of commodities, for they made little attempt to mask the circumstances of their production in the way that Reddy describes. Book announcements, far from erasing the intellectual labor that went into making a book, made this labor a central selling point. "I have now come so far with my seven-year-long examination of the intestines of animal bodies," Götze announced, "that I am ready to present my findings to the public through print." Orders for the work would be accepted through Michaelmas 1782.[109] Similarly, when the Frankfurt publishers Varrentrapp and Wenner announced in 1789 that they were publishing a Latin and German translation of Miller's *Illustration of the Sexual System of Linnaeus*, they went into great detail about the origins of the work. The idea for the translation had first come from a Kriegsrath Merk in Darmstadt; he had put the publisher in contact with Herr Weiß in Rothenburg. Weiß, the publishers noted, was "distinguished among the botanists of Germany for his eager efforts to make the learning of botany easier and to further its useful application." Weiß had put great care into completing the translation, had added many things to the text, and had included a useful foreword for botanical beginners. This was not just a translation, the publishers said; it was in many ways an entirely new work.[110]

A translation project is a good place to look at this period's assumptions about what gave a book its value. Unauthorized reprints, translations, and bowdlerizations were common in the late eighteenth century, and as a result so were complaints about booksellers whose books were nothing more than mechanical reprints or cut-and-paste compilations thrown onto the printing press. Booksellers and printers who behaved in this way, critics argued, deserved contempt. They did not want to serve the public good or further learned debate; their only desire was for profit, and they cheated authors of the just desserts of their labor. By the early nineteenth century, such concerns would lead to the first copyright laws, which both enshrined the author's intellectual work as the most important ingredient in the making of a book and accorded him special financial protections on that basis.[111]

The exact distribution of legal and financial credit for a book's production was still an open question in the late eighteenth century. Nonetheless, in the Varrentrapp and Wenner advertisement cited above, we can see that the dual labor of the volume's instigator, Kriegsrath Merk, and the book's translator and author, Herr Weiß, were presented as a key constituent of the book's value. The publishers did not just emphasize the utility of the book's content; they described the hard work involved in its production, and the stature and learnedness of the men involved. In emphasizing the intellectual labor that went into their translation, the publishers preemptively answered the charge that their book was devoid of novelty, just a mechanical reproduction of something already in existence. In other words, the value of their book depended on its *not* appearing before the public as a commodity of the kind that Reddy described. If potential buyers could not see signs of the specific labor that produced the book, the local circumstances of its creation, its worth was significantly compromised. Similarly, when

Johann Bernoulli wanted to convince people to buy his edition of Lambert's *Nachlaß*, one of the ways he chose to describe the value of the manuscripts he was editing was to show the eminent hands through which these "learned treasures" had passed. After Lambert's death, his manuscripts had been bought by Johann Georg Sulzer ("also taken from us much too soon," Bernoulli noted), then purchased "for a large sum" by the Prussian Academy of Sciences. Here, Bernoulli offered dual criteria to convince potential buyers that Lambert's manuscripts contained things worth their time and money—he mentioned the high price the academy had paid for them and also the eminent learned man who had considered them worth saving.[112]

Anke te Heesen and Staffan Müller-Wille have shown how naturalists who sent one another specimens treated these objects as gifts; even after the specimen changed hands, the identity of the giver remained relevant to the object's value. In an exchange of commodities, the relationship between seller and the commodity ends when the cash changes hands; a gift, in contrast, still carries with it traces of the giver after it leaves her possession. It constitutes a bond between giver and receiver that persists past the moment of exchange. Many eighteenth-century naturalists honored this bond by carefully marking the source of the specimens in their collection; Linnaeus named species after people who sent him novel specimens.[113] Natural history specimens retained associations with their previous owners even after they had been passed on. Following a similar logic, Bernoulli's advertisement presented information about the previous owners of Lambert's papers as relevant features of their present value.

In a variety of ways, learned authors kept many of their specific social markers even when they stepped before the public in print. They typically listed their state offices and academic titles on their cover pages, and often their memberships in learned societies as well. A reader picking up Karl Gustav Jablonsky's *Natursystem aller bekannten in- und ausländischen Insekten* knew immediately that this was a work written by the personal secretary of Her Majesty the Queen in Prussia; the author also was a member of the Nature-Researching Society in Halle.[114] The title page of Franz von Paula Schrank's *Allgemeine Anleitung, die Naturgeschichte zu studieren* informed the reader that the author was a doctor of theology, the holder of a church office in the Palatinate, the director of the Economic Society of Burghausen, and a member of learned societies in Munich, Erfurt, Berlin, Lund, Leipzig, Rome, and Görlitz.[115] All of these social facts were relevant to the public authority of a natural researcher.

CHAPTER 1

The public persona of a learned author had folded within it copious references to the certifying networks of the republic of letters and to the specifics of his social station. As we saw above, in some cases authors might distribute their work through their learned correspondents; they might invoke the urgings or approval of their friends in their prefaces. Other marks of honor were readily displayed, too. They listed their academic degrees and their public offices; in subscriber lists and dedications, they advertised their connections to the powerful and noble. Jablonsky dedicated his entomological work to a Prussian princess who had since married into the House of Orange; E. A. W. Zimmermann dedicated his 1791 translation of William Smellie's *Philosophy of Natural History* to the Holy Roman Emperor Leopold II.[116] All of these specific social facts, which were still readily on view in the published works of late eighteenth-century learned authors, were things that a complete process of commodification along the lines Reddy described would have erased; all of these things were essential to stabilizing the public value of an author's work.

There is one practice that was common within Enlightenment public culture that would seem to cut decisively against the argument I have just made above. Book reviews were often published anonymously in the eighteenth century, and this would seem to be a perfect instantiation of how enlightened public discourse produced a subject-position that spoke with the voice of universal Reason rather than as any specific, socially defined person.[117] If we consider the status accorded the *collective forums* in which anonymous book reviews appeared, however, we might be able to better understand how this discursive conflation of an individual author with Reason per se could have had such plausibility. To late eighteenth-century educated Germans, periodical journals were thought of in associational terms. Under the Prussian General Law Code of 1794, journal editorships and readerships could also be considered private societies, and scientific journals were often the product of fairly well-integrated communities, groups of colleagues who saw themselves as part of a collective project.[118] One of Germany's most famous early review journals, the *Göttingischen Gelehrten Anzeigen*, was the mouthpiece of the Göttingen faculty. The reputation of Friedrich Nicolai's *Allgemeine Deutsche Bibliothek* was built on his renown as a man with extraordinarily good connections within the world of German letters. Nicolai intended his journal to bring together the learned and the wider public into closer relation; when he went searching for reviewers, however, he looked to the learned world.[119] Through his extensive network of correspondents, he sought (and sometimes failed to get) reviews from prominent men with established learned reputations.[120]

The early periodicals of the seventeenth century had emerged out of learned correspondence networks, and the journals of the late eighteenth century were in significant ways still sustained by them. Within a journal like Nicolai's, in other words, the voice of the anonymous reviewer was a voice that had been vetted before it spoke. To use Klopstock's terms, the reviewer had been recognized as a citizen of the republic of letters rather than one of the mob. In writing anonymously, the individual author subsumed his personal voice into the voice of a collective entity, the journal, that gave additional force to his own judgments.[121]

Furthermore, for an author to appear before the scholarly world as the *mere* purveyor of a commodity would have been to sacrifice all claims to learned legitimacy. As La Vopa has shown in his study of theology students, eighteenth-century German culture was still deeply uneasy about the disruptive effects of excessive personal ambition, and an overweening desire for profit was a particularly base and disreputable kind of ambition.[122] The publishers of learned books disavowed all such motives. The Frankfurt publishers of *Miller's Illustration of the Sexual System of Linnaeus* hoped that "the true learned public would support this enterprise, which is certainly not motivated by a desire for profit."[123] Johann Bernoulli hoped that his potential customers would see that he had decided to edit Johann Heinrich Lambert's papers only out of admiration for the deceased mathematician and "a desire for self-instruction."[124] A reviewer in the *Allgemeine Literatur-Zeitung* praised Albrecht Höpfner's *Magazin für die Naturkunde Helveticus* as a "truly patriotic organ, free from all ambition for fame or profit."[125]

Profit motive was seen to threaten learned communication in a variety of ways. It could keep learned men from being able to examine valuable material. Given the commercial value of whale bodies, Johann Gottlob Schneider noted that the "curiosity of the natural researcher" generally had to take a backseat to the profit motives of whalers, and it was rare to get to dissect a whole specimen.[126] Similarly, learned writers described people who saw natural history specimens just as profit-generating commodities as untrustworthy cheats. "Human hands motivated by profit [*gewinnsüchtige Menschenhände*]," according to Bernhard von Nau and Georg Zinner, created fake specimens that were then falsely perceived as natural wonders. Joseph Bergmann agreed; people motivated by profit (*gewinnsüchtige Leute*) used clever artifice to get collectors to waste their money on counterfeit natural rarities.[127] Learned authors, in contrast, described themselves as providing gifts and performing services by publishing their work (even as they offered it for sale). Johann Jacob Volkmann described his translations of recent Italian works on natural history as a "service" to the learned public.[128] The anonymous reviewer of a new edition of Linnaeus's *Systema vegetabilium* thought the work was "a more pleasant gift to the learned public" than a similar earlier edition by a Herr Planer. Another reviewer wrote that Casimir Christian Schmidel had "obliged himself to the botanical public" through his classificatory efforts.[129]

Against the backdrop of the eighteenth century's burgeoning consumer culture, it might be tempting to read such language as so much ad copy, merely a strategic mask for transparently commercial motives. Learned books were, after all, gifts and services that were up for sale. But booksellers, too, had honor to preserve, as Pamela Selwyn has shown in her study of Friedrich Nicolai.[130] Learned authors remained embedded within networks in which other kinds of exchanges were not just possible, but essential to the constitution of a learned reputation. Among the learned, gifts, services, and specimens traded hands; people described themselves as friends bound by mutual obligation and also behaved accordingly. Membership in this learned public was not something that could be purchased. It had to be earned. In order to make sure their special status was clear to their readers, learned authors did a great deal to graft their specific, complex social identities onto their printed works. They crowded their title pages with lists of academic degrees, official titles, and the names of societies that had honored them with memberships; they signaled their connections to powerful patrons or statesmen, and mentioned the promptings they received from their learned friends. In turn, the critical public to whom they submitted their work was full of "friends" and fellow connoisseurs.

When learned natural knowledge was described as a patriotic activity that contributed to the public weal, it was sometimes also presented as a contribution to *"bürgerliche Gesellschaft,"* or civil society. In conceiving of himself as part of civil society, however, the learned man did not need to abandon his learned status. The study of natural history and natural philosophy, Johann Daniel Curio wrote in 1786, made "useful members of the state and of civil society." Curio also noted that "many learned and insightful men" had praised his periodical, which was intended to introduce young readers to these valuable fields.[131] References to learnedness remained invaluable as a way to talk about expertise, something enlightened Germans thought civil society could not do without.

Boundary Markers

Two kinds of boundary markers were often evoked to indicate the line that separated the learned and the educated public. The first of these boundary markers was gender; the second was Latinity. Though neither was applied with complete consistency, these two boundary markers offer another way to see how the learned public adapted itself to the broader enlightened values of taste and public service central to Germany's expanding vernacular public sphere.

Advocates of polite sociability and civic patriotism both looked down their noses at the *Stubengelehrte*, the unsociable, argumentative traditional learned man who spent his time closed in his study, locked away from nature and from life, the sources of all real knowledge. According to much eighteenth-century satire, this kind of Latinate learned man was as useless as he was awkward, his mental energy wasted on empty word games in a dead language. His projected opposite was the enlightened and cultivated savant.[132] In some versions, the latter carried the sheen of polished and courtly *politesse*; in others, he donned the more earnest bourgeois garb of the patriot. In either case, rejecting the mantle of old-fashioned pedantry came with a commitment to putting the fruits of learning to work in the wider world. This critique, however, did not ban the learned man from a public place of honor; it served to create an updated version of learned identity.

In crafting their public personas, learned friends of nature made ready use of the other resources available in late eighteenth-century culture. One of these was the language of polite sociability, still often identified in the 1770s and 1780s with Parisian salon culture and the French-speaking German courts. The circles around Frederick the Great might be taken as the apogee of this Francophilic vein within German intellectual life. Under Frederick's reign, the Prussian Academy of Sciences imported stylish, salon-tested natural philosophers like Pierre Louis Moreau de Maupertius and published its proceedings in French.[133] Frederick's court was a predominantly masculine affair; German learned men were well aware, however, that the standard French model of sociability gave a prominent role to women.[134] They knew of the power that aristocratic women exercised in the salon-centered world of Parisian letters. According to Schlosser, for example, it was the French who had first hit upon the idea that the best judge of a learned man's worth was his ability to appeal to ladies (Schlosser himself did not think very highly of this standard).[135]

According to Londa Schiebinger, many of the early modern structures that had allowed women access to the sciences, such as guild ties or noble status, had largely eroded by the late eighteenth century. At the same time, the countervailing processes of scientific professionalization and privatization of the family had supposedly progressively excluded women from public scientific activity.[136] But the cultural innovations of the second half of the eighteenth century also provided women of a certain social background with novel opportunities for participation in intellectual life. Women often participated informally in the reading societies of the period, which typically met in homes and built on more diffuse patterns of "open house" sociability in which women acted as hostesses.[137] Women, usually noblewomen, sometimes appeared as honorary members in learned or patriotic-economic societies. Halle's Nature-Researching Society admitted a local unmarried woman as a member in the early nineteenth century. Louise Corthum, the daughter of local merchant and gardener Johann Carl Corthum, became a member of the society in 1805.[138] Her participation was even more noteworthy given that her father was not a member of the society.

On the whole, however, the networks of natural knowledge were networks of men, and it is worth pointing out that the two most powerful structures that kept women at the margins of learned sociability were not new to the late eighteenth century. The masculine culture of the universities and the monopoly men held on official state offices meant that both learnedness and service to the state were heavily gendered. Though these structural constraints did not settle the question of women's participation beyond dispute, they did stack the deck in a very serious way.[139]

Indeed, learned men often used women's presence as a way to mark the difference between the learned and the educated reading public. Among naturalists, the interest of female readers was sometimes noted positively as a mark of broader public success.[140] On the other side of the fence, when commentators wanted to denigrate excessive enlightened enthusiasm for the spread of knowledge, they often made sarcastic references to reasoning ladies, as we have already seen with Schlosser.[141]

The jauntily named *Galanterie-Mineraologie* offers another example of how gender could be used to mark the boundary between learned and polite readers. The author of this work, C. F. von Arenswald, anticipated two objections to his text. The learned would mock him for speaking of such subjects before ladies. The ladies, in turn, might greet him with skepticism, expecting to be bored. He dismissed the former critics as pedants and promised his female readers that he would not be dull; he was "not that sort of learned man." It was true, Arenswald acknowledged, that mineralogy was a learned subject, but he planned to present it in a way appropriate to his audience. His book would enlighten the genteel female reader on a subject already close to her heart—her jewelry. What could be more interesting than that?[142]

The ideal of refined, feminized, French-accented sociability was an object of both desire and jealousy for ambitious German men of common birth. The young Johann Gottfried Herder went to France hoping to immerse himself in the revivifying waters of *sociabilité*; he returned disgruntled and with a renewed commitment to the power of the German vernacular. For Herder and for many of his contemporaries, the cultivation of German as a literary language offered a counterweight to the world of decadent, French-speaking courtiers, with their accompanying ladies. German was the language of the newly expanded reading public and of enlightened associational life, and the flowering of a standardized national literary language was often framed as a *"bürgerliche"* affair, even when members of the nobility signed up in support of the cause.[143]

Like their literary counterparts, natural philosophical authors also aspired to measure up to these new German linguistic ideals. Johann Friedrich Blumenbach, for example, made a point of informing the readers of his natural history handbook that he had avoided using regional dialects in giving the German names for plants, animals, and minerals.[144] Friedrich Ehrhardt, a man with less impressive educational credentials than Blumenbach, noted defensively that he might not always write in the purest German, and apologized in advance to his reader.[145] In a review of recent works on butterflies, Ferdinand Ochsenheimer assessed not only the content but also the style of the books he discussed. One handbook by a J. F. Stopp was dismissed as a "highly mediocre little work." "The style is poor and incorrect" was all Ochsenheimer felt the need to say beyond that, and he criticized several other works for their bad German.[146]

Even with the growing dominance of the vernacular, however, Latinity remained an important distinguishing mark of learned status within the burgeoning enlightened public sphere. Latin's share of the German book market declined significantly over the last third of the eighteenth century, but it did not disappear entirely.[147] In the natural sciences, the percentage of Latin books contained in the Leipzig *Messe* catalogs went from 30.44 percent in 1770 to 16.28 percent in 1800.[148]

In the preface to the first edition of his *Handbuch der Naturgeschichte*, Johann Blumenbach positioned his work right on the border between a learned audience and a wider readership of casual dilettantes, and Blumenbach's preface illustrates well the linguistic complexities of straddling that line. To pitch his book, he called on both Latinate erudition and ideals of polite cultivation; both sets of references jostled next to each other with neither gaining the upper hand. Blumenbach's *Handbuch* was perhaps the most successful work of its kind in this period. It went through twelve sanctioned editions and several unsanctioned printings before the eminent naturalist's death in 1840. As Blumenbach noted in later editions, the book "found its public." In the first edition of 1779, however, the young professor took care to specify the kinds of readers he hoped to attract. The work could be used for introductory academic lectures in natural history, Blumenbach explained. He presented it as a "compendium" in Bacon's sense of the word, a memorial to the current state of a science. But it was also, he wrote, a book for "dilettantes" who wanted to study a little natural history in the free hours left to them by the practice of their profession.[149]

Blumenbach's preface was sprinkled with Latin words, but also with Germanized versions of French terms, and this linguistic heterogeneity was part of a joint appeal both to traditions of Latinate learning and to French-inflected ideals of polite sociability. In courting a non-Latinate audience, he used words borrowed from French; he promised *"Dilettanten"* that they should not allow themselves to be bored (*ennuyirt*) by the brief Latin descriptions that appeared in the book. They could simply pass over them. He assured his reader that he had chosen only the most entertaining and interesting facts for inclusion, and had exerted himself to avoid the "insuperably boring prolixity" of many natural history handbooks. Other authors were likely so boring, he noted, because of their constrained social experience, "domestic constriction" in his formulation, a phrase that evoked standard criticisms of the unsociable traditional pedant. But Blumenbach also suggested that the supposedly dull tone of much natural history writing could be blamed on the shortcomings of unlettered authors, *"illiteratis"* (whom, he added to cover his bases, still deserved credit for their goodwill). In another instance of mixed lineages, he used a French label to describe the scholarly reader, noting that "people of *metier*" would of course understand Latin. Blumenbach also warned that natural history, though a useful and entertaining study for "all educated people [*alle gebildeten Menschen*]," was no replacement for a solid foundation in the *Humaniora*, the humanist curriculum.[150] Latinity marked the boundary between the serious scholarly reader and the mere dilettante.

Conclusion

The Regensburg pastor Jacob Schäffer can stand as a typical example of someone who counted as a serious student of nature during the Enlightenment. Schäffer's intellectual interests were many and varied. Among his published works one could find studies on electricity, lens grinding, birds, insects, and mushrooms; he also drafted plans for a mechanical washing machine, experimented with furnace design, and tested new techniques for making paper.[151] The loose field that Schäffer sometimes (though by no means always) referred to as "*Naturwissenschaft*" plausibly included all of these activities.[152] Alongside his publications, Schäffer's learned correspondence and his personal collections also lent weight to his intellectual reputation. His extensive cabinet was a requisite stop for a learned man traveling through Regensburg (Goethe dropped by in 1786). Schäffer's correspondents included the Academy of Sciences in Paris and the leading naturalists of his day, men like Carl Linnaeus and René Antoine Ferchault de Réaumur. Numerous learned societies, among them the Botanical Society in Florence, the Royal Society of London, the Society of Nature-Researching Friends in Berlin, and the Academy of Sciences in Munich, accepted him as a member.[153]

In past discussions of the relationship between science and the public sphere, historians have pointed to a seeming paradox. Science is difficult and highly technical; it is also, at least in theory, supposed to be publicly accessible knowledge, open to all.[154] To the late eighteenth century, this looked less like a paradox and more like a division of labor. The republic of letters had been Central Europe's *Öffentlichkeit* (public) long before the late eighteenth-century expansion of a large vernacular reading public, and in the world of eighteenth-century nature study, the "public" could refer to multiple things. It could be the anonymous, commodity-driven public described by T. C. W. Blanning, which expressed its collective taste through the consumption of cultural products. Or it could be the knowledgeable public, the republic of letters, made up of people whose interrelationships were far more complex.

In the final third of the eighteenth century, "the learned world" was deeply intertwined with the broader literate and sociable public, but it was also still distinct from it in various ways, too. Rather than thinking of the old republic of letters as something that was swept away around 1750, or simply as another name for the enlightened public as a whole, we need to pay attention to the ways that these two kinds of publics evolved alongside each other. When authors submitted their work to the learned public, the faith they expressed in this community of critics came in part from the mutual goodwill and vetted expertise that was supposed to characterize members of the learned world. This goodwill was not just a disembodied abstraction; it still took on concrete expression through learned correspondence and personal exchange. Friedrich Martini's faith in the knowledge produced within the Society of Nature-Researching Friends was faith in a community of learned friends. By the late eighteenth century, this faith had been linked with, without being totally absorbed into, a more general discourse of publicity. As Thomas Broman has put it, a crucial task in explaining the prestige of science in the Enlightenment is to explain not just why individual experts inspired trust but why *expertise* as such did.[155] A better understanding of the structures of the "learned public" can move us closer to an answer to this key question.

All of these points are central to the history of *"Naturwissenschaft"* because they help us understand the community of common interest that existed around the study of nature in the eighteenth century. *Naturforscher* thought of themselves as sharing common tastes, common passions, and a common desire to serve the public good. They also shared a common kind of prestige as members of the learned public. In this chapter, the learned public has appeared mostly in the highly idealized garb in which its members typically clothed it. In practice, the learned world was rife with controversy, bad blood, and competition; any study of the period's many scientific disputes could be cited to illustrate as much.[156] The tensions within the republic that are of greatest interest in this book, however, are the ones that appeared along its external edges, at that border zone where the learned *Naturforscher* met and mingled with the broader public of "the educated estates." Mapping out that contact zone is the subject of the following chapter.

It would be premature at this stage to draw too sharp a contrast between developments in German-speaking Central Europe and those elsewhere, but there are ways in which one can already see some differences emerging. In looking at the eighteenth-century Lit and Phil movement in Britain, Roy Porter has argued that these groups were primarily interested in reproducing the cultural life of London in the provinces. These societies were more interested in promoting taste and enlightenment, Porter says, than in cultivating useful knowledge.[157] These two motives are harder to disentangle in the German case, where good taste and utility often went hand in hand. To a much greater extent than in Britain, state officials constituted the personnel of the German enlightened public; to a much greater extent than in Britain, the nascent bureaucracies of the German states also made room for men with expertise in natural knowledge.[158] Being able to argue that one was, at least in some general way, benefiting the public weal was a powerful source of distinction in its own right. Furthermore, the study of nature was linked with a distinct tradition within the bureaucracy, the cameral sciences, that stood in competition with the more venerable faculty of law. As a result, not all learned men were equally convinced that studying nature was useful to the state.[159] As a result of these three factors, the pursuit of useful natural knowledge loomed somewhat larger as a defining common cause within the landscape of German enlightened associational life than it did elsewhere.[160] A British Lit and Phil society, as its name suggested, was supposed to further all the arts and the sciences together; French provincial academies were designed to do the same.[161]

In the German states, enlightened "utility [*Gemeinnützigkeit*]" was broadly defined; the study of nature was seen as useful for moral and spiritual reasons, not just material ones. German enlightened interest in practical knowledge, however, was much more than just window dressing. One can question whether or not eighteenth-century learned patriots actually managed to be as useful as they claimed. The avalanche of practical observations and experiments published in the last third of the eighteenth century, however, can leave little doubt that they were trying very hard.[162]

TWO

The Expanding Ranks of Nature's Friends

Andreas Ficinus came from a long line of learned men. The famous Renaissance humanist Marsilius Ficinus, head of the academy at Cosimo de Medici's court in Florence, was one of his forbearers. Andreas's father had been a jurist; his older brother also studied law. Ficinus himself had planned to study theology, but the Seven Years' War ruined his prospects for a university education. The upheavals of the 1750s and 1760s cost him his parents and his source of financial support, and instead of a minister, he became an apothecary. After fifteen years working for others, he acquired his own pharmacy in Dresden in 1777.[1]

Through the relationships he built in the Saxon capital, Ficinus partially recovered the learned status that European politics and bad luck had taken from him as a young man. Like his ancestor Marsilius before him, Ficinus benefited from the patronage circles around an ambitious court. The apothecary's facility for analyzing the chemical composition of medicines won him the attention of the Saxon royal family's two personal doctors, Johann Ehrenfried Pohl and Johann Gottfried Leonhardi. Their support gained Ficinus admittance to chemical lectures held in the home of the well-connected nobleman Joseph Friedrich von Racknitz. In 1795, Ficinus became a member of the Leipzig Economic Society, a group with strong ties to Saxon officialdom. He went on to publish chemical analyses of several Saxon mineral springs, a topic of interest to those who hoped to add to the kingdom's wealth by attracting visitors seeking water cures. By the end of his life, Ficinus could claim a minor place in the German learned world; he received a brief entry in Hamburger and Meusel's *Gelehrtes Teutschland*.[2]

Ficinus's life history nicely illustrates a key point: belonging to the learned world was not the same thing as belonging to the learned estate. The members of the traditional learned professions provided the core of Germany's republic of letters, but they were not its only members. Indeed, learned networks of natural knowledge could be decidedly socially heterogeneous, and Ficinus's perambulations capture this fact well. In these networks, learned physicians, fashionable noblemen, and men from more practical backgrounds all had their place.

Late eighteenth-century Germans sometimes wrote as if the learned estate and the learned world were one and the same, but particularly in the case of natural knowledge, this had not been true for quite a while. As recent research on the sixteenth and seventeenth centuries has shown, people of varied social stations contributed to the early modern period's philosophical revolutions. Natural history flourished in the cabinets and gardens of small-town universities, at the courts of German rulers, and in the homes of wealthy merchants.[3] The new experimental, chemical, and mathematical philosophies made the experiences of artisans, artists, and empirics relevant to learned men.[4] When Johann Burkhard Mencke bragged in 1715 that a reader could find over 240 "learned farmers, artisans, merchants, and also learned soldiers" in the pages of his *Gelehrten-Lexicon*, early modern natural knowledge could claim significant responsibility for this multifaceted republic of letters.[5]

The previous chapter focused on the forms of communication typical of learned exchange; this chapter looks more closely at the social composition of the learned world. The boundaries of the republic of letters were flexible, but only to a point. Particularly in normative discussions of who *ought* to be counted among the learned, educational background still had a strong delimiting function. Several other features of the republic of letters, however, made this delimiting function much less effective than the structures of later professionalized science. The pursuit of useful natural knowledge invited a wider range of people into natural historical and natural philosophical conversations, though not necessarily on equal terms. In addition, members of the learned world often described their activities as a form of private leisure; this, too, could make it hard to see exactly where the boundaries of the learned world ought to be set. This ambiguity only became more acute by the early nineteenth century, when the burgeoning leisure

CHAPTER 2

and consumer culture of Germany's major urban centers introduced an ever-larger circle of people to the charms of studying nature as a form of rational leisure.[6]

To be seen to best effect, the late eighteenth century's expanding leisure culture needs to be viewed in a local setting. A learned reputation, too, was still based in important ways on physically localized resources, things like the possession of a cabinet. A man secured his reputation as a learned expert through his connections to a cosmopolitan and dispersed community, but also through his activities within the local social world in which he lived. Andreas Ficinus, like numerous other apothecary-chemists of the period, built his reputation from the equipment found in the laboratory of his apothecary shop; he forged ties with local learned physicians and intellectually active noblemen.

In order to explore one local setting in sufficient detail, this chapter, like many later ones, spends a great deal of time in Ficinus's hometown, the Saxon court city of Dresden. Given the great diversity of urban settings within German-speaking Europe, no one city can stand as a "typical" place in which to examine late eighteenth-century science in a local context. But Dresden shared numerous things with other court and administrative centers of the period, and its peculiarities also have heuristic value. From the late seventeenth century forward, Saxony's rulers had been particularly enthusiastic importers of French ideals of polite sociability and royal display; they built up an extensive set of collections, which included both natural objects and physical instruments. The Wettin dynasty also tied its image to Saxon strengths in mining and luxury manufacturing.[7] One outcome of these interests was the famous mining academy in Freiberg; the famous porcelain manufactory in Meissen was another. Dresden, in other words, provides an obvious counterexample to R. Steven Turner's assertion that, since Germany lacked a tradition of polite noble patronage of the sciences, the learned world and the learned estate were one and the same.[8]

Dresden was shaped by its court, but not cravenly so; it also possessed an affluent and assertive Lutheran citizenry who in 1722 had decided to start construction on their monumental, elaborately decorated Church of Our Lady (*Frauenkirche*) just a few blocks away from the royal family's Catholic court church. Dresden had an active cultural life beyond the circles of the court; it was a place that, according to one late eighteenth-century visitor, "included more manufacturers and useful artisans than all of Bavaria" and where "knowledge of the things that belong to civil society and polite life" was so widespread among the citizens that even a Frenchman would feel at home.⁹ In the city's rich mix of diversions, natural philosophical and natural historical amusements held a respectable place.

In the late eighteenth century, the taste for cultivated amusements earned a person the right to be called *"gebildet."* In thinking about what this label meant, we should not turn too quickly to a closed story about "the transformation of the learned estate into the modern educated middle class." When late eighteenth-century authors used *"Bildung"* as a criterion to divide up the social world, they did not cut along the lines that would later mark out the *Bildungsbürgertum*.¹⁰ Their concept of "the educated estates [*die gebildeten Stände*]" extended upward to include the nobility (or at least the curious and intellectually cultivated members of the same); it extended out beyond the learned professions to include more affluent and cosmopolitan kinds of urban burghers, the booksellers of Leipzig or the textile magnates of Zurich. Highly skilled men from artisanal backgrounds could also fall within this category. Members of the educated estates, in other words, were people who belonged to Mack Walker's category of "movers and doers," a group that cut promiscuously across the divisions of the three-estate model of society, and did not simply prefigure some later nineteenth-century social group.¹¹

The Social Heterogeneity of the Learned World

In 1809, Gottfried Haymann published a lengthy and rambling lexicon he had been working on for most of his adult life. His book, *Dresdens Schriftsteller*, was a monument to the "learned authors" of his native city. After years of collecting stories and references, Haymann had written a work over five hundred pages long, one that offered a conscientious accounting of the writers, collectors, skilled artists, and artisans who had lived in the Saxon capital in the second half of the eighteenth century. Haymann himself held no academic degree. Just like Andreas Ficinus, he had had his young life disrupted by the Seven Years' War. This troubled time, he wrote, had robbed him of the chance to go and "sit at the feet of the greatest teachers in Leipzig." But, as he told his reader, he had worked his whole life to make up for this missed opportunity and had later traveled to Leipzig to visit the learned men of that city and make them "personally useful to himself." His father's worn and annotated copy of Georg Henning Behrens's *Curiöser Harz-Wald* had first drawn him to the charms of that venerable learned genre, the *Historia literaria*.¹²

Haymann's book paints a detailed picture of Dresden's late eighteenth-century cultural life. In three different sections on physical, economic, and mathematical authors, he listed 106 men who were either currently living or had lived in the Saxon capital; he listed another 18 names in his section on medical authors.¹³ Roughly 70 percent of these authors were the usual suspects of the German Enlightenment, members of the learned professions and state officials. In the section labeled "physical authors" (Haymann used this term to include nonmathematical natural philosophy, chemistry, and natural history), medical doctors were the largest group; 27 percent of the entries in this category were learned physicians. Another sizable percentage held a state or court post; this group included university-educated commoners and noblemen (many of the latter had also spent at least some time at a university).¹⁴ Within the period's economic and learned societies, one can find a similar social breakdown, though the constitution of a given society varied according to the social, political, and economic complexion of the city in which it was located. Zurich's Physical Society, for example, was an alliance between its urban patriciate, many of whom had made their fortunes in textiles, and the educated professions of the city, joined by a smaller percentage of artisans.¹⁵ In most places, such societies drew the bulk of their membership from the university-educated professions and the state bureaucracy, and their memberships included officials of noble as well as bourgeois origin. In Prussian societies, for example, medical doctors and professors of varying ranks represented the largest group, at 43.6 percent. State officials formed about 30.8 percent of the membership; trade and agriculture contributed a small but not insignificant 12.4 percent. In Saxony's economic and learned societies, state officials were the largest group, at 32.7 percent; doctors and professors composed 27 percent of society members; trade and agriculture, 16.7 percent.¹⁶

As a raw statistical label, the category "trade and agriculture" is somewhat opaque. In Haymann's book, the kinds of men behind this label emerge with greater clarity. The people who fell under "trade," at least in Dresden, were a set of highly skilled men involved in the eighteenth century's most high-tech economic pursuits. Among Dresden's published authors one finds several men who benefited from Saxon royal interest in promoting mining and certain specialized kinds of manufacturing. The skilled machine builder Johann Gotthelf Studer, for example, had studied at the well-respected Freiberg mining academy. The author of a work on mathematical instruments, he held the title of court machinist. The autodidact Christian Glieb Pötzsch worked in Meissen at the royal porcelain manufactory. Pötzsch had interests in mineralogy and meteorology; he was a member of Berlin's Society of Nature-Researching Friends and Jena's Mineralogical Society.[17]

The representatives of "agriculture" in Haymann's book belonged to an emerging group of skilled estate managers and agricultural experts. In the final third of the eighteenth century, noble landowners often hired managers to run their estates in their absence.[18] One such man, Johann Heinrich Rosenhayn, published a flora of Saxony and numerous works on agriculture; he ran an estate in nearby Sachsa. Johann Martin Fleischmann trained in the court gardens and went on to manage the royal vineyards; he published prodigiously on botany, silk manufacturing, and viniculture.[19]

The Saxon royal family's collections and elaborate households also supported a reservoir of people with natural philosophical and natural historical interests. We have already met the court machinist; the curators of the Saxon royal collections and the keepers of their gardens also sometimes became published authors. Several court gardeners—Johann Gottfried Hübler, Johann Heinrich Seidel, and Friedrich Traugott Pursch, Seidel's assistant—published botanical works in the late eighteenth century. Within the royal household, the position of court gardener had existed since the sixteenth century. Dresden had long participated in the broader European trade in both plants and gardeners, and the court's interest in exotic plants was already well established in the early modern period. In the late seventeenth century, a Saxon court gardener had been sent on a collecting expedition to Siam with the Dutch East India Company. These positions ran in families, often passed down from father to son, and court gardeners enjoyed a much higher status than their more modest local counterparts, the fruit and vegetable growers who made up the rank and file of the gardening trade. In 1719, the court gardener had even been granted the right to wear a small sword, a key sartorial privilege and an important mark of status.[20] By the late eighteenth century, learned ornaments also augmented the court gardener's position. Johann Heinrich Seidel's garden, the Herzogin Garten, was often called the "Botanical Garden" (a title typically used for university gardens) because of the diversity of species it contained. Seidel printed several catalogs of the garden's contents, ordered according to the Linnaean system.[21]

Apothecaries formed another large group among Haymann's "physical authors." Ursula Klein has argued that many eighteenth-century apothecaries enjoyed a hybrid status that combined elements of learned and artisanal identity. Many apothecaries attended at least a few lectures at a university as part of their training, and their laboratories and practical skills gave them excellent resources to contribute to chemical discussions. They published chemical works and enjoyed local prestige as the analysts and certifiers of chemical products. In Lorenz Crell's chemical journal, the most important of the period, almost half of the contributors were apothecaries. Three apothecaries in a row held the chemistry chair in the Prussian Academy of Sciences.[22] One sees a similar pattern in Dresden with men like Andreas Ficinus. Another local apothecary, Carl Gottfried Bünger, the owner of the *Löwenapotheke*, gave lectures to local audiences on experimental chemistry; he also published in Gilbert's *Annalen der Physik* and other journals. During his apprentice years, he attended lectures at Halle.[23]

Local noblemen, too, figured prominently in Haymann's book. They owned some of the city's largest natural history collections. The versatile Freiherr von Racknitz composed piano music, wrote art criticism, and collected insects and mineralogical specimens. He published a natural historical description of the fashionable spa town of Karlsbad. The Count von Hoffmannsegg's estate near Dresden had one of the richest private botanical gardens of the period.[24] One of Dresden's most enthusiastic entomologists, Peter Ludwig Heinrich von Block, amassed fifteen thousand specimens over the course of his life (he also had an extensive collection of Saxon shoes). He published a 1797 work on the insects of the Plauen Gorge, along with numerous journal articles on literary and natural historical topics.

Heinrich von Block's life illustrates how a nobleman's early training for state service could also serve to build up learned contacts. The young Baron von Block had first been slated to follow his father into military service. After his father's death, he decided to enter civil service instead and to this end joined the household of a maternal relative who was a state official in Wurzburg. While in Wurzburg, Block attended lectures at the town's university; when he returned home, he continued his studies under the physician Carl Heinrich Titius, curator of the royal natural history collection. The baron next entered the household of a young Russian nobleman, the Prince Bariatinsky, and accompanied the prince on his travels through Germany, northern Italy, and Switzerland in 1790. During these travels, Block collected insects, socialized with exiled French aristocrats, and sought out connections with eminent learned men. In 1793, he returned home to Dresden to enter state service and shortly thereafter was elected to membership in numerous learned societies, including groups in Halle, Siena, Leipzig, and Florence.[25]

In this period, a small number of farmers also became famous for their learned activities. One of them happened to live in a village just outside of Dresden. The "astronomer-farmer" Johann Georg Palizsch, described in his day as a "model to his estate," corresponded with the academies in Paris, London, and Saint Petersburg. He made several widely celebrated astronomical observations; he had a cabinet of physical instruments, minerals, zoological specimens, and shells. His garden was famous for its botanical diversity, and he was the first man in the Elbe valley to grow the potato.[26] The autodidact Swiss farmer Jacob Guyer, an early advocate of stall feeding, was another rural man who became a European celebrity. Hans Hirzel's account of Guyer's life appeared in multiple French, English, German, and Italian editions through the 1760s, 1770s, 1780s, and 1790s.[27]

Through their individual biographies, we can also see these different kinds of *Naturkenner* interacting with one another. The apothecary Ficinus attended lectures in Joseph Friedrich von Rackowitz's home; the Baron von Block learned natural history from Dr. Titius in the royal natural history collection. To better understand the status issues involved in these interactions, it would be helpful to look more closely at organized scientific sociability, where such relationships were formalized. Since Dresden had no learned society for natural research in the late eighteenth century, we will have to leave the Saxon capital to look at how these issues played out in other locations.

Like the Freemasons' lodges of the period, learned societies often combined men of noble and common birth, men who often also worked together as officials of the state. Both of these enlightened social forums cut against the divide between noble and nonnoble elites but did not erase this distinction completely. For example, Katrin Böhme has shown that the early Society of Nature-Researching Friends in Berlin shared a number of characteristics with Freemasonry. The society's seal incorporated Masonic symbols, and several of its founding members were also in local Freemasons' lodges.[28] Its statutes specified that "general equality . . . without consideration of birth, rank, or standing" was the rule of the group, so much so that "even the slightest attempt to assert any privilege was as good as forfeiting all the rights of membership."[29] In its published *Beschäftigungen*, the society gave only authors' names but not their *Amt*, or official title, a policy it took the trouble to mention explicitly in the introduction to its first published volume.[30] Each member's status in the society was supposed to come from his standing as a man of enlightened goodwill, a quality that was evidenced in his willingness to participate in useful collective projects. In and of itself, a serious interest in the study of nature spoke well of a person's character. "I assume from the outset," Martini wrote in his initial proposal for the group, "that rational collectors of natural rarities are also men of good taste, and consequently also people of pleasant manners and sociable friends of humanity."[31]

The desire to suspend differences in status and estate created potential problems, particularly when the society printed its membership lists, which contained men and a few women of high social rank. As we have seen, it was by no means standard practice for learned authors to erase marks of their social status when they appeared before the public in print. The Society of Nature-Researching Friends went to extraordinary trouble to ensure that no hint of rank made it into these lists, but it also explained its procedures to avoid offense. The twelve local members were listed in an order determined by lot, "without giving consideration to birth, rank or other privileges." The four additional local honorary members were named "in the order of their acceptance as members." The society listed its nonlocal honorary members alphabetically, a procedure it followed "for well-considered reasons." The widely differing social status of the members within this category, which included high-ranking aristocrats as well as modest village pastors, no doubt made this a diplomatic choice.[32]

Berlin's Society of Nature-Researching Friends, which had a closed inner core of twelve "ordinary members," combined aspects of Freemasonry with the habits of academies. It was a hybrid of intimacy and publicity, a "closed society [*geschlossene Gesellschaft*]" whose conversations were often shared with the wider world. Many of the lectures given at society meetings were printed in the local papers.[33] In addition, the society also functioned as a nonlocal publishing collective, organizing a vast system of correspondence and exchange. Significantly, the group's published proceedings documented the activities of *all* members, not just the local core. This delocalized quality was characteristic of many such publications. The first volume of the Halle Naturforschende Gesellschaft's *Abhandlungen* included twenty different contributions from twelve different authors; only two of these authors were from Halle.[34]

Indeed, the Berlin society's carefully orchestrated fictive equality seems to have worked best among its dispersed correspondents. Initially, its local core crossed a significant religious divide (one of its founding members was Jewish), but it did not cross the divide between noble and commoner. This was a distinction that still mattered a great deal within Prussian state service in the late eighteenth century, and many of the Berlin society's members were common-born state officials. The membership patterns of the society's early years suggest that social distinctions were more easily leveled from a distance than in local, face-to-face interactions. Martini sought out only men of the middle classes to serve as the society's local core, claiming that the nobility lacked the desired sense of public spirit necessary in such a collective enterprise.[35] The group's nonlocal membership suggests that this explanation was disingenuous, for plenty of noblemen appeared in that context. A better explanation lies with the casual, intimate face-to-face socializing that characterized the society's local meetings, a style of sociability that offered none of the complex ritual that helped create a protected space for contact between nobles and commoners in the period's Freemason lodges. For the first decade and a half, the society's nonlocal members were considerably more socially diverse than its local members. All of the local members were of bourgeois origin until 1790, most of them nonnoble state officials serving in the Prussian capital. Between 1773 and 1800, twenty-four of its thirty-five local members held state posts (68.6 percent), and the society had only four noble members. The nonlocal members, in contrast, already contained plenty of nobles in the 1770s.[36]

Most of the women who appear on the rolls of enlightened learned societies were noblewomen, and this is another good reason to think that the fictive equality of society memberships was in fact often used to forge what could still be understood as patronage relationships between people of different status.[37] Of the approximately one hundred women who belonged to German learned societies in this period, most were from the nobility and the wives of important public officials.[38] Two of the Society of Nature-Researching Friends' three female members were high-ranking noblewomen, Princess Christiane Louise von Hohenlohe-Langenburg-Kirchberg and Princess Katharina Romanowa Daschkowa, and neither was resident in Berlin. Two other noblewomen (the Princess Sapiehar von Jablonowska and Duchess von Podewils) were mentioned as possible candidates in the society's protocols but were never entered into the membership lists. The one female member of bourgeois origin was also the most celebrated German-speaking female naturalist of the period. Catharina Helena Dörrien was a gifted flower painter and author of a well-received flora of Nassau.[39]

As we have seen, the networks of natural knowledge also extended downward (or sideways) as well as upward, to artisans, gardeners, and other kinds of practical experts. How were these relationships managed? "Natural history is well known as that branch of knowledge which, more than any other in our current day, is universally loved and almost as universally practiced and cultivated," wrote the Regensburg pastor Jacob Christian Schäffer in 1764. Emperors, kings, and princes, all manner of learned men, even farmers and artisans—according to Schäffer, all were beginning to recognize the benefit of "occupying themselves with the observation, exploration, and treatment of nature."[40] Similarly, Johann Scheuchzer argued that natural philosophy could be valuable to "all intelligent [*vernüftigen*] people of both sexes," from the housewife in her kitchen to the artisan in his workshop and the farmer in his field. "Even a farmer [*Bauer*] can be a *Naturforscher* in his own sphere," wrote F. W. von Leysser in 1774.[41] Johann Samuel Schröter called natural history "the favorite study of the learned and the unlearned."[42] By the 1790s, people also referred to chemistry as a *Lieblingswissenschaft*, a favorite science, of the educated public.[43]

Such statements might make it seem as if the doors of natural research were wide open in the late eighteenth century, particularly when we juxtapose these descriptions with the diverse cast of characters Haymann recorded in his book on learned Dresden. Learned naturalists and natural philosophers often asserted that members of every estate could gain from the study of nature. In this regard, the promotion of natural history fit well with the period's other efforts at popular enlightenment, and the enlightened public's fascination with figures like Jacob Guyer suggest that the thought of "philosophical farmers" was genuinely exciting to many.[44] Daniel Jenisch called the scientific study of agriculture, forestry, mining, and manufacturing one of the greatest accomplishments of the past century; such efforts, to his mind, were central to the "perfection of civil society," the improvement of the moral and material conditions under which people lived.[45] But there was a strong counterstrain within enlightened discussion that was more skeptical of what the social expansion of knowledge would mean. Enlightenment discourse was full of dismissive references to the undereducated who imposed themselves on the public's attention.[46] As we saw in the preceding chapter, when Johann Schlosser looked at the efforts of enlightened popularizers, he saw not the spread of knowledge but its trivialization and degradation.

Furthermore, when Schäffer and Leysser wrote that "even farmers" could do natural research, they did not imagine a world in which the spread of knowledge would erase the difference between the estates. As Anthony La Vopa has shown, this period's discussions of merit and mobility remained well grounded within reference points provided by corporate values.⁴⁷ Schäffer, for example, was quite explicit about the division of labor he imagined: "The learned man must collect, observe, check, test, describe, identify, analyze, and put in order. The unlearned man must search things out, bring them to the learned man, and put them in his hands."⁴⁸ Johann Röhling described the agenda of his 1796 German flora in similar terms. He had written in German, he explained, in order to make the work accessible to beginners and collectors; these lesser enthusiasts could help (Latinate) learned botanists finish the project of creating a complete German flora.⁴⁹

Men like Schäffer and Leysser imagined a future where everyone still worked within their allotted sphere, but worked more effectively. Like the period's cameralist-trained officials, they wanted to perfect a social order that they saw as already largely well formed.⁵⁰ When Schäffer proposed in 1764 that farmers ought to take up natural history, he imagined that this would create novel exchanges between himself and members of other estates, but these novel exchanges would be set within stabilizing relationships of deference, adjusting the social order, not reordering it.

The anticipated deference was not always forthcoming. Learned cameralists had always faced challenges from practical critics, men who questioned whether or not the university educated, for all their impressive titles and book learning, really had any decent advice to offer farmers or miners or craftsmen. Andre Wakefield's unflattering picture of the many failures of cameralist officials suggests that these practical critics often had a point.⁵¹ From this perspective, the Latinate, university-educated man was just one species of *Naturkenner*, not the defining type. Some critics claimed that traditionally learned men, with their heads stuffed full of dead languages, could even hinder attempts to make natural knowledge useful.⁵² An 1805 article from the proceedings of the provincial Westphalian Society of Nature-Researching Friends offers one such example. The society announced that its proceedings contained articles chosen for their utility; its readers would surely find these works more pleasing than ones that were "purely speculative."⁵³ Among these useful contributions was an article bemoaning the continued use of Latin for treatises about nature. The people who most needed to hear about new discoveries, the anonymous author of this article wrote, were those who would use these discoveries for "production, artisanal industry, and cooperation in the maintenance of the state," and it was precisely these people who were often unable to comb through "dead languages" to find new information.⁵⁴

The anonymous critic suggested that extensive linguistic training might even be a handicap. The human soul could only hold so much knowledge, and men who had spent time learning dead languages might well have already used up a good portion of their available spiritual powers. Men whose minds were uncluttered with such antique paraphernalia were likely more innovative. Their "spiritual powers still [had] room" for new discoveries. "*Selbstdenker*," independent thinkers, were often more useful than men of erudition ("*Sprach-Gelehrten*"). "We have a thousand useful inventions that we owe to unlearned burghers, artisans and farmers," he wrote, "without which the erudite learned world [*die sprachgelehrte Welt*] would starve and freeze." Elegant Latinity did not necessarily make a man a good *Naturforscher*. "The amount of linguistic knowledge is not the only thing that defines the useful learned man." The useful learned man proved his worth through practical discoveries about natural objects and their powers, and a classical education did not necessarily aid in this cause.[55]

The Westphalian society's criticisms point to the dangers that the new public forums of the late eighteenth century posed to university-educated *Naturforscher*. Enlightened men had argued that farmers and artisans could be *Naturforscher*, too, in their own limited way. From that starting point, it became possible to claim, as the anonymous Westphalian critic did, that practical men who worked with nature every day might contribute more to the storehouse of natural knowledge than the traditional learned estate.

Natural Knowledge as Leisure

In the 1770s, figures like von Leysser and Schäffer had promoted natural history and natural philosophy as a form of rational leisure, and the success of that project also created new boundary problems for the learned world. To understand why this was true, we need to remember that in this period men with serious scholarly interests often described their natural philosophical or natural historical work as the product of their "*Mußenstunden*," their leisure time, since their scientific activity was something done outside the demands of their official state office or profession. The author of a natural historical bibliography from the 1780s explained that he had first become interested in natural history through the purchase of some shells, which then inspired him to study natural history "in his hours of leisure."[56] Friedrich Ehrhart bragged that he had devoted "those hours which many people use to laze around" to the study of botany.[57] Christian Pötzsch, in the introduction to his

Vom Granit und Siemit, said that he had devoted "the leisure allowed me by my official duties . . . to the study of mineralogy," and he flattered himself that he had come to a correct understanding of the subject as a consequence of his efforts.[58] The Society of Nature-Researching Friends in Berlin noted that its member Bernhard Feldmann had devoted "his hours of relaxation to the most useful natural philosophical activities" (Feldmann had been an avid collector of shells and coral).[59] The society's statutes also specified that it would not ask any of its members in "public offices" to do anything they could not do "in [their] spare time [*bloßen Erhohlungstunden*]."[60] Johann Reuß described his contemporaries as practicing natural history as a "pleasant and useful leisure activity."[61] Describing science as a form of leisure activity did not indicate a clear social or epistemological line, as it would later in the nineteenth century. Indeed, well into the nineteenth century, even people with serious interests in the natural sciences received a great deal of their training through family friends, colleagues, and independent reading.[62]

When F. W. von Leysser praised natural history as a useful leisure pursuit, he held this to be true for a variety of reasons. The study of nature drew one closer to God, and it was a distraction from less morally salutary ways one might spend one's free time. Beyond the benefits to an individual's character, however, the study of nature also put one's leisure to work for the general good.[63] Not everyone who picked up a natural history handbook deserved the notice of the wider learned public, of course, but this initial act of self-cultivation was part of a continuum that might lead to publication and a recognized place among *Kenner*, or experts. Useful leisure could provide the raw material for serious intellectual activity.

That was the guiding assumption, in any case, behind many introductory scientific handbooks. Handbooks offered to initiate new readers into the basics of natural philosophy, chemistry, and natural history. Unlike other genres of popularizing scientific writing, their aim was to introduce their readers to scientific activity, not just provide her or him with information. They offered to guide readers through the early stages of what might later become a serious learned avocation. These kinds of works multiplied rapidly in the last third of the eighteenth century. For example, the number of botanical textbooks published in German-speaking cities grew every decade between 1770 and 1810. Eighteen appeared in the 1770s, twenty-eight in the 1780s, thirty in the 1790s, and forty-seven in the first decade of the nineteenth century; this steady growth is even more striking considering that war and invasion disrupted the German book market in the 1790s and 1800s.

Between 1770 and 1800, the primary language of these works also shifted from Latin to German.[64]

Local, regional, and national floras were also often written to be accessible to beginners. Many of the German national floras published in this period offered both catalogs of German plant life and introductions to botany as such, and their title pages and prefaces proclaimed the desire to reach readers from a variety of social stations.[65] Linnaeus's sexual system of classification, which offered a quick, easily learned introduction to the plant world, proved a useful tool in this regard. It was a popular choice for botanical handbooks, even when authors recognized the philosophical superiority of the more complex natural systems proposed by Antoine Laurent de Jussieu and others.[66] In addition to introductory handbooks and floras, beginning students also had access to other kinds of products to aid in their education. The Hanoverian botanist Friedrich G. Ehrhart offered a herbaria of preidentified plants, touting it as "one of the first and best" of the "diverse aids for familiarizing oneself quickly and reliably with plants."[67] Beyond the growing scale of the book market, social developments also expanded the reach of printed material. Through the proliferation of reading circles and lending libraries, books also became less difficult to acquire for a wider range of people.[68]

Natural history and natural philosophy's heightened prominence in print culture offers the clearest metric of their expanding public presence. As the German book market grew in the final third of the eighteenth century, the natural and practical sciences not only expanded apace, they gained a larger market share.[69] In 1740, 3.31 percent of the titles offered at the Leipzig book fair dealt with the study of nature. This percentage had grown to 6.20 percent in 1770 and to 7.12 percent in 1800.[70] The periodical landscape was similarly rich. In the period between 1766 and 1780, German-speaking Europe had 236 different periodical publications dealing with the natural sciences, medicine, and the practical sciences. The sheer number of titles can be somewhat misleading as a sign of health and expansion; one reason the number of titles was so high was because most journals were fairly short lived. Even adjusting for this fact, however, German-speaking Europe still boasted a more active field of scientific and technical periodicals than France or England.[71] The natural sciences also had an important place in the Enlightenment's more general periodicals, representing about 25 percent of their content by one estimate.[72]

Within natural historical disciplines, some knowledge of Latin remained important, since it was the standard language used for scientific descriptions of plants and animals (and still is today).[73] The need for Latin placed certain obstacles on the social expansion of natural history, but many authors argued that these obstacles could be at least partially overcome.[74] For those who had missed out on childhood training in the classical languages, published aids offered to help bridge this educational gap. At least six different Latin-German botanical dictionaries appeared between 1780 and 1820.[75] Many botanical handbooks came with a terminological introduction, instructing their readers on the proper Latin terms for different plant characteristics, from "veined" or "hairy" to "blue" and "bald." J. J. J. Koch's botanical handbook advertised in its lengthy title that it offered "the correct pronunciation of the Latin plant names," protecting the non-Latinate novice from embarrassing mispronunciations.[76]

Learned natural knowledge also had other connections to the period's emerging culture of leisure and consumption. Alongside things like public concerts, natural philosophical lectures offered diversion to the late eighteenth-century urban public. Scientific lecturing has been much better studied in France and Britain, but one can find similar activities going on in the German states.[77] In Dresden, a young doctor of theology named Andreas Tauber held mineralogical lectures. The physician Wilhelm Friedrich Gerresheim "satisfied the great fondness of the inhabitants for natural history, physics, and chemistry through lectures, for which he put together a beautiful natural history cabinet." The previously mentioned owner of the *Löwenapotheke*, apothecary Carl Gottfried Bünger, lectured on experimental chemistry. Some of Dresden's local lecturers even published textbooks based on their lectures, following a habit common among university professors. A local preacher named Winkler published a textbook on experimental philosophy, and another on geography and natural history, both based on lectures he gave in Dresden. Carl August Blöde, a state official who studied in Freiberg and Leipzig, published his lectures on Franz Joseph Gall's theory of the brain.[78]

Itinerant performers also came through town hawking natural philosophical amusement. Chemical and electrical displays were particularly popular in German-speaking Europe, as they were elsewhere, and a handful of people seemed to have been able to earn a living presenting philosophical marvels to the urban public.[79] In the early 1810s, Gottfried Reichard, who would later settle on the outskirts of Dresden, offered chemical lectures in the fashionable Hotel de Pologne. Reichard and his wife, Wilhelmine, also staged demonstrations of hot air balloons, one of the most popular spectacles of the period. The Reichards'

CHAPTER 2

Dresden ascent was one of the young couple's most dramatic. They arrived in the city after celebrated appearances elsewhere but had to cancel two performances due to bad weather. On their third attempt, the Reichards faced a restless audience and another windy day. Wilhelmine braved an ascent despite the inauspicious weather and came tumbling back to earth just outside of Tharandt (her injuries were luckily not serious).[80] Alongside hot air balloons, complex automata were also popular objects for public display. A number of craftsmen exhibited machines in the Saxon capital in the later eighteenth century, and these demonstrations of technical ingenuity were often billed as philosophical or *"wissenschaftliche"* events.[81]

For elite experimental philosophers like Georg Lichtenberg, these public lecturers were dangerous frauds; they compromised the dignity of natural philosophy by catering to the credulity of the ticket-buying public. At least based on the evidence from Dresden, there is little to suggest that complaints like Lichtenberg's did much to dampen public interest in the traveling demonstrations of chemical and electrical lecturers. The aforementioned Reichard delivered his chemical lectures at one of the most fashionable addresses in town, and the hefty entry fee charged suggested that his public was a well-heeled one.[82]

In addition to events held within city walls, natural history's popularity also stemmed from the countryside's rising importance as a site of urban leisure. By the early nineteenth century, the countryside around Germany's larger towns had taken on new importance for urban residents, and the practices of natural history—collecting and identifying natural objects—helped structure the use of the countryside as a space for relaxation and cultivated amusement.

Late eighteenth- and early nineteenth-century educated Germans saw the experience of nature as both emotionally and morally salutary, and this belief infused meaning into time spent in the countryside. The value placed on contact with "nature" in this period came in part from Enlightenment pedagogical writings, especially the Rousseauian tradition that idealized rural life as healthy and virtuous.[83] The popularity of the English landscape garden also served as a conduit for these ideals, as did landscape painting, a once low-status genre whose prestige had risen a great deal by the early nineteenth century.[84] All of these activities belonged to the formative experiences central to cultivated or *"gebildete"* selfhood. Contact with "natural" spaces (which were in fact often carefully cultivated) provided a stage set for socializing that was cast as emotionally authentic and unforced, free of the formality of life within the city. The capacity to appreciate nature was one of the things that separated the cultivated or educated from the rest of the *Volk*. Proper sensibility for nature also crafted emotional bonds among the cultivated and became a stereotypical foundation for friendship. When two people faced the beauty or sublimity of the natural world and experienced common feelings, it was a widely recognized mark of supposed spiritual kinship.[85] Natural history gained additional meanings against this broader backdrop, but it also imparted emotional and moral meaning in its own right. Among the most ardent friends of nature, after all, were naturalists themselves.

Several sorts of activities brought city dwellers into the countryside with greater regularity by the early nineteenth century. The most important of these was gardening. Between 1750 and 1830, it became common for more affluent middle-class families to own a garden, a small plot of land at the city outskirts where the family might spend part of the summer months.[86] Regular walking also became part of middle-class self-care in the early nineteenth century, and the city perimeter was frequently the forum in which this activity took place. In addition to its salutary benefits, a daily stroll offered a chance to socialize, to see and be seen. The removal of town walls, a project already under way in some places during this period, freed up spaces for promenades around urban centers.[87] Regional tourism, centered in particular in the striking mountainous landscapes of areas like the Harz and the various pseudo-Switzerlands of the Germanies (the Fränkische Schweiz, the Sächsische Schweiz), was also becoming more common.[88]

Dresden's reputation was particularly closely linked with its surrounding countryside. A quick look at the city's immediate rural hinterlands illustrates the degree to which new forms of leisure drew urban residents out into the landscapes surrounding the city.[89] Beginning in the late eighteenth century, images of Dresden typically included the surrounding countryside as a backdrop. Benjamin Gottfried Weinart's 1777 guide to the city described Dresden as "cradled" by its countryside. "All the attractions of Nature have taken this city into their lap," he wrote. "Forests, at an appropriate distance, mountains, some near and some far, surround the entire area and offer the eye a charming amphitheater."[90] By the early nineteenth century, travel guides to Dresden usually included the nearby countryside, as well as information on the slightly farther afield Sächsische Schweiz.[91] While the royal art collections were perhaps the city's most famous attraction, a stop to admire the landscapes around the nearby village of Tharandt was also a standard part of a visit to Dresden. On his last free afternoon in the city, the young Heinrich von Kleist, having spent the morning at the famous picture gallery, chose to go to Tharandt instead of visiting other standard sites within the city.[92] The Plauen Gorge, famous for its cliffs, was a well-established destination for city natives and visitors alike.[93]

As it became a frequented social space, the local countryside also became the muse of a staggering amount of cultural production. Between the years of 1770 and 1830, over one thousand works of visual art took the Plauen Gorge as their subject.[94] The early Romantics celebrated the emotional effect of the nature around Dresden, where Friedrich Schlegel claimed to have "seen Nature in a more beautiful form for the first time."[95] To the northeast of the city lay the Dresden Heath, the subject of several works by the Romantic landscape painter Caspar David Friedrich in the 1820s.[96] On a less monumental scale, other local artists such as Carl August Richter, Adrian Zingg, and Johann Friedrich Wizani churned out prints of Dresden's surroundings for display in middle-class homes. These modest etchings showed a countryside interpenetrated by court and town. Rural laborers tended animals, harvested crops, and gathered fruit, undisturbed by the strolling city dwellers who crossed their paths. Urban men and women walked along country lanes, admiring striking views; the men pointed out items of interest to their female companions, who listened intently.[97]

The sharp line between city and countryside continued to blur in the early nineteenth century. Dresden began to remove its fortifications in 1810, creating more space for socializing around its perimeter. The walls were partially rebuilt in 1813, and then the final round of removal began again in 1817. Much of the resulting space was used for gardens and walking paths.[98] The *Grosse Garten*, a large tract of land that had originally been designed as a royal hunting park, was another possible destination.[99] There were also a number of gardens inside the city. An 1829 guide devoted an entire section to Dresden's gardens, listing eight large public gardens as well as a number of smaller private ones.[100]

Many of the basic practices of natural history—identifying individual natural objects and perhaps also collecting and arranging them—became part of the perceptual habits that guided an educated urban dweller's interaction with the countryside. Gardening offers perhaps the most direct example of this relationship. In the early nineteenth century, gardening came to be informed by what Abigail Lustig has called a "botanical aesthetic." Whereas the English landscape garden had emphasized the expansive sweep of a view, the connoisseurship of individual plants played a much greater role in nineteenth-century gardens. The flow of new species from the Americas, Africa, and Asia into European gardens added to this fascination with botanical variety. For the educated viewer of this period, the multitude and diversity of the plants on display became one of the criteria by which one evaluated the beauty and value of a garden.[101] As Dresden emerged as an important commercial gardening center over the first half of the nineteenth century, its commercial gardens offered an unprecedented variety of species, one far in excess of what would be available by the end of the century.[102] The ability to appreciate this diversity had been carefully cultivated in the gardens of the preceding decades. Handbooks on garden design sometimes included directions for setting up a natural history cabinet; later memoirs often also recalled childhood natural historical instruction in the family garden.[103]

One can also find traces of heightened natural historical perception in landscape painting, as well. The Dresden Academy of the Arts, the leading center for landscape painting in Germany at the time, took its students into the countryside to sketch from nature. This practice had been part of standard artistic practice for a while, but a new concern for the specific characteristics of plants appeared in the 1780s, when academy professors like Adrian Zingg directed their students' attention to botanical details.[104] Paintings depicted natural particulars with greater precision than had been common earlier, paying more attention to things like the specific shapes of tree leaves.[105]

The union of scientific and aesthetic perception was a cause célèbre of the early Romantics, but in a less sophisticated form, it was also a cultural commonplace by the early nineteenth century. As we saw in the preceding chapter, naturalists had long accorded an aesthetic dimension to their activities. Natural historical texts about Dresden and its surrounding countryside often assumed that naturalists had a heightened appreciation for the beauty of nature. One guidebook described the Plauen Gorge as "without a doubt the most noteworthy and striking of all the many natural beauties" around Dresden. It was an "enchanting place for poets, artists, natural researchers—in short for any friend of the potent charms and graces of the beautiful and the sublime."[106] The *Naturforscher*, the poet, and the artist all shared a sensitivity to the beautiful and the sublime in nature, and catalogs of *naturalia* sometimes appeared side by side with more lyrical descriptions of natural landscapes. In his work *Der Plauische Grund bei Dresden, mit Hinsicht auf Naturgeschichte und schöne Gartenkunst*, Wilhelm Gottlieb Becker called on a variety of resources to describe the Plauen Gorge. The first half of the volume contained illustrations and verbal descriptions of Plauen's most famous views; a description of the area's geology and catalogs of plants and insects filled the rest of the work.[107]

CHAPTER 2

Collecting plants, insects, or minerals could be part of an excursion, whether a daily stroll or a lengthier hike, with these natural objects serving as mementos of a pleasurable trip. One published description of the Plauen Gorge, for example, promised to instruct "friends of nature" on finding "pleasant mementos" to carry home from their visit.[108] Making botanizing or mineralizing part of a tour through the countryside was not a universal habit. It was, however, not a rare pursuit. The first travel guide for Saxon Switzerland, Wilhelm Lebrecht Götzinger's *Beschreibung der Sächsischen Schweiz*, also included natural historical information, alerting travelers to the insects, plants, and minerals that could be found in the region.[109] Part topographical description, part guidebook, Götzinger's work was first published in 1804, with an expanded second edition in 1812. In the introduction to the second edition, Götzinger noted that he had made special efforts in his revisions to make the work useful for "friends of nature and history, and especially for botanists, entomologists, and mineralogists."[110]

Natural historical collections and gardens were also standard attractions for a traveler. In Freiberg, home to the Saxon mining academy, the visitor could see the late professor A. G. Werner's collections. At the Wettin royal palace of Pillnitz, the "botanical treasures" of its gardens were an important draw. Travel guides also noted interesting geological features. The area around the town of Meissen was described as noteworthy for its coal deposits, for example.[111] Travelers could also take *naturalia* home as mementos even when they did not collect them themselves. A local bureau in Freiberg offered visitors fossils and preorganized collections for purchase.[112] Mineralogical specimens could also be bought in Dresden, where for a while E. F. Liebenroth assembled suites of minerals for sale.[113] One of the royal gardens in Dresden also sold its rare plants, publishing a yearly catalog of its offerings.[114]

The subscription list for Wilhelm Becker's book on the Plauen Gorge, a large and lavishly illustrated volume, provides some insight into the mixed noble and middle-class group at the upper end of the market for the period's travel guides and topographical descriptions. The sixty-one advance subscribers listed in Becker's preface included three women (all aristocrats, a princess and two duchesses). Close to half of the subscribers were nobles (twenty-seven out of sixty-one, or 44 percent). State officials, both bourgeois and noble, were also well represented, as they had been in the rolls of learned and patriotic societies. A little over a third of Becker's subscribers had some court or higher administrative title (twenty-two out of sixty-one, or 36 percent). Most of the other subscribers were not identified by occupation; among those whose profession was listed, there were (not surprisingly) six book merchants, as well as an apothecary and a landscape painter.[115]

Books that appeared in smaller, cheaper formats like the aforementioned guide by Götzinger had a wider audience. With its modest dimensions, Götzinger's work was meant to be carried along on a trip, and he offered a number of practical tips for travelers venturing into Saxon Switzerland. Though regional leisure travel was less common than it would become later in the nineteenth century, by the 1810s and 1820s the Saxon countryside already had a regular influx of urban visitors.

The young drawing instructor Carl August Friedrich Harzer took regular trips into Saxon Switzerland in the early nineteenth century, and his personal journal provides an example of how natural historical knowledge informed common understandings of the countryside, even for less affluent members of Dresden's reading public. Harzer had been orphaned at age fourteen, and after attending the Dresden Academy of Art, he made a living as an illustrator and teacher. He began taking trips to Saxon Switzerland in his early twenties. There is no indication that Harzer had serious natural historical interests as a young man; nonetheless, his journal notes recorded a general awareness of what a naturalist or a cameralist would have found interesting in the landscapes through which he passed. In his particular case, this more casual natural historical eye eventually evolved into something more serious. By the 1820s, Harzer had developed an interest in entomology, which then became part of the ritual of his trips; in the 1830s, he founded a natural historical society in Dresden, discussed in chapter 5. His early journals, however, offer an example of the more diffuse ways in which natural historical knowledge was interwoven into early nineteenth-century culture.

Harzer took his first trip to Saxon Switzerland with several young male friends—another drawing instructor, a court musician, and the sons of several local petty bureaucrats. Many of his journal entries recorded comic episodes; part of the fun of such a trip was making loud noises in places with good echoes and laughing at your friends when they tripped or ended up with blisters. At one point, overindulgence in schnapps left all the young men nauseous and with horrible headaches.[116]

The young drawing instructor also stopped periodically to sketch views of natural or cultural landmarks, and he walked through the Saxon countryside with a keen eye for points of natural historical and economic interest. He took notice of the clover fields he passed, which were the fruits of the last few decades' agricultural improvements.[117] He also knew to comment on the basalt near the town of Stolpen (basalt had been the focus of intense learned controversy between rival mineralogical camps in the late 1790s).[118]

Harzer was well schooled in the appropriate emotional formulas for socializing in the countryside. After an encounter with another party of travelers (initially strangers), he wrote that they did not remain strangers long, since they had "the same purpose and the same feelings." "Soon all courtly [*höfische*] and urban [*städtische*] etiquette was banished and gone." Together in nature, people met one another with natural ease. The group of young Dresden men encountered similar parties throughout their trip—acquaintances from Dresden, a group singing an aria through a gorge, another party of both men and women they joined on a boat ride to Schandau.[119]

Harzer's journal depicted a countryside adapting to provide services to this new influx of urban tourists. He paid local guides to lead him and his friends on hikes, using one local man, Gottlieb Petzschke, on more than one occasion over the course of several years.[120] Harzer noticed with pleasure when rural children met his activities with incomprehension, remarking in his journal on a "sharp girl, of whom we asked the way," who "did not want to believe that we did not know it, and [who] said she was sure 'we would find the way soon enough with our glasses.'" A local boy who had accompanied him on a hike, Harzer wrote, seemed to think that the Dresden teacher "had fallen from the moon." The boy told him "among other little scientific [*wissenschaftlichen*] tid bits" that "the bilberry vines had to grow before they could get big."[121]

The amusement Harzer took in such episodes certainly came in part from the cultural distance they helped to mark, the differences between the local, limited world of rural people and his own sense of his cultivated mobility. His trips to Saxon Switzerland were important landmarks in his life, and he took at least six between 1808 and 1817. He described them as something that he "worked, saved and looked forward to for the entire year," and he understood these excursions as indications of his status as a *"gebildeter"* man. This conviction was expressed most clearly on the one occasion in which nature failed to provide the promised harmony among his travel companions. On his 1817 trip, Harzer got into an argument with his friends and had to go home alone, and he described his fellow travelers' bad humor as the result of their lack of *Bildung*. He bemoaned having his pleasure ruined "through the ingratitude of uncultivated [*wenig gebildeter*] people."[122]

At the beginning of the nineteenth century, part of being *"gebildet"* was knowing about basalt and clover, possessing the ability to see the countryside with an eye informed by works of natural history and topographical description. When educated Germans looked at rural landscapes, elements of the natural and cameral sciences directed their attention, providing common perceptual reference points.

Conclusion

In 1821, the medical professor Heinrich Ficinus, the son of the Andreas Ficinus with whom we started this chapter, published a flora of Dresden. In its preface, he discussed his local botanical predecessors, listing several published precursors to his flora, including a 1799 work on the flora of the Plauen Gorge and an 1806 flora by a local surgeon. He also mentioned a local collector who had explored the area thoroughly but never published his findings. This local military officer, a Captain Löber, had been "very experienced in the mathematical and philological sciences," and his passion for the study of nature had eventually led him to give up his military post and live instead "in philosophical quiet and independence" on the resources of his small pension. Löber had explored the Dresden area and collected specimens of the plants he found there, which he then shared with a number of "nonlocal friends [auswärtigen Freunde]." The one thing Löber had not always done, however, was share his findings with other local botanists or publish his results, and as a consequence, many of his discoveries had not become more widely known.[123]

Not to share your results with a broader public violated one of the central values of Enlightenment culture; nonetheless, it is notable that Ficinus could still write in 1821 that Captain Löber "should certainly not be missed from the history of Dresden botanists." Amid the late eighteenth century's impressive expansion in published exchange, it is easy to overlook the ways that more localized, embodied trappings of learned distinction continued to have force, even in the absence of publication. Of course, Captain Löber's knowledge of local plants was hardly enough to earn him immortal fame in the annals of learned Europe. But neither was his collecting activity and extensive correspondence completely without purchase, at least in a local context. Löber's correspondence and collecting still gave him some right to memorialization, and Ficinus compared the retired military officer to an early eighteenth-century botanist, "the famous Ruppius," the author of a flora of Jena whose life had been portrayed by Albrecht von Haller.[124]

Though Ficinus was perhaps generous to include such a man in his introduction, even in 1821 the history of botany in Dresden was not necessarily just a history of knowledge that had ended up in print. Expertise that had been tended through collecting and broadcast through networks of correspondence still seemed worth mentioning. By the early nineteenth century, the republic of letters could swell to an impressive size if one employed these older standards of learned distinction. The figure of Captain Löber serves as a useful reminder that as the community of authors had expanded, other marks of learnedness had spread as well. Not only was published authorship a more common achievement; other, more physically localized marks of learnedness (serious collecting, for example) were also much more widely distributed. Keeping these local, embodied aspects of learned reputation in view is essential to understanding the tensions that plagued the learned world around 1800; it is also essential to understanding the solutions the coming generation proposed to these tensions.

As long as learned credit could be issued for activities like collecting, the growing popularity of the natural sciences—the explosions of handbooks, of casual butterfly collections, and public chemistry lectures—had very real implications for the learned world. Learned naturalists had always complained about the bad work done by *"Halbkenner,"* or half-educated pretenders. By the early nineteenth century, these complaints and anxieties had become even more acute. Two publications on entomology can illustrate the change in tone. In a 1773 article in the Berlin journal *Mannigfaltigkeiten*, Johann Samuel Schröter noted that "one finds many collectors [of butterflies] here, people who otherwise do not collect anything," but who arranged butterflies under glass as decoration for a room. Schröter noted that such people often arranged their collections badly, and he offered them some friendly learned advice on how to correct their mistakes.[125] Three decades later, Ferdinand Ochsenheimer noted similar activity in Saxony with less sympathy. Ochsenheimer was a famous stage actor, but he had also spent lots of time in the countryside chasing butterflies. He was not alone. "Entire armies [of butterflies] are captured and murdered in Saxony every year," Ochsenheimer wrote, "without serving anything other than the demands of a passing fancy [*Liebhaberey*]." Most collections were nothing more than "a bright array of colors" without scientific order, their unscholarly arrangements a reflection of the partial and haphazard education of their owners.[126]

Given his reference to "passing fancy," one might assume that Ochsenheimer was criticizing people like those Schröter had mentioned in the 1770s, collectors whose interest in butterflies was limited to the insects' decorative charm. But Ochsenheimer went on to complain that many of these ill-informed collectors had also published books on entomology. Where learned work ended and mere *Liebhaberey* began was by no means always clear. Ochsenheimer thought that many people who had authored entomological work had no business sharing their thoughts with the broader public, and the actor presented himself as a reformer and a guide to this motley crew of collectors and bad authors.

Despite the sharp tone he used to describe entomological dabblers, Ochsenheimer nonetheless did not write as an expert speaking to a lower class of permanent amateurs, and it would be a mistake to understand his complaints simply as an example of nascent professional consciousness. He was an actor, after all, and he constructed the stage from which he spoke out of his claims to learned expertise, claims based on his own personal history of careful study, wide reading, and assiduous collecting. He spoke from a stage, in other words, constructed out of the materials available through the republic of letters. Ochsenheimer wanted to use his expertise, he claimed, to reform the way that other, less well-trained entomologists went about their work. His book was inexpensively produced, he said, and included full descriptions of all the species found in Saxony. This was to aid those who could not afford an extensive library, ensuring that they struck out on the right path in their studies. As a reformer, he hoped to bring Saxony's wandering butterfly enthusiasts back to the straight path of genuine scholarship.[127]

Ochsenheimer's work illustrates two things: the learned public was increasingly crowded by the early nineteenth century, and under considerable strain; at the same time, many of its fundamental assumptions remained unchallenged. The persistence of the learned world, as both an organizing concept and a set of social practices, speaks against seeing 1800 as a sharp break between a specializing science and public culture more broadly. Nor is it satisfying to frame this relationship in terms of paradox, a sleight of hand where seemingly publicly available knowledge was in practice anything but. Generally speaking, eighteenth-century Germans saw no paradox at all. In natural history and natural philosophy, the *learned* public, not the public in general, had the right to judge a work's quality. The truly thorny question was a somewhat different one: who belonged to the learned public? The next generation's answers to this question are the focus of the following three chapters.

THREE

Defending Learned Dignity

In the late eighteenth century, a man could join a group like Halle's Nature-Researching Society and list his specialty as "mining." By 1830, this was no longer possible.[1] A few decades into the nineteenth century, learned societies for natural research had a much more distant relationship with practical economic improvement than they had had a generation earlier. Under the broad enlightened cause of useful natural knowledge, new systems for assessing the value of agricultural land got discussed in the same forum as newly discovered foreign species of frogs. Farmland and frogs were both "treasures" to be explored. Improving farming was a clear practical good, while newly discovered frogs offered yet another example of the variety and intrinsic interest of nature.[2] In the nineteenth century, this enlightened amalgam of concerns broke apart. During the Enlightenment, *Naturforscher* had already drawn a *formal* distinction between natural history and natural philosophy on the one hand and the practical sciences of nature on the other. In the nineteenth century, that formal distinction took on flesh and blood. It became a principle used in the social organization of knowledge, not just an intellectual divide.

In place of useful natural knowledge, elite *Naturforscher* now rallied around a new collective aim: the creation of a general science of nature. This general science of nature was still ultimately relevant to practical concerns, but *Naturforscher* insisted that if narrow practical aims called the tune to which general science danced, the latter would never reach its full potential. The natural sciences also needed to be pursued together, as a unified collective enterprise. Different scientific disciplines had to remain in conversation with one another, because a scientific [*wissenschaftlich*] understanding of nature depended on keeping more general horizons in view. When looking at any one part of nature, the learned *Naturforscher* always needed to keep an eye on the whole.

One could give a short and simple intellectual explanation for the changes I just described. In the early nineteenth century, German natural researchers joined other educated men in celebrating the ideal of pure *Wissenschaft*, which was deeply disdainful of enlightened utilitarianism. The Romantic generation also had an avid taste for unifying systems; little wonder, then, that unity became a major theme in the organization of intellectual life.[3]

Both of these statements offer a partial explanation for these changes, but not a completely satisfying one. Recent work on the history of German science has often questioned just how powerful the ideal of "pure science" actually was in shaping scientific institutions and has often accorded it very little motive force in other settings.[4] An older historiography once portrayed *Naturphilosophie* as the dominant force in Germany around 1800, but we now have a much more detailed picture of the complex philosophical commitments and empirical practices that shaped German science in these years. German researchers certainly did not share one (or even two) versions of what a "general science of nature" ought to be.[5]

This chapter and the following two examine why this idea became a compelling collective cause nonetheless. The unifying dreams of the *Naturphilosophen* and other Romantic *Naturforscher* played a role in shaping this collective cause, but they cannot take full credit for its widespread success. Granted, some of the most ambitious philosophical system builders of this period were also its most ambitious social organizers. The image of the lonely Romantic genius, misunderstood by the philistine masses, was as popular among *Naturforscher* as it was with their literary counterparts, but in practice, *"Einsamkeit und Freiheit"* ("solitude and freedom") was hardly the rule of the day. Most leading figures in German Romantic science were tirelessly social.[6] Henrik Steffens led the Natural Scientific Section of Breslau's Silesian Society for the Culture of the Fatherland (Schlesische Gesellschaft für vaterländische Cultur).[7] Carl Gustav Carus played a prominent role in the local scientific societies of his native Dresden. Lorenz Oken, famously, was the guiding force behind the Society of German Natural Researchers and Doctors.[8] His fellow *Naturphilosoph* Christian Nees von Esenbeck initiated an effort to revive the Leopoldina, the official academy of the now-defunct Holy Roman Empire.[9]

There are good reasons to be cautious, however, about assuming that a commitment to a general science of nature required individual researchers to subscribe to any one particular brand of the same. Men like Oken and Nees von Esenbeck saw themselves as members of a scientific avant-garde. By their own assessments, they had only marginal success in winning large numbers of long-term converts to their ambitious intellectual projects.[10] The new *Naturphilosophie* in particular, with its heavily metaphorical language and ambitious synthetic scope, was not easily disseminated, and its advocates sometimes openly acknowledged as much. In his introduction to his *Grundzüge der philosophischen Naturwissenschaft*, for example, Henrik Steffens made no apologies for the esoteric, aphoristic quality of the book. "It goes without saying that a work such as this cannot be generally accessible [*gemeinverständlich*]."[11] In another context, he described his local intellectual circle as "the Freemasonry of our day." Unlike the Freemasons, however, the group possessed secrets that, because of their difficulty, no longer had to be protected by "external machinations." The intellectual elite would not be understood by the masses even when it spoke in public. "We can present [our secrets] completely openly and to everyone; they will still remain secret."[12]

Despite their fervent rhetoric about the synthesizing power of the great individual intellect, men like Carl Gustav Carus and Lorenz Oken did not actually craft intellectually consistent communities, infused with the spirit of their new philosophical natural science. If one looks at the learned societies of the period, one sees that Romantic geniuses spent a lot of time talking with pedants. In Dresden, for example, Carl Gustav Carus joined with local medical professors who had little sympathy for his Goethean brand of morphology (shortly after coming to Dresden, Carus had only just prevented one of his fellow professors from devoting his keynote speech at the medical academy's opening to a critique of Goethe's views on natural science).[13] Oken's national Society of German Natural Researchers and Doctors can hardly be described as an institutional vehicle for the spread of his *Naturphilosophie*.[14]

However contested and controversial a general science of nature might have been in any of its particular forms, as a social project it worked well. To further understand why, we need to look more closely at how a broad set of shared status anxieties drew together researchers with divergent philosophical and disciplinary commitments into a single, roughly articulated common front. As we saw in the preceding chapter, the learned world had become quite wide and unwieldy by 1820. In rallying under the new banner of a general natural science, university-educated researchers sought to clarify what separated the true learned expert from the half-informed enthusiast, and what separated true natural knowledge from a mere isolated bit of practical know-how.

This chapter explores the second of these two issues; it examines why separating theoretical from practical intellectual forums came to make so much sense to university-educated German researchers in the first third of the nineteenth century. Over these decades, learned societies and economic societies, so closely intertwined in the Enlightenment, largely went their separate ways. The break between these two kinds of groups was not complete (economic societies still often named professors as honorary members, for example), but by 1830 their divergent trajectories were clearly laid out.

The most important cause of this split was the roaring success of the Enlightenment's practical sciences, and their continued consolidation in the post-Napoleonic era. In referring to the "success" of enlightened practical science, I am using a crude measure: not the actual impact that these fields might have had on economic activity but their formidable presence within print culture and associational life.[15] These fields also continued to attract state support in the early nineteenth century, as teaching in the cameral and practical sciences continued to expand at both universities and more specialized academies.[16]

By 1830, two previously interwoven projects—general *Naturforschung* (natural research) and the pursuit of practical natural knowledge—had became distinct social enterprises that drew on two differently constituted publics. When enlightened patriotic societies survived into the nineteenth century, they generally evolved into an alliance between improvement-minded bureaucrats and various men of practice—landowners, estate managers, and a miscellaneous group of men from other skilled occupations. University-educated men from outside of the bureaucracy, while not entirely absent, no longer made up a sizable minority within these forums. Nineteenth-century agricultural societies still owed a clear debt to learned traditions: they published proceedings and their members presented papers. But they became sites where people with a direct stake in agricultural matters met to wrestle with practical concerns. And these groups spread like wildfire. In addition to the handful of enlightened societies that survived the Napoleonic era, there were numerous new societies founded in the 1800s and 1810s; by 1820, Prussia alone had 15 agricultural societies; by 1852, it had 361.[17]

In the face of this massive increase in intellectual activity on the part of other social groups, many elite researchers asserted that what *they* really cared about was a general science of nature rather than anything so mundane as petty material benefit. Set in context, this claim looks less like a noble renunciation and more like elite *Naturforscher*'s acceptance of their limited power to shape practical economic discussion. But that would be to set too modest a cast on the aim of such rhetoric. A general study of nature might be pursued for its own sake, but it was supposedly essential to practical life all the same. The only way to gain a true understanding of nature, elite natural researchers argued, was to study it in its entirety. This was hence the only *real* way forward in the improvement of practice as well. Practical men who studied practical problems knew only bits and pieces of nature; only the learned man knew nature as a whole.

One can point to numerous examples of how the tumultuous political developments of the early nineteenth century helped drive forward reforms within the republic of letters during these decades. The humiliations of the Napoleonic Wars inspired Lorenz Oken to give a rousing speech on how natural history could contribute to the renewal of the German nation. The Prussian educational reforms prompted the Society of Nature-Researching Friends in Berlin to revise its statutes. In the late 1810s, a group of professors began planning to bring back the Leopoldina, the scientific academy of the Holy Roman Empire; their conversation was one of many about how Germany could move forward in the face of the empire's demise. The upheavals of war and territorial consolidation prompted significant internal political reforms throughout the German states, and these same developments helped focus attention on the need for reform within the learned world. The specific reforms that German *Naturforscher* suggested, however, were primarily responses to tensions peculiar to the learned world itself, not reflections of more generalized kinds of social or political anxieties. At least in its general and broadly shared outlines, this period's concept of a unified natural science did not emerge along tracks parallel to any single set of political developments. It was not the product of the Reform Era, nor a tool of reaction. It was an attempt to come to terms with problems that had troubled the learned world with increasing frequency since the late eighteenth century, and it rose to prominence as part of a defense of learned dignity. In an age where practical men giving practical advice appeared to be popping up faster than mushrooms, and gaining the ear of high-ranking officials to boot, learned dignity seemed in dire need of defense.

Learned Dignity and Practical Expertise

In 1809, the young naturalist Lorenz Oken marked the beginning of his zoology professorship at the University of Jena with a lecture entitled "On the value of natural history, especially for the education of the Germans." In celebrating the value of natural history, Oken followed in a long line of Enlightenment commentators. He rejected, however, the preceding generation's favorite justification for the pursuit of natural knowledge:

> To speak about the uses of natural history may be in our times useless talk, but the more people have praised its uses, the more they have degraded it, and the less they have recognized its true worth—when I look around at writings that praise natural history, all that I find are lowly assessments of the profits that it brings when the artisan brings along a few pieces of natural historical knowledge to his workshop.[18]

Jena had been the scene of a devastating Prussian defeat just three years earlier, and Oken spoke with an urgency fueled by the political and military blows of the Napoleonic Wars. Natural history offered a path to German national renewal, he thought, but only if cultivated in the right form. If proper hierarchies of value were not reestablished—if natural researchers did not again make the production of truth, rather than the advancement of material prosperity, their first aim—all that would be left, he warned, would be a "dyer's, forester's, and manufacturer's natural history." Earlier naturalists' love of utility, and their tendency to "rip out fragments of natural history and treat them as independent sciences," represented "the flight of all truly learned sensibility." "The learned man does not exist," Oken pronounced, "only to be demoted to an artisan."[19]

When Oken warned that "general" natural history might sink under the weight of its various technical subbranches, the image was not an entirely unreasonable one. The Enlightenment had produced an enthusiastic audience for practical knowledge, a public that continued to grow throughout the first third of the nineteenth century. Natural history and chemistry were popular sciences; their practical cousins were more popular still. A complete bibliography of eighteenth-century works in the various branches of cameralist science includes over fourteen thousand publications, and the pace of publication on these topics did not slacken in the first decades of the nineteenth century, it quickened.[20] By the 1830s and 1840s, individual branches of the practical sciences boasted impressive specialized bibliographies of their own. Wilhelm Engelmann's 438-page *Bibliotheca oeconomica* covered literature on agriculture, animal husbandry, cooking, and common rural industries like the distilling of spirits.[21] C. P. Laurop's more specialized 1830 *Handbuch der Forst- und Jagdliteratur* was 432 pages long and included 5,648 entries.[22] Johann Heinrich Dierbach's 1831 handbook covering all of "general botany," in comparison, filled only 251 pages with a similar-sized font.[23]

Given the many material and political pressures that the German states faced in the 1800s and 1810s, these practical fields in many ways held even greater allure than they had in the preceding century. Across the German states, educational support for the practical sciences grew on a variety of fronts. Both the agricultural sciences and forestry experienced significant institutional gains in the first two decades of the nineteenth century.[24] Prussian officials, struggling on multiple fronts to raise agricultural productivity, funded Albrecht Thaer's famous institute in Möglin.[25] Several older state academies also received new statutes in this period, which sometimes emphasized the need for academies to make themselves useful to the state. When the king of Bavaria gave his Academy of Sciences a new constitution in 1809, he noted that, while he did not want to place chains on the "spirit of research," he would reserve his highest gratitude for academicians who turned their minds to problems "of direct practical benefit to the fatherland."[26] Erfurt's Academy of Useful Sciences (Akademie gemeinnütziger Wissenschaften), another Enlightenment society that received new statutes in the mid-1810s, got a similar mandate to concern itself with "those subjects of human knowledge that have *direct* influence on civic life [emphasis in original]." The academy should promote these fields of interest to "*practical* life [emphasis in original]."[27] As members, the society desired "not just learned men, but also skilled *artisans* and enlightened, thinking *agriculturalists* [*Landwirthe*] [emphasis in original]."[28] As it turned out, the venerable old academy still ended up with many more learned men in its ranks than artisans; by the late 1840s, the academy's members were mostly state bureaucrats, teachers, ministers, medical doctors, or academics.[29]

In the membership lists of private agricultural societies, however, one begins to see a public assembled that differed significantly from its eighteenth-century enlightened predecessor. For example, the Leipzig Economic Society, one of the Enlightenment's largest and most successful patriotic-economic societies, underwent a thorough reform in the early nineteenth century and decided that henceforth it would focus more or less exclusively on practical agricultural improvement. The society's new statutes were issued in 1817, right in the midst of the severe agricultural crisis of 1816–17, and just a few years after Saxony had lost two-thirds of its territory and about half of its population to Prussia as punishment for its ill-fated alliance with Napoleon. The society had always had close ties to Saxon court officials; in its post-Napoleonic guise, the association developed an even more intimate relationship with the Saxon state. In 1816, the group moved its yearly meeting, along with the society's collections, from Leipzig to the political capital of Dresden and took the awkward new name "The Economic Society in the Kingdom of Saxony." The move was the final stage in a slow drift toward the political capital that had begun very early in the history of the group; the society's governing officer had been a Dresden resident since the 1770s.[30] The Economic Society fashioned itself as a central clearing house for several other agricultural societies that had recently been founded elsewhere in Saxony.[31]

The new officers of the Economic Society overlapped heavily with the group of men who led the Saxon state through the first third of the nineteenth century. State officials held all of the ten governing positions in the society, and the society's director, Count Detlev von Einsiedel, was the highest official at the Saxon court. The men on the society's governing board were the same ones who participated in the heated political debates over Saxon agriculture reform and who argued bitterly about the abolition of feudal rights and duties during the 1820s.

While these men clashed on questions of political reform, the promotion of technical agricultural innovation was one project on which they could agree, and the society's governing board drew from opposing parties within Saxon officialdom. Both the state's reform and conservative factions were represented in its officers, though its director was Saxony's most influential conservative.[32] A few years before the society's reorganization, the Count von Einsiedel had led the successful campaign against constitutional reform; he remained the highest-ranking official at the Saxon court until he was driven from office by the uprisings of 1830–31.[33] His fall from political grace also led to his resignation as the director of the Economic Society, suggesting that the two posts were viewed as interconnected.[34]

The revised statutes of the Economic Society made it clear that the group had little interest in discussing general natural philosophical or natural historical issues; they also emphasized the group's particular desire to recruit members with real practical knowledge of farming. The

CHAPTER 3

Saxon Economic Society solicited direct reports from men of practice—its newly revised statutes stated that it welcomed reports from "farmers, artists, artisans, and so on," and its membership requirements stipulated that "the society would admit anyone as a member who has distinguished himself through patriotic sensibility and knowledge [*Kenntnisse*] or experiences [*Erfahrungen*] in those areas that the society aims to promote."[35] The body of theoretical knowledge that the society now considered most relevant to its activities was the "theory" of agriculture, and it expressed little interest in other areas of natural research. In the first volume of its new *Schriften*, the society asked its readers to send in "true and exact descriptions of actually existing farms." Because Albrecht Thaer's work *Grundzüge der rationellen Landwirthschaft* "seemed to have exhausted the theory of [agriculture], the current aim should be the practical use of this theory, the comparative examination of the proposed laws."[36]

In 1817, the Economic Society's membership composition did not yet reflect its new, narrower focus on agriculture. It did, however, reveal a well-established pattern of choosing members who had practical forms of expertise. In this respect, the 1817 statutes' emphasis on practice over theory ratified a preexisting pattern in the recruitment of new members. In the eighteenth century, Saxony's various economic and scientific societies had drawn only about 32.7 percent of their membership from state officials, with another 16.7 percent coming from trade and agriculture. Professors and medical doctors made up another 27 percent.[37] By the early nineteenth century, the Economic Society showed a marked shift in its membership composition. In 1818, 76 percent of the society's 151 working members had some form of administrative or court title; this category included very high-ranking officials within the state, but also a large number of men who had other kinds of court titles, skilled artisans who had been honored as the official providers of products to the Saxon royal family. These were men, in other words, whom the royal household recognized as skilled and practically useful in the most direct possible way; these were the men who provided them with cutlery, clothing, and beer. The remaining members were landowners or men involved in skilled trades like machine making or in other kinds of manufacturing. Medical doctors and university professors were almost entirely missing from its working membership; only 6 out of the society's 151 regular members were professors or doctors (4 percent).[38] Weighted so heavily toward the upper levels of the Saxon bureaucracy, the society also had a higher percentage of noble members than was standard for middle German civic associations;

36 percent of the society's members were nobles (the average within associational life in general in the middle German states, at least in the eighteenth century, had been around 25 percent).[39]

In the 1770s, Johann Beckmann had claimed that everyone interested in natural philosophy was also interested in agriculture.[40] By the 1820s, the relationship between these two fields was no longer nearly as cozy. The emphasis the Economic Society placed on descriptions of "actually existing farms" and topics of *direct* relevance to practical life signaled a certain sympathy for views of practical critics who claimed that learned theorizing was so much wasted paper. The Economic Society, however, did not take things quite that far. Indeed, it placed clear value on learned knowledge as a source of cultural prestige. Learned men still had an ornamental function within the group. A large number of university professors were named as honorary members, but they were no longer part of its working core. The society also highlighted its allegiance to Albrecht Thaer's new program of rational agriculture. It invited correspondents to "test" Thaer's laws. But since the society also said that Thaer had already written all there was to write about the theory of agriculture, it seems that the group's officials were not, in fact, proposing the kinds of tests a *Naturforscher* might carry out. The point was not to see if Thaer's laws were true; that was taken as read. The point was to collect information about "actually working farms." This suggests a fairly truncated view of agricultural science; Thaer appeared more as a talisman than as a partner in an evolving conversation about the general principles that should guide agricultural practice. With the crisis of 1816–17 just behind them, the Saxon officials who had taken charge of the Economic Society made it their first priority to learn more from people who had actual experience running productive farms. Thaer had already done all the theorizing that needed to be done; it was time to focus on practice.

Enlightened learned authors had often claimed the right to act as mentors and guides to practical men; learned natural knowledge was supposed to be a wellspring whose waters flowed out to every corner of practical life. In referencing Thaer's rational agriculture and appointing professors as honorary members, the Economic Society continued to pay homage to that claim. But as a working forum, the society pulled in a somewhat different direction. Aristocratic officials, artisans, and noble estate landowners now talked among themselves about practical topics, largely without the mediating presence of members of the learned estate. The Saxon Economic Society was also not unusual; in its basic ambitions and aims, it was a fairly typical example of the agricultural societies coming into existence in the early nineteenth century. Its statutes indicate the general direction in which agricultural associational life would develop in later decades. Agricultural associations typically drew on state officials, landowners, and estate managers as their core constituencies; by midcentury, a smaller number of prosperous peasant farmers were also joining their ranks. Practical agricultural improvement, often coupled with only a thin nod to the academic natural sciences, remained their stock-in-trade. Both the Prussian and Bavarian governments encouraged similar groups within their states, and beginning in 1837, a national Association of German Agriculturalists met every year. The next section examines how learned societies changed in the early nineteenth century, once the universe of "useful natural knowledge" had begun to grow too large for learned governance.

Defending Learned Dignity

To understand the strategies that university-educated men chose to defend learned dignity, we should return to Lorenz Oken, whose rousing condemnation of enlightened utilitarianism opened the last section. Oken's speech on natural history and the German nation contained several other interesting features. The first half began with a strident critique of how utilitarian values wreaked havoc on proper "learned sensibility." An overweening emphasis on practical utility degraded the learned to the level of artisans. The second half of the speech then argued that the weakness of the German nation came from the deep division that existed between the learned and the rest of the *Volk*—in short, the Germans' lack of a shared national *Bildung*, or education. The natural sciences, Oken argued, could provide this common *Bildung*. A natural scientific education was far superior, in this regard, to a classical one. A classical education isolated the learned from the rest of the *Volk*; a natural scientific education could bind the *Volk* together. Other nations (one surmises the French and the British) had made the natural sciences the centerpiece of their national *Bildung*, and they were much stronger than the Germans as a result.[41]

It has long been common to see the early nineteenth-century language of *Wissenschaft* and *Bildung*, which Oken used heavily in this speech, as a replacement for the early modern period's language of learnedness. The problem with this view, of course, is that Oken drew on the latter just as heavily as the former. If we think of *Bildung* as a value that replaced learnedness or stood in competition with it, his arguments become incoherent. Oken wanted the whole nation to become *gebildet*, but the learned were also to keep their special dignity. Perhaps we could throw up our hands and simply label him a "transitional" figure, a man with one foot within the old world of estate society, another in the new future community of liberal national citizens.[42]

Oken's references to learnedness are better understood, however, as a continuation of the eighteenth-century habit of separating the learned public from both the educated public and civil society more generally. Oken described the relationship between the learned and everyone else as follows:

Every man needs information about nature; the working man needs to be presented with additional, particular information that is necessary to his work. But to the learned man, whatever estate he is from, the merely particular is a plague. He is there to watch over the sanctuary of the sciences; this sanctuary is constituted only through its integrity, its wholeness, not through individual instances of specialized refinement, nor in the service of the profiteer or of clever ignorants.[43]

The learned could only help everyone else achieve the right kind of moral self-cultivation if they kept the sanctuary of the sciences pure, if they pursued the sciences in all their majestic generality. Building on the "universal *Bildung* that is fitting for the learned," one could then further the "masculine *Bildung* of our fellow citizens [*Mitbürger*]." Oken's speech is yet another excellent example of the ways that the Romantic generation continued to try to answer the central questions that had occupied their enlightened predecessors, albeit with the use of some novel intellectual resources.[44]

The key concept Oken used to defend learned dignity, the idea of *Wissenschaft* pursued for its own sake, is a familiar landmark in the intellectual history of the period around 1800. This ideal was once assigned a great deal of power; supposedly it emanated from its first institutional home, the University of Berlin, to place a decisive stamp on nineteenth-century German intellectual life.[45] Historians have grown skeptical of this story for a number of reasons. The neohumanist ideals of *Wissenschaft*, Charles E. McClelland observed, left "a certain residue" on the universities but did not do much to determine the actual form that they took.[46] Historians of science have concurred, pointing out that there was little in the general ideal of *Wissenschaft* that offered much guidance for the particular ways in which natural science should be taught. Mundane considerations like the training of secondary teachers and competent bureaucrats determined the structure of the university curriculum, and practical, material concerns continued to shape the sciences throughout the entirety of the nineteenth century.[47] Nineteenth-century universities became so important precisely because they did not become the bastions of pure learning that early nineteenth-century neohumanists had envisioned.[48] The founding of the University of Berlin also no longer seems the definitive watershed it once did.[49]

With these qualifications duly noted, it is nonetheless true that the advent of a new rhetoric of pure *Wissenschaft* around 1800 did effect a permanent change in the vocabulary that educated Germans used to describe academic pursuits, consolidating trends that had begun in the eighteenth century. Once this new ideal of *Wissenschaft* rose to prominence, older descriptions of the learned "*Wissenschaften und Künsten*" came to seem quaint and dated, and men in the leading circles of German intellectual life spoke of the relationship between theory and practice in different terms. This broad lexical change demands further explanation. In the early nineteenth century, *Wissenschaft* finally shed its more expansive early modern meaning. It was no longer used to designate just "knowledge" (both academic and nonacademic) in general; more important, it lost its early modern partner, the learned "*Künste*," a term that had also sometimes functioned as its synonym. Under the Enlightenment's broad understanding of useful natural knowledge, natural philosophy and natural history had been discussed in the same forums as practical topics like agricultural improvement; by the early nineteenth century, this unifying rubric had lost a great deal of ground.[50]

In stabilizing the division between the learned guardians of the sciences and Oken's so-called "profiteers and clever ignorants," local scientific sociability played a crucial role. Learned societies and academies have been widely recognized as central to German Enlightenment science, but in histories of the nineteenth century, they have typically been accorded only an ancillary role. After the transformations of the Reform Era, supposedly, the "age of academies" gave way to the "age of universities."[51] There is little actual evidence, however, to suggest that the leading figures of the early nineteenth century thought that the age of academies was over. In both their formal discussions of scholarly institutions and their social practice, university-educated Germans—and in particular university professors—continued to place great value on learned societies.

Many of the best-known writers on university reform also enthusiastically endorsed the importance of learned societies, seeing these two kinds of institutions as both complementary and interdependent. If the university passed on *Wissenschaft* to the next generation, the society or academy was the place where the accomplished researcher met his peers. These were the forums in which "the masters of *Wissenschaft* [came] together," to quote Friedrich Schleiermacher. As Henrik Steffens wrote, it was in these institutions that one cultivated "the deepest foundation of one or more sciences."[52]

Putting learned societies back where they belong, at the heart of the intellectual politics of the early nineteenth century, does more than just correct a lacuna in the institutional history of science. In the changing patterns of associational life, one can see the wider pressures that made the rhetoric of pure science an appealing resource for Germany's university-educated elite.[53] The rise of pure *Wissenschaft* has often been associated with the so-called revolt of the philosophical faculty against the traditional higher-ranking faculties of law, medicine, and theology. It has also been explained as a defense of the university over and against the new practical academies of the Enlightenment. Both of these contexts were certainly important, but in the hands of leading natural researchers, the ideal of pure *Wissenschaft* also spoke to concerns that stretched beyond educational reform or the relative status of the philosophical faculty. Significantly, it was often the utilitarian forums of the Enlightenment that received the brunt of these commentators' ire rather than the traditional higher faculties of law, medicine, and theology. Friedrich Jacobi, for example, spoke derisively of "Societies for the Conservation of Wood, the Finding of Coal and Peat, the Draining of Fens"—societies that were interested in nothing more than uncovering resources of practical economic value.[54]

Many natural researchers who praised the pursuit of pure *Wissenschaft* had little direct stake in the internal politics of university faculties; some of them also had ties to specialized academies. Carl Gustav Carus was a professor at the medical academy in Dresden, a post he left to become court physician to the Saxon royal family. Friedrich Heinrich Jacobi also never held a university chair. Gustav Wucherer was instrumental in saving the University of Freiburg from closure in the late 1810s, but his allegiance to the university and to pure *Wissenschaft* did not mean he was opposed to technical academies. He founded a polytechnic school in Freiburg and also authored a mathematical textbook for technical school students. Several years later, he left Freiburg to teach at a school in the court city of Karlsruhe. At stake for these men was not just a set of institutional interests but the broader question of how learned dignity could be squared with the demands of practical utility. Their central concern lay with the future of learned identity, which in their eyes seemed unstable and unsure against the backdrop of flourishing, socially diverse exchanges about practical natural knowledge.

In championing the cause of pure *Wissenschaft*, leading natural researchers evoked images of a virtuous but declining learned estate, threatened by masses of second-rate, petty, and materialistic laborers. These images turned the values and aims of the preceding generation on their head.[55] The positive adjectives that Enlightenment societies had often affixed to natural knowledge, words like "useful" or "public-spirited," were the particular object of scorn. In 1813, Gustav Wucherer delivered a speech at the University of Freiburg that was full of references to learned dignity and complaints about the degrading position of servitude that utilitarian demands placed on the sciences.[56] The president of the Bavarian Academy of Sciences, Jacobi, also complained that the sciences were only valued when they exhibited a "competency for handicraft." "Everyone is supposed to explain to which guild or branch of industry they belong," he noted sarcastically.[57] To be assigned to a guild just like an artisan, or to be forced to serve the demands of the productive estate (*Nahrungsstand*)—this was more than learned dignity, the advocates of pure *Wissenschaft* suggested, should be willing to bear.[58]

Between 1800 and 1830, eighteen new local societies for natural research were founded within German-speaking Europe. Most of these groups had different social profiles and narrower intellectual horizons than their eighteenth-century predecessors, horizons that aligned, at least roughly, with the new ideal of a general science of nature. A number of private societies died around 1800—Leipzig's Linnaean Society faded out of existence in the first years of the nineteenth century, as did Jena's Nature-Researching Society.[59] The Revolutionary and Napoleonic Wars disrupted the regular rhythms of intellectual sociability in many cities, but there are also signs that the old enlightened cause of "useful natural knowledge" had ceased to work effectively as a unifying force. For example, the previously vibrant Economic Section of the Zurich Physical Society was disbanded in the early nineteenth century, and the group continued on with a narrower range of activities.[60]

Societies that survived into the nineteenth century did so in altered form, often rewriting their statutes. Berlin's Society of Nature-Researching Friends revised its statutes in 1810, and the society's plans linked up with broader Prussian reform efforts. The group donated its collections to the new University of Berlin, and it also built strong ties with the faculty of this new institution.⁶¹ The Halle Nature-Researching Society, too, became more strongly connected to the local university as it entered the nineteenth century. These ties had not been absent earlier, but they became much denser after about 1810. The society became, essentially, a forum for the university faculty.

By 1830, the majority of German university towns had local societies for natural research or, alternately, societies that joined together natural research with its traditional learned sponsor, medicine. Most were of fairly recent origin. Nine out of the eleven societies in university towns had been founded since 1800 (Bonn, Breslau, Erlangen, Würzburg, Heildelberg, Freiburg, Leipzig, Marburg, and Rostock), while older societies survived in Berlin, Halle, and Königsberg.

Many of the new societies founded in the first third of the nineteenth century were devoted solely to *Naturforschung*; others chose an older learned ally, medicine. In the case of these jointly medical and natural scientific societies, medicine and "general natural science" were sometimes divided into separate sections. In Dresden, for example, only the medical doctors in the society attended the special "medical" meetings, but everyone in the society came to the "general meetings" devoted to the natural sciences as a whole.⁶²

The typical early nineteenth-century nature-researching society (*Naturforschende Gesellschaft*) was founded in association with a university (or in a town like Frankfurt or Dresden with a medical academy), with the professoriat taking a leading role. The active members of Saxony's two new natural scientific societies, for example, were dominated by university-educated commoners. In Leipzig's Nature-Researching Society, founded in 1818, 78 percent of the fifty-seven active members were university-educated men of bourgeois origin, most of them either university professors or medical doctors; the only aristocrat was one of the few society members who was also a university student.⁶³ The natural scientific-medical society founded the same year in Dresden had a similar complexion. Its officers were generally professors from the city's recently founded medical academy. Over the first decade and a half of its existence, the society had a total of sixty-five members. Fifty-three percent of these were university-educated physicians; another 15 percent were from the next rung down in the medical profession, the surgeons or military doctors that the Dresden medical academy had been created to train. Reflecting its seat in a political and military center, the society also included a handful of military officers (five, or 6 percent), and state officials (fourteen, or 21.5 percent). The state officials who joined the Dresden Society for Natural and Medical Knowledge

(Gesellschaft für Natur- und Heilkunde) were generally technical bureaucrats from areas like the mining administration, however, and not the men with broad responsibilities who made up the leadership of the Economic Society. Only two members were from the nobility (a paltry 3 percent of the total).[64]

While it did include a small number of men from the lower ranks of the medical professions (surgeons and military doctors), in seeking out local members, the Society for Natural and Medical Knowledge of Dresden skipped over a number of non-university-educated local men who might have had plausible claims for inclusion, especially given the fact that plenty of the society's members had little publication record. Heinrich Ficinus, in his 1821 *Flora* of Dresden, thanked local colleagues who had "made [his] work easier through their love of science [*Wissenschaft*] and through kind sharing." The list included a number of apothecaries whose names did not appear in the rolls of the city's local scientific society.[65] Several honorary members of the Dresden Economic Society were also passed over by the Society for Natural and Medical Knowledge. These included an apothecary, a mathematics teacher from the Dresden Ritterakademie, a local machinist (*Mechanicus*), a chemist (*Chemicus*), and all of the city's numerous court and commercial gardeners. These were men who, in the eighteenth century's more catholic understanding of *Naturforschung*, might well have belonged to a society devoted to natural research.

The city's preeminent family of court gardeners, the Seidels, represented one of the most notable exclusions from the new society. Around 1800, Johann Heinrich Seidel had been a noted figure in Dresden's intellectual landscape. The well-respected court gardener had published a scientific catalog of the royal garden he managed, he had an extensive library of botanical works in several languages, and he was widely admired as an accomplished autodidact; Goethe visited him to chat about botany when he came to the Saxon capital.[66] Seidel had been a member of the Economic Society, and there were signs that the Seidel family's connections to local university-educated men were beginning to take on more than just an intellectual character by the late eighteenth century. Most of Seidel's daughters married other gardeners, but one married the son of the court tutor.[67] Johann Heinrich died in 1815, too early to be included in the Society for Natural and Medical Knowledge, but his sons, who followed in their father's footsteps, were never asked to join the new scientific society, despite the fact that they also published on botanical topics. The Seidel brothers remained in the city as court and commercial gardeners, presiding over gardens of impressive botanical variety, but they were not integrated into the new forums of early nineteenth-century learned discussion.

Between Saxony's Economic Society and its new scientific societies, there was a clear asymmetry in the assignment of honorary memberships. University professors and medical men still had a symbolic place in the Economic Society, though they no longer participated in large numbers in its active membership. The converse was not true; a contribution to *Oekonomie* was not automatically a contribution to *Naturforschung*. While most of Leipzig's university professors and Dresden's medical academy professors were offered honorary memberships in the Economic Society, the scientific societies did not return the favor. The Leipzig society's honorary memberships were split between university-educated men of bourgeois origin (mostly professors at other German universities) and Leipzig's local commercial elite. The officers of the Economic Society were not offered honorary memberships in either the Dresden or the Leipzig society; only two (possibly three) members of the Economic Society had been named honorary members of the Society for Natural and Medical Knowledge by the late 1820s.[68]

Even when a man had an extensive public reputation, the language of *Wissenschaft* could be used to deny men of artisanal background the title of *Naturforscher*, as Myles Jackson has shown in his work on Joseph Fraunhofer, the most famous artisan-*Naturforscher* of the early nineteenth century. Fraunhofer received a great deal of recognition for his work in optics, including an honorary doctorate, but there was still significant debate over whether his discovery of the dark lines of the solar spectrum deserved to be credited as a true "scientific" discovery or merely an example of artisanal skill. Fraunhofer's policy of secrecy regarding his glassmaking techniques exacerbated the problem, but his lack of a university education emerged as a key point in the debates over his work. When his name was put up as an ordinary visiting member to the Royal Academy of Sciences in Munich, one of the opponents of his membership, the Prussian mining official Joseph von Baader, argued against his admittance on the grounds that "the Academy must not become a corporation of artistes [*Künstler*], factory owners [*Fabrikanten*], and artisans [*Handwerker*]." Fraunhofer was eventually admitted to the academy at a slightly lower rank, as an extraordinary visiting member rather than as an ordinary one.[69]

In other words, increasing intellectual sophistication alone cannot explain the new patterns that emerged within local scientific sociability in the early nineteenth century. Fraunhofer, despite the fact that his work (both retrospectively and from the perspective of many at the

CHAPTER 3

time) was some of the most important of the period, still raised a classificatory problem for learned organizations, newly intent as they were on guarding the line between scientific accomplishment and artisanal skill. The boundaries of early nineteenth-century science's many nascent epistemic cultures were not coterminous with the boundaries of local scientific societies, and men who participated in discussions that took place in print culture might not have access to local associational life.

Scientific societies were not entirely closed, however, to the non–university educated. Frankfurt's Senckenberg Society, as Ayako Sakurai has shown in a recent study, was a joint effort between that city's powerful commercial elite and its learned professionals. But Sakurai makes the valuable point that the *learned* remained a distinct group within the Senckenberg; the fact that the urban patriciate and members of the learned professions joined together to form a natural scientific society did not mean that they dissolved into a single group.[70] In Leipzig, another city with a well-organized and culturally prominent commercial middle class, merchants and bankers also appeared in the Nature-Researching Society, though they were typically honorary members, not active ones (these were two formally distinct categories of members). The society also included the owners of several local printing houses, important allies to cultivate for those whose reputation depended on their access to the printed word.[71] The world of commerce, especially the book trade, was well represented in the list of the society's local honorary members, supporters of the society who were not expected to present scientific work. Forty-five percent of Leipzig's local honorary members were university-educated men; another 40 percent were involved in commerce, printing, or finance. Their contemporaries in manufacturing and craft fields other than printing (von Baader's dreaded *Künstler, Handwerker,* and *Fabrikanten*) gained the same status much less frequently.[72]

The advocates of pure *Wissenschaft* did not want to sever the relationship between theory and practice, however; they wanted to renegotiate it. When these commentators discussed how *Wissenschaft* related to practical, productive activity, the two fields were assigned clearly asymmetrical values, with technical creativity presented as the offspring of general theoretical insights. Handicraft appeared as the passive object of *Wissenschaft*'s animating spirit. Practical knowledge was mechanical and mindless; only *Wissenschaft* was creative and generative of novelty. If this assertion sounded implausible to their readers, the advocates of pure *Wissenschaft* promised that it was only cultural amnesia that kept

Wissenschaft's unique role in producing technical novelty from being self-evident. Jacobi claimed that people had merely forgotten the intellectual origin of common technical amenities. Once great thoughts became materialized in the mechanical gestures of artisans and in their tools, they passed out of sight for all but the most careful observers.[73]

Wissenschaft was not just generative of technical innovation. It also contained folded within itself the essence of all practical knowledge. According to Oken, the worker (*der Arbeitende*) only needed as much natural knowledge as was necessary to carry out his particular form of work (the word Oken used here, significantly, was "*Naturkenntniß*," and not "*Naturwissenschaft*"). The learned man, on the other hand, carried within himself the full complement of human knowledge, "out of which he independently brings forth that which the artisan only imitates."[74] In uncomprehending mimicry, artisans repeated the gestures of a materialized fragment of *Wissenschaft*. This new, more aggressive hierarchy between pure *Wissenschaft* and practical knowledge involved a demotion of the Enlightenment's practical sciences. Lorenz Oken, for example, did not describe *Oekonomie* and *Technologie* as independent sciences in their own right, despite the fact that he included them in his list of natural sciences. Instead, he described these fields as belonging (though only nascently) to "the lists of scientifically [*wissenschaftlich*] grounded activities."[75]

Similarly, Dresden's Society for Natural and Medical Knowledge did not completely abstain from references to utility, but it positioned itself very differently from its local counterpart, the Economic Society. The society's statutes dictated that new members should be chosen "with care" and in accordance with the goals of a "learned society." They needed to be men of "*wissenschaftliche Bildung*."[76] In a circular released in 1823, the society announced to the public that it wanted to "support the work of educated [*gebildeter*] workers and skilled artists and artisans." Unlike the Economic Society, which asked for direct contributions from artisans (and sometimes offered such men memberships), the Society for Natural and Medical Knowledge presented itself as a potential advisory board. Its members were "more than willing" to assist artists and artisans who might benefit from greater contact with the natural sciences. Such men should contact the society should they want "to inform themselves about new discoveries" or draw on "the advice of a member of the society in reference to some work they have undertaken."[77] The society fashioned itself as a source of useful expertise; men who sought members' advice on practical questions, however, did so without becoming part of the community of learned experts themselves. If the society's surviving correspondence is an accurate record, its offer of aid did not garner much response.

The Tricky Business of Separating Theory from Practice

In the early nineteenth century, practical economic societies and learned scientific societies largely went their separate ways. That should not be taken to mean, however, that science and practice severed all ties. The natural sciences remained densely interconnected with economic and practical concerns in numerous ways, and at the level of the individual researcher, these ties could be very strong. It would be analytically counterproductive to take distinctions between "pure" and "practical" knowledge at face value. The categories of pure and practical knowledge, even once reified, were hardly stable. In physics and chemistry, for example, it was only in the 1820s that separate printed forums emerged for the "theoretical" and "practical" sides of these disciplines (in the form of the *Journal für praktische Chemie*, for example), and even then the separation was not complete.[78] Indeed, specialized chemistry journals of all stripes succeeded in the nineteenth century because of the audience of "practical men" (including apothecaries and industrialists) who subscribed to them, read them, and sometimes contributed to them.[79] Conversely, leading *Naturforscher* with university chairs continued to take an interest in practical fields throughout the nineteenth century.[80]

Although natural research did not actually sever ties with practical and economic concerns, the distinction between *Wissenschaft* and practice did take on new relevance in organizing local communities of discussion. Societies dedicated to improving fields of productive activity (agricultural societies, and also later industrial societies, which are discussed in chapter 7) pulled away from learned societies, drawing on a wider range of occupational groups for their members. These groups used many of the communicative behaviors typical of early modern learned societies (published proceedings, protocoled meetings), but become separate forums and at least partially distinct communities.

These distinctions also took on new force within print culture, though it is difficult to generalize across specialized disciplines. In the case of the natural historical sciences, the new distance between theory and practice can be seen in textbook formats and classificatory aids. In botany, for example, textbooks offering readers introductions to both theoretical and practical botany in one volume no longer appeared in the nineteenth century. Natural historical bibliographies also slowly came to exclude practical works. The Heidelberg professor Johann Heinrich Dierbach's *Repertorium botanicum*, published in 1831, started with "general botany" and promised to continue with a second volume on "medical-pharmaceutical and chemical botany," and another on "economic, technical, and forest botany," but he never completed the second and third volumes, and these practical sciences now boasted their own separate bibliographies.[81] By 1841, M. S. Kruger's *Bibliographia botanica* felt comfortable excluding "practical" botany (i.e., agriculture, forestry, and gardening) without comment. "Botanical literature" now covered only what Dierbach had called "general botany" and what the late eighteenth century would have called "theoretical botany."[82]

To see in more detail how this new division worked at a local level, and how it bisected activities that were joined together at the level of individual biography, it would be useful to take a closer look at the activities of one of the few men who was a regular member of both Dresden's Society for Natural and Medical Knowledge and its Economic Society, the entrepreneur and medical doctor Friedrich August Struve. In the late 1810s and 1820s, Dresden became the home to a number of new manufacturing concerns, enterprises that extended the density and reach of the city's economic ties beyond local or regional markets. One of the most important of these new businesses was Friedrich Struve's mineral water manufactory.[83] In 1818, Struve, a Dresden pharmacist and doctor, developed a new technique for producing artificial mineral water, a highly profitable innovation given the importance of mineral water cures in aristocratic and upper-middle-class culture.[84] Using his new techniques, Struve could fabricate mineral waters that mimicked those found at famous spas like Karlsbad, Ems, and Marienbad, and in 1820, he built a factory in Dresden to produce these waters on a commercial scale. A year later, he opened a spa where people could come and "take the waters" produced in his factory. In 1823, he started a second facility in Leipzig, and from there, Struve's enterprise expanded quickly to European prominence, with branches in Brighton, Königsberg, Warsaw, Moscow, Petersburg, and Kiev.[85]

The protocols of the Society for Natural and Medical Knowledge recorded the early years of this medical and commercial enterprise but in a truncated, partial form. Struve was not in the original group of men who founded the scientific society in 1818, but his name came up for membership during the first year of the society's existence, in the spring of 1819. There is some evidence that he had a somewhat elevated status within the society from the start; in the vote for the first board of directors, taken very shortly after he became a member, he finished just out of the running to be one of the four directors, receiving the sixth highest number of votes.[86] Struve's first presentation to the society on his artificial mineral water, entitled "Thoughts on the Relationship between Natural and Artificial Mineral Water," took place in 1820, the same year that his manufactory opened.

In the society protocols, Struve's paper was recorded purely as a chemical one, with no reference to its economic and medical implications. He had been asked to present at one of the society's "general meetings" (those dedicated to *"allgemeine Naturwissenschaft* [general natural science]") rather than one of the medical meetings, and nothing in the protocols suggested that his talk addressed questions that were of more than theoretical interest. His presentation, as recorded in the official society protocols, focused on the question of whether or not substances produced through different processes could be truly identical. "The consideration of the electrical forces that can emerge from chemical bodies," the summary began, "have led many to conclude that originating forces maintain themselves in their products." This was not the case with his artificial water, Struve argued; it was not significantly different from waters that had been produced naturally. In particular, he dismissed the claim that artificial mineral water always cooled faster than natural mineral water, an assertion that had been used to support the opposing position that the natural and the artificial were somehow ontologically distinct. This apparent difference, he claimed, appeared only because past researchers paid insufficient attention to the chemical mixtures involved. If truly chemically identical, naturally and artificially produced mineral waters were exactly the same.[87]

Perhaps the actual conversation that day was more wide-ranging. The secretary Carl Gustav Carus chose to record Struve's presentation, however, only as an answer to a question about chemical ontology, taking no note of the extraordinary medical and economic implications of Struve's ability to replicate naturally occurring mineral waters. The learned society's records offered no clue that Struve was at the time opening a manufactory in Dresden to produce these medically useful waters on a commercial scale. In other words, in documenting the talk in the Society for Natural and Medical Knowledge protocols, Carus failed to record precisely those aspects of Struve's activity that would have been of greatest importance to a late eighteenth-century society.

Struve's new mineral water enterprises were clearly of great interest to his fellow society members; in fact, his lecture occurred when it did because the board of directors had taken the somewhat unusual step of requesting that he present something about his work.[88] Part of their interest was certainly intellectual; some of the chemical issues addressed in Struve's presentation had appeared in society discussions before. In the meeting in which Struve's name first came up for membership, the medical professor Heinrich Ficinus had given a talk "on the progress of the newer chemistry in the last ten years." More specifically, the society also had heard a paper the preceding fall on Jöns Jacob Berzelius's theory of combustion, and members' discussion of this paper had branched out into a more general conversation about the nature of light and heat and the role of electrical forces in chemical transformations.[89]

Struve's artificial mineral water was not interesting just for its philosophical implications, however. His new factory was an undertaking with particular significance for the medical community of Dresden, both economically and therapeutically. At Struve's spa, visitors to the court city could have the medical benefits of Central Europe's great spas in one location. Over the next decade, members of the city's medical community, all of them also members of the Society for Natural and Medical Knowledge, helped promote Struve's mineral water to the European public. Struve's own 1824 book about his mineral water had a foreword by Friedrich Ludwig Kreyssig, a professor at the Dresden medical academy.[90] Kreyssig also authored a guidebook to German spas for a French readership that hawked Struve's products. Friedrich August von Ammon wrote a guide to spa cures that included a pitch for Struve's spa and his mail-order service in its preface.[91] Collectively, Dresden's learned society may have addressed only questions of chemical ontology in its meeting, but as soon as its members stepped outside of this carefully segregated space, Struve's findings again became facts that had medical, economic, and philosophical significance all at once.

Latinity, Erudition, and Learned Identity

Nineteenth-century ideals of pure *Wissenschaft* and *Bildung* had particularly strong roots within classical philology.[92] Historians have sometimes described the natural sciences' use of these concepts as a sign of their cultural subordination to the hegemonic power of classics.[93] This is misleading in two ways. As we have seen, university-educated *Naturforscher* had their own reasons for arguing that knowledge ought to be pursued for its own sake rather than for mere practical ends.

They also had good reasons to hold on to Latinity as a mark of learned status. Latinity was not just generally prestigious; it had operational value within the networks of natural knowledge. The skills imparted in a classical education distinguished the university educated from the broader crowd of unlettered natural researchers, and this was a distinction the university educated were increasingly eager to reinforce in the early nineteenth century.

As it had been in the eighteenth century, Latinity was still useful as a status marker among *Naturforscher*, and it could still be used as a tool (either for good or for ill, depending on the author's perspective) for establishing hierarchies among different groups with an interest in studying nature. The practically oriented Erfurt Academy of Useful Sciences, which wanted to reach out to artisans and farmers, took the trouble to state explicitly in its statutes that "the language of the academy will be German; the protocols will be taken and the diplomas filled out in this language."[94] The more self-consciously "learned" Leopoldina society began its yearly journal with a Latin preface.

The Leopoldina society, German-speaking Europe's oldest existing scientific academy, offers perhaps the clearest example of Latin's continued relevance to the cultural politics of early nineteenth-century natural philosophy. Founded by a group of doctors as the Academia Naturae Curiosorum in 1652, the home base of this Baroque natural philosophical and medical society migrated along with its presidency throughout the eighteenth century, and its members were spread throughout the Holy Roman Empire. Its colloquial name, the Leopoldina, came from the expanded version of its name used after it received official imperial recognition—the Academia Caesarea Leopoldino-Carolina Naturae Curiosorum. The Leopoldina had gone into steep decline in the late eighteenth century. The imperial society remained in existence in name, but it published its last volume of *Acta* in 1783. A popular legend claimed that the only reason that the academy had not been abolished along with the other institutions of the Holy Roman Empire was because Napoleon forgot about it, and this explanation, even if false, captures the modest stature of the society around 1800 well.[95]

The Leopoldina was revived as a national institution in the early nineteenth century primarily through the efforts of the botanist Christian Nees von Esenbeck, the man who assumed its presidency in 1818. In its renovated form, the new Leopoldina continued to employ older marks of learned status.[96] The eighteenth-century publications of the academy had been in Latin, and though its nineteenth-century *Nova acta physico-medica* contained plenty of vernacular articles, its Latinate roots were still readily visible. Each volume opened with a Latin preface, and some of the contributions were in Latin (these dropped off markedly without completely disappearing by the 1840s).

In ceremonial contexts like the Leopoldina's yearly prefaces, knowledge of Latin was employed primarily as a cultural ornament. But a classical education could still be useful for more substantive kinds of intellectual communication. By 1800, the majority of natural philosophical discussion took place in the vernacular, but not all of it did.[97] Though German was clearly dominant, Latin's importance as a scholarly language was not entirely negligible.[98] Latin also had some usefulness as an administrative language in Central Europe, particularly for Prussian administrators who dealt with the kingdom's Polish territories or for Habsburg state servants communicating with the emperor's Hungarian subjects.[99] The language was still used for university dissertations, and it also retained a place in medical writing, though its prevalence was fading.[100]

A commitment to Latin as an active scholarly language did not mean that an author was a Baroque relic who had somehow accidentally survived into the nineteenth century. Germany's most important early nineteenth-century mathematician, Carl Friedrich Gauss, was also perhaps its most committed natural philosophical Latinist. He published work in Latin and even kept his personal mathematical journal entirely in Latin.[101] At least one seminal nineteenth-century scientific discovery was initially published in Latin—Hans Christian Ørsted first circulated a brief description of his 1820 experiments on electricity and magnetism in a short Latin pamphlet, a format that allowed him to address men from a variety of European countries through the same document.[102] In fields like botany, a number of important works, particularly in systematics, were written in Latin.[103]

Though the vast majority of articles published in early nineteenth-century scientific periodicals were written in the vernacular, not all were. For example, the first volume of D. F. L. von Schlechtendal's journal *Linnaea*, which appeared in 1826, contained articles in Latin as well as in German. The first German journal to publish Ørsted's Latin treatise, the influential *Journal für Chemie und Physik*, published a short editorial note encouraging submissions in Latin after Ørsted's piece appeared. The editor's note came at the bottom of the first page of another Latin contribution, Karl Schrader's dissertation (also on electromagnetism), and the journal's editor, J. S. C. Schweigger, mentioned in this note that Ørsted's use of Latin had received "the approval of a number of learned men" and that the journal would welcome future essays in that language since Latin offered advantages for communicating with foreigners.[104]

Furthermore, like their neohumanist colleagues, early nineteenth-century natural researchers cared about the style and correctness of their Latin. When Schweigger decided to include Latin articles in his *Journal für Chemie und Physik*, he indicated that the journal would only take articles in which the Latin was "pure."[105] Along similar lines, Ludwig Reichenbach began his *Catechismus der Botanik* with questions about the proper words for different aspects of botanical science. Terms that had been adapted without proper attention to their Greek or Latin etymology he dismissed in acerbic footnotes. Reichenbach ridiculed "glossologia" (a synonym for "the science of descriptive terms") as "an unneeded word, which emerged out of an ignorance of the Greek language."[106] Philological correctness had been a touchy point of honor for university-educated natural researchers in the preceding century, and these anxieties continued. In the preface to his 1815 botanical handbook, Johannes Christian Mössler expressed the hope that his work would help lead to the "complete extermination of that deleterious, though still widely spread prejudice . . . that botany, the loveliest of all the sciences, simply consists of a dry artificial language and the knowledge of a bunch of barbaric names."[107]

Several university professors of natural science also published work in classical philology or history in this period. Most of these men, significantly, had been born into learned families. Many were ministers' sons, for whom the mastery of classical languages represented part of their patrimony. The botany and chemistry professor Heinrich Friedrich Link, the son of a Lutheran minister, published a new scholarly edition of Theophrastus, the leading ancient philosopher to write on botany (Leipzig: Vogel, 1818–21), and he also authored a jointly historical and natural historical work, *Die Urwelt und das Alterthum, erläutert durch die Naturkunde* (Berlin, 1821–22; 2nd ed., 1834). Johann Jacob Paul Moldenhawer, another minister's son who held a chair in botany and pomology in Kiel, also lectured in classics there and, like Link, published work on Theophrastus.[108] The professor of botany in Halle, Kurt Polycarp Joachim Sprengel, produced a new edition of Dioscorides's *De materia medica* (Leipzig: Cnoblauch, 1829–30), the most important classical work on pharmacology.

Philologists also occasionally drafted naturalists to help with their own editing projects. August Friedrich Schweigger, the son of a theology professor at the University of Erlangen and himself professor of botany at the University of Königsberg, was recruited to help with the Prussian Academy of Sciences' new edition of Aristotle. In 1821, Schweigger left Germany for southern Europe to travel through Sicily and Greece, in part to conduct natural historical research for August Immanuel Bekker's massive new edition of the Aristotelian corpus. (Schweigger did not return from the trip; he was murdered while botanizing in the Sicilian countryside.)[109] Schweigger's brother, the Halle physicist J. S. C. Schweigger, authored a book that was perhaps the nineteenth century's most unusual combination of philology and natural science, the 1836 *Einleitung in die Mythologie auf dem Standpunkte der Naturwissenschaft*, which argued that certain myth cycles and works of classical literature, most important the *Iliad*, were actually encoded descriptions of electromagnetic phenomena.[110] Particularly for the sons of university-educated fathers, such erudite activities were an outgrowth of an early introduction, often under paternal tutelage, to the world of classical learning. The mastery of classical languages continued to provide a genuine source of learned solidarity across the disciplines and a meaningful status marker that separated the university-educated *Naturforscher* from his less well-heeled brethren.[111]

Conclusion

By distinguishing between merely "useful" and truly "learned" knowledge, elite *Naturforscher* neatly exempted themselves from thorny, complicated questions about their practical relevance. They also shored up distinctions between university-educated men and other "friends of nature," defending a social boundary that Enlightenment societies had guarded much less carefully. In contrast, the early nineteenth century's practical societies welcomed a much broader range of participants and were eager to emphasize their direct connections to economic concerns.

In breaking with the Enlightenment cause of useful natural knowledge, the founders of early nineteenth-century learned societies also narrowed the range of activities that fell under the label "*Naturforschung*," drawing a firmer line between the concerns of learned science and the demands of practice. The period's new (or newly reformed) learned societies provided powerful spaces for preserving this distinction once it was asserted. They helped to reify the categories they used, making the new, stronger differentiation between *Oekonomie* and *Naturforschung* seem intuitive and obvious. The differences between these categories ceased to be merely philosophical—the differentiation was now built into the social lives of university-educated natural researchers. The new classificatory boundaries produced through these changed patterns of discussion made natural research's thick connections to practical knowledge less visible, even as such relationships persisted. In the process, local habits of intellectual sociability helped transform the categories of abstract philosophical reflection into the commonplaces of later nineteenth-century public discussion.

FOUR

Nature in a Local Microcosm

In 1824, the category *"Naturwissenschaft"* appeared for the first time in the widely read Brockhaus lexicon. The entry began, "Nature is mirrored in the spirit of the cultivated [*gebildeten*] person, and this reflection, this ideal image of nature, is natural science."[1] With the brevity to which encyclopedias aspire, the Brockhaus glossed over one of the most difficult philosophical issues of the early nineteenth century in a single sentence. In the wake of Immanuel Kant and Friedrich Wilhelm Joseph Schelling, German thinkers had struggled mightily to make the triad this sentence evoked—knowledge, the self, and the external world—hold together. For a generation immersed in these epistemological debates, there was no theory of knowledge without a theory of the subject, and the major figures of this period wrestled valiantly with the tricky problem of bringing the self and the world into meaningful relation, without ever coming to general agreement about a solution.[2]

The Brockhaus's description of *"Naturwissenschaft"* is interesting in several ways. First of all, it shows the heavy debt the concept owed to the philosophical and intellectual ferment of post-Kantian German intellectual life. From the 1790s forward, proposals for a new, unified philosophical account of the natural world branched out from the systems of Kant and Schelling and intertwined with ongoing debates within natural philosophy and natural history. The proposals put on the table for a general science of nature were many and varied. Through his morphology, Goethe thought he had shown how natural researchers could come to a concrete intuition of the whole of nature through the examination of its parts.³ Alexander von Humboldt crafted a "master physics of the earth" that tried to map the global play of forces through precise measurement.⁴ Researchers working on chemistry and electromagnetism hoped that they were zeroing in on the most primary and basic of nature's forces, which seemed to be polar in nature.⁵ The works of the *Naturphilosophen* offered some of the most audacious synthetic efforts of the period; figures such as Lorenz Oken and Henrik Steffens produced sweeping and controversial new general accounts of the natural world, systems that presented complex, highly metaphorical histories of a dynamic, organically unified Nature.⁶

The precise architecture of this new general science of nature was the subject of heated debate; it was even difficult for German *Naturforscher* to agree on what this new thing ought to be called. There were a number of competing labels in circulation. Gustav Wucherer spoke of *"die allgemeine Naturlehre* [general natural philosophy]"⁷ Carl Gustav Carus wrote of *"die Naturwissenschaft im allgemein* [natural science in general]"; he then divided *Naturwissenschaft* into "sensory [*sinnliche*]" and "philosophical" sides.⁸ Oken, in contrast, considered *Naturphilosophie* and *die Naturwissenschaften* as interrelated but separate entities. Henrik Steffens and Heinrich Link wrote works proposing the criteria for a *"philosophische Naturwissenschaft"* (philosophical natural science) and a *"philosophische Naturkunde"* (philosophical natural knowledge), respectively.⁹ Among the new learned societies of the period, one finds a similar lack of consensus in the business of naming. "*Naturforschende Gesellschaft*," already a popular choice in the eighteenth century, remained a common label; several other groups used "*Naturkunde*" in their names. Only one society before 1830 used "*Naturwissenschaft*," and it did so with an additional adjective that suggests the need to explain the significance of the word—the society in Marburg was a society "for the promotion of all of the natural sciences" ("*zur Beförderung der gesammten Naturwissenschaften*").

Here we see why it is important not to conflate the history of a word with the history of a thing. The basic idea that all of the natural sciences formed a unity had become an often-asserted cultural commonplace by the 1820s, though there was still significant disagreement about what exactly bound the different fields of natural knowledge together, and similar disagreement about what a new, general science of nature ought to be called.

Once again, the Brockhaus entry helps cut through seemingly hopeless complexity to reveal a broad general pattern. Its strategy for sorting out the tangle of this period's complicated scientific and philosophical debates was both simple and elegant. The unity of natural science, the lexicon said, was grounded in a particular social persona. Natural science was the image that a "cultivated" person formed of the natural world. The unity of natural knowledge was realized through a certain form of selfhood, the integrated personality of the *"Gebildeten."*[10] In the thinking of the period, the *"gebildete"* individual was never one-sided (*einseitig*); his or her personality was pleasingly harmonious and balanced. As a result, nature perceived only in separate fragments had little value in the process of *Bildung*. Nature needed to be grasped in her entirety if she were to lend her own harmony and unity to the minds of those who understood her. As Carl Gustav Carus put it, understanding the deep interconnectedness of nature, her beauty and her lawfulness, helped to bring a person's "inner life into similar harmony and clarity."[11]

Already from the preceding chapter, we should be able to see that the Brockhaus lexicon was onto something. The general study of nature was indeed linked to a particular social persona, the learned *Naturforscher*, now also described as a man of *"wissenschaftlicher Bildung."* In the previous chapter, we saw that declaring allegiance to a general science of nature was part of the way that an elite *Naturforscher* could set himself off from the expanding crowd of people interested in more directly useful natural knowledge. This chapter looks at how local learned societies also helped Germany's university-educated elite set themselves apart from the growing crowd of new authors and collectors that we left behind at the end of chapter 2. As we have seen, Germany's burgeoning print and consumer culture had made the raw material for a learned reputation much more widely available by the early nineteenth century. The new elite learned societies of the early nineteenth century were in part founded to deal with the ambiguities these circumstances created. Within these new, more elite learned societies, university-educated *Naturforscher* marked themselves off from a growing public of more casual collectors and dilettantes. The butterfly collectors that Ochsenheimer complained about so bitterly in 1806 were not much in evidence in the Dresden Society for Natural and Medical Knowledge or in Halle's Nature-Researching Society. These groups did not just exist to guard the lower border of the learned world, however; as we shall see, their benefits went far beyond that.

These societies were the organs of an emerging elite science; this elite science, however, was not yet professional in the later nineteenth-century sense. Its internal structure is still best understood as a fur-

CHAPTER 4

ther development of networks of the eighteenth-century learned world, and its social logic is easiest to see when looking forward from 1780, not backward from the late nineteenth century. As a result, university professors, though they were very important, should not be allowed to push other kinds of elite natural researchers out of view. Before we turn to look in more detail at the social practices of this period's learned societies, however, there are two other general issues that we need to examine first: the relationship between unity and specialization, and the relationship among local, national, and cosmopolitan settings for scientific discussion.

Specialization and Unity

The Romantic era's enthusiasm for the unity of natural knowledge has sometimes been described as a form of nostalgia, a futile final attempt to hold on to older ideals of learned breadth in an age where increasing specialization was the order of the day.[12] Historians are often quick to note, for example, that the generalist GDNA divided itself up into individual disciplinary sections shortly after its founding, hence making itself relevant to emerging patterns of elite discussion. If one looks more closely at early nineteenth-century intellectual life, however, unity and specialization were rarely discussed as mutually exclusive choices. As we will see in later sections, the many general, multidisciplinary scientific forums founded in this period are best understood as complements to specialized disciplinary discussion; they were organizations that proliferated as part of strategies to manage the tensions and opportunities that specialization presented, not attempts to hold specialization at bay.

This reciprocal relationship between general collective forums and individual specialization was also not novel in the nineteenth century. Berlin's Society of Nature-Researching Friends, for example, had also been intended to facilitate intellectual specialization. "Every lover or connoisseur of Nature has his own favorite subject," Friedrich Martini proffered, and it was best to expend the most effort in this preferred area.[13] The society was supposed to allow each individual member to focus on his favorite area of natural history; each member would then instruct his fellow members in his specialty. Working within the context of the society, each person could excel in one narrow field instead of diluting his interests over the entirety of nature. Specialization was in many ways a defining feature of the group, despite its general horizons. The society included a topical index in the first volume of its proceedings, suggesting that it did not expect its audience to read the journal from cover to cover, but rather to seek out articles of particular interest.[14] From the early modern period forward, people had praised learned academies and societies because they allowed their members to do collectively what no man could do alone. This justification for intellectual sociability took on heightened importance in the rapidly specializing landscape of early nineteenth-century German science.

The German academic elite, the same men who were at the heart of the period's specializing disciplines, also put a great deal of effort into creating generalized forums where these disciplines came together. All but one of the university seminars founded in the first half of the nineteenth century covered all of the natural sciences together, at least in their formal administrative design.[15] As we saw in the last chapter, generalist societies for natural research proliferated, too.

In past work on the history of German science, there has been a strong tendency to present general forums as early stages of an institutional process that leads toward more specialization.[16] Specialized groups, it has often been assumed, were further along the path to modern professional science than generalist ones. As a result, it has been easy to see the local natural scientific societies of the early nineteenth century as a phenomenon largely external to the history of elite science, the vehicles of a divergent amateur scientific culture.

As it turns out, if one assumes that specialized societies were in the academic vanguard, and that generalist groups housed second-tier figures, one ends up with a strange map of German intellectual geography. Many of the "specialized" societies founded in the first half of the nineteenth century were actually quite modest in stature. The Entomological Society in the small town of Stettin, for example, cuts a poor figure when cast as the vehicle of a new professionalizing scientific elite; so does the Ornithological Society of Görlitz. The *Flora*, the journal of the Regensburg Botanical Society, was full of just the sort of newsy natural history notes that made elite German botanists cringe. In contrast, Leipzig's generalist Nature-Researching Society was packed to the gills with university professors, and its list of honorary members reads like a Who's Who of German academic life. As Katrin Böhme-Kaßler has shown, Berlin's Society of Nature-Researching Friends was just the same.[17]

Some degree of specialization might be the trademark of a serious *individual researcher*, but when Germans were deciding whether or not a particular *place* was an important location on the map of the learned world, the collective breadth of its local learned community mattered a great deal. When Lorenz Oken evaluated Dresden's standing as a natural scientific center, for example, comprehensiveness was his first criterion. He thought that if the city acquired a good zoologist (the one thing missing, in his view), its claims to be an important scientific place would be strong.[18] The move from a specialized focus to a more general one also sometimes coincided with a local scientific society's attempts to claim greater publicity. Several early nineteenth-century learned societies described their origins this way. The Marburg Society for the Pursuit of All the Natural Sciences wrote that it could trace its origins to "an earlier society that existed for many years here in town, but which was more private, and just devoted to one branch of natural history, namely ornithology."[19] The Nature-Researching Society of Görlitz had a similar genealogy. It had grown from "a group of businessmen" who wanted to devote "the hours of their leisure to a collective discussion of ornithological topics"—that is, who wanted to get together to talk about their pet birds. The society had now become "almost exclusively a society of real learned men," and had broadened its intellectual horizons accordingly. In both of these cases, the collective move from specialization to generality was seen as the intellectually ambitious one, and not the reverse.[20]

Dilettante businessmen gathered to talk about narrow topics, like the care of their pet birds. Men with serious learned ambitions had more important—and more general—things to discuss together. Elite researchers wrote for specialized journals, but they also founded general local learned societies. We should not gloss over this latter kind of activity, which was a central feature of elite German scientific life for decades, merely as the result of an incomplete transition to disciplinarity or a product of nostalgia. This phenomenon deserves to be explained on its own terms. To do so completely, however, requires us to look more closely at the cultural politics of *place* in this time period.

Setting the Stage: The National, the Local and the Cosmopolitan

In looking at later figures like Rudolf Virchow and Emil Du Bois-Reymond, previous historians have argued that German efforts to create a unified science were directly linked to their efforts to build a unified nation.[21] One could point to similar connections in the early nineteenth century, too. Oken's Society of German Natural Researchers and Doctors (GDNA) has been seen (with good reason) as a project that joined nation building to the promotion of science.[22] In his 1809 speech on natural history and the German nation, Oken began with an argument about the need for a general natural history and then moved on to argue that the natural sciences as a whole could provide the necessary foundation for a unified German national culture.

It would be a mistake, however, to see the desire for a unified science of nature simply as a projection of the desire for a unified nation. To start with, learned dignity was the first thing that Oken's general science of nature was designed to defend, and his anxieties on this point were widely shared. Furthermore, cosmopolitan, regional, and local allegiances played a central role in shaping German scientific life in this period; the national context was not the only thing that mattered. As Glenn Penny has argued for the later nineteenth century, placing too strong an emphasis on nation building can lead us to miss many other important things that shaped the development of science in the polycentric German states.[23]

In order to develop this point further, we should pay a visit to the national GDNA. In 1830, the Scotsman James Johnston attended the annual meeting of this organization and penned a chatty and detailed account of the proceedings once he got back home. The 1830 meeting of the society was held in the northern German trading center of Hamburg, and Johnston began his account of the event with a description of the conference's opening days. All the arriving *Naturforscher* came first to Hamburg's city hall, and men from different cities arrived together as group, massed like delegations at a diplomatic conference. The *Naturforscher* already assembled watched the procession of later arrivals with interest, noting the size or eminence of the delegations that had come in from each place.[24]

Johnston's description of the GDNA can be used to make several points. First of all, it illustrates the way in which national forums provided a stage set for representing more local kinds of identity. He described the groups from different German towns, particularly from German university towns, arriving together as the collective representatives of their place. These men were joining one another as German *Naturforscher*, but their allegiance to the German fatherland did not erase other, more local ties. They cared about the reputations of their respective towns and their respective dynastic states, too. As Johnston makes clear, when the congress debated where its next meeting ought to be held, the competing delegations vied to win that "honor" for their state or city. Recent work on political identity within German-speaking

Europe has made a similar point about the multivalent character of political loyalties in this period. The nation could be a significant object of allegiance, particularly for many educated Germans, but dynastic states or cities commanded allegiance, too, and these local and regional identities did not necessarily wane as national loyalties grew. Indeed, after 1815, many German ruling princes worked hard to cultivate a greater sense of dynastic loyalty among their subjects, and did so with significant success. Germany's different states also competed with one another on a variety of cultural fronts.[25]

The effort that *Naturforscher* put into local scientific societies should be seen against this broader backdrop of cultural competition on a national stage. A man's reputation was linked in no small way to the reputation of the place where he lived, and local scientific societies, as the following section will explore, helped make a given city's local scientific culture more visible to the rest of the republic of letters. It pulled together the diverse activities of individuals and broadcast them outward to advertise the intellectual vitality of a particular place. This was a project that had special appeal in academic centers like Berlin, Leipzig, and Halle, but the strong link between science and the politics of place in this period is also visible from outposts like Frankfurt am Main, one of the few city-states left standing after the territorial consolidations of the early nineteenth century. As Ayako Sakurai has shown, the Frankfurt patriciate used its new scientific society, the Senckenberg, to advertise its distinctive republican traditions within a German Confederation dominated by larger, aggressively centralizing territorial states. The Senckenberg Society was designed to advertise both the knowledge of the city's learned men and the civic virtue of its urban patriciate.[26]

In the more politically concentrated landscape of post-Napoleonic Germany, cultural authority was now more intimately linked with place than it had been before. The university reforms of the late eighteenth and early nineteenth centuries were a part of the process of creating new, more powerful central places, and local learned societies provided a corollary to this state-driven process in the form of private cultural initiatives.[27] A local society helped a city mark its place on the map of learned Germany, and university professors were particularly eager to take advantage of this tool. As the Göttingen Academy of Sciences had been in the mid-eighteenth century, early nineteenth-century private societies were complementary institutions to universities and academies, resources for professors seeking to raise their visibility within Germany as a whole.[28]

A national audience was not the only target of these efforts, however. The broader European world of science and learning still mattered, too. The horizons of the nineteenth-century republic of letters were still cosmopolitan in significant ways. Johnston's account of the 1830 meeting makes this point nicely, too. When the Scotsman walked into the conference's reception hall, he described it as an entry into a linguistic Babel. As the assembled *Naturforscher* came together, "the mixing up and compounding of languages where so many are spoken, and so few can speak them all" caused a great deal of mirth and confusion. One might, in quick succession, encounter a Dane, then a Swede, then a German, then an Englishman, and next a Pole. Everyone stumbled through as best they could with the words they had in each tongue, and when all else failed, people spoke French. The king of Denmark had sent a delegation to try to bring the meeting to Copenhagen in the coming year, and the proposal was seriously discussed, though eventually rejected.[29] This captures well the multivalent complexity of "Germany" around 1830, and also the continued openness of this national forum to cosmopolitan guests.[30]

At this national meeting, it was also possible to go too far with nationalist enthusiasms. Johnston reported that when Friedrich Georg Wilhelm Struve gave a speech on the history of astronomy that praised the Germans at the expense of other nations, there was much grumbling among his listeners. When Friedrich Sigismund Leuckart complained that two Englishmen had been elected to chair the zoology section, his remarks caused general embarrassment in the audience, and the next speaker criticized Professor Leuckart's misguided rudeness.[31]

Helmut Walser Smith and Lorraine Daston are certainly right to point to the revolutionary and Napoleonic Wars as an important turning point in the history of the cosmopolitan republic of letters, but one should be careful not to overstate this case.[32] Early German theorists of *Wissenschaft* were quite clear about the fact that science was a cosmopolitan project, even if it might be made to serve the cause of national revival.[33] For Henrik Steffens, the nation was an important community, but one that represented only a single dialectic moment between the poles of pure individuality and pure cosmopolitanism, opposing principles that any true practitioner of *Wissenschaft* would jointly embody. Friedrich Heinrich Jacobi openly scoffed at the idea that *Wissenschaft* could belong to any individual state or nation; such views were only for crude utilitarians.[34] Even in the GDNA, where spontaneous choruses of Ernst Moritz Arndt's popular anthem "Was ist des deutschen Vaterland?" (What belongs to the German fatherland?) could break out at any moment, crossing too far into an anticosmopolitan chauvinism could still be a mark of bad taste. Local learned societies gained meaning from this broader cosmopolitan frame, too; their members were well aware of their place within a network of societies that stretched from London to Philadelphia, Florence, Berlin, and beyond.

Print Culture and the Crisis of Learned Identity

"Literature has become an exercise in loneliness." In putting these words into the mouth of his narrator, the novelist Karl Immermann offered his diagnosis of Germany's supposed state of artistic malaise. "There is no public anymore," his anonymous narrator complained. "That word presumes a number of receptive listeners. But who still listens, and who wants to receive?" Literature had become a lonely enterprise, not because there were too few authors but because there were now too many. The glut of writers, it seemed, had produced a dearth of careful, receptive readers, and the fruits of an author's intellectual labor were easily lost in this overpopulated void. Among such an "excess of intellects," Immermann noted grimly, each man stood alone.[35]

The novelist's scholarly and scientific contemporaries lodged similar complaints of overcrowding. In the early nineteenth century, biographical handbooks struggled to keep track of the large number of people of diverse social origin and variable renown who participated in print culture as authors. While in 1750 the Leipzig professor Christian Gottlieb Jöcher could claim that his biographical encyclopedia included "scholars of all estates who have lived since the beginning of time" in a mere two volumes, by the early nineteenth century, guides of similar scale existed for different dynastic states (*Das gelehrte Baiern*, *Das gelehrte Sachsen*) and even for individual cities (*Das gelehrte Erfurt* and *Das gelehrte Berlin*).[36] The Napoleonic Wars disrupted the German book market for a while in the first two decades of the nineteenth century, with a low point of production coming in 1813. The market returned to prewar levels by 1821, however, and was soon growing again at a steady clip. In 1826, 5,168 books were published in German-speaking Europe; by 1837, that number had almost doubled (10,118).[37]

For the sciences of nature, one of the most significant developments of the decades before 1800 had been the rapid expansion of periodicals. Not surprisingly, patterns of journal publication in Germany were less centralized and more diverse than in France or Britain. In the eighteenth century, German-speaking Europe had produced twenty-seven different general review journals (admittedly of varying success and longevity). In comparison, France had three and Britain four.[38] The Germans also had many more natural philosophical journals than other European countries—thirty-four to France's two and Britain's seven. If the German journals were greater in number, they were also often shorter in life span than their foreign counterparts.[39]

The decentralized, multicentered nature of German learned publishing was a source of both pride and concern to early nineteenth-century *Naturforscher*. J. S. C. Schweigger thought the lack of a dominant cultural center had a generally salutary effect on German *Wissenschaft*, but this situation came with the hazard, he noted, that publications sometimes failed to find their proper audience. Significant discoveries often languished in obscurity due to the lack of a "scientific point of unity."[40]

Indeed, by the early nineteenth century, the very ubiquity of books and periodicals had greatly devalued authorship's worth as a mark of intellectual status. Distinguishing the truly learned from the ever-expanding crowd of mere authors seemed an essential but difficult task. The learned handbooks of the period wrestled with this issue, often without coming to definitive conclusions. As the ranks of authors swelled, bibliographers debated whether all published authors deserved a place in compilations of the "learned." Earlier scholars had complained of the tediousness and difficulty of compiling guides to the learned world; now bibliographers wondered whether such exhausting industry was even worthwhile.

A comprehensive handbook, bibliographers worried, immortalized the names and achievements of men who had only a dubious claim to posterity's attention. Klement Alois Baader remarked that the title of his work, *Das gelehrte Baiern* (Learned Bavaria), was misleading. "Everyone knows that what one really means by the learned," he wrote, "is actually authors." He noted acerbically that there were "many writers and bookmakers who are not learned."[41] Baader's counterpart for the region of Schwabia, the preacher Johann Jacob Gradmann, mentioned that many people had advised him only to list "the most remarkable learned and worthy Schwabian writers." His book was, however, "not a *critical*, but a literary" work. Despite its name, it was not a reference work only about learned men, but a *Schriftsteller-Lexikon*, an encyclopedia of authors. Gradmann was more sanguine about the usefulness of compiling lists of second-tier authors. Even people who had only published short pieces in periodicals were included in his book, he informed his reader, "because news of even mediocre and lesser known authors can also be interesting in some respects, even should it only serve to provide knowledge of the literary industry of this or that city, region, party, and so on."[42]

Though Baader and Gradmann followed the bibliographic precedent of their predecessors, their introductory prefaces revealed a growing unwillingness to grant participants in the world of print the title of *"Gelehrte"* without qualification. Jöcher's 1750 lexicon, by contrast, had been explicit about its expansive use of the term; the work listed "learned people of all estates, both male and female." By 1800, bibliographers were more careful to distinguish between the truly learned and other kinds of authors, even as they still aspired to compile complete lists. This commitment to bibliographic completeness itself helped produce the sense that the learned world had become unwieldy and overpopulated. Information overload was not just a simple fact; the conscientiousness with which scholars pursued their projects of compilation helped make the expanding number of authors a source of social and epistemological concern. Attempts to concentrate, centralize, and manage the world of print made the new flood of published writings even more visible as a problem.

The increasing volume of print culture had other implications for the structure of learned identity. Well into the eighteenth century, extensive correspondence networks had still been part of the backbone of the learned world, but these patterns of communication faded into senescence as print culture expanded. The archives of learned societies capture this transformation clearly. In the eighteenth century, Berlin's Society of Nature-Researching Friends had expended much time and effort on handwritten correspondence with its geographically dispersed membership.[43] The correspondence files of Dresden's Society for Natural and Medical Knowledge, founded in 1818, make for much less interesting reading. After 1800, the letters that learned societies exchanged with nonlocal members were typically perfunctory. Societies sent letters offering honorary memberships, and they received back short, gracious, formulaic replies. The original statutes of the Berlin society had included extensive guidelines about how letters should be written, sorted, and sent. One of the first acts of the Dresden association, by contrast, was to order numerous scientific and medical journals for its members' collective use.[44]

Of course, personal correspondence continued to have a place in intellectual life in the nineteenth century, and printed periodicals already had an important role in learned exchange in the late seventeenth century. What had changed was the relative importance of each of these media, not the complete replacement of one by the other. It is also worth noting that this change did not represent a replacement of personal or private modes of exchange with public ones. Writing a letter was also a publicly relevant act in the republic of letters. The world of handwritten exchange had also not necessarily been an intimate one; important figures often had massive numbers of correspondents. But these relationships were still mediated through the language of friendship, and information was brokered through complex rituals that involved the repeated assertions of mutual *personal* goodwill.[45] The slow decline of correspondence networks (or, perhaps better, their increased privatization) did away with these reassuring communicative gestures; no similar individualized exchanges took place between the editors of journals and their readers.[46]

Some observers lamented the fact that other learned activities no longer commanded the respect they once had, now that publication had become so important. Early modern learned reputations had been built in part through printed exchange, but also through letter writing, teaching, and collecting. Even in the most self-consciously modern of forums, the declining value of these other sources of cultural status could cause anxiety and regret. In 1816, for example, F. A. Brockhaus, the printing house that produced the most popular German encyclopedia of the nineteenth century, published the first volume of a new reference series, a collection of biographical sketches entitled *Zeitgenossen*. The series, Brockhaus explained, would present the life stories of men and women who were "already known to the world," and who were "important and influential in a larger circle." To qualify for the series, a person must have led "a *public* life" and have had "a visibly historical relationship to their time." *Zeitgenossen* would chronicle the lives of great statesmen and military leaders, extraordinary businessmen, and "extraordinary masters of art and *Wissenschaft*." In the case of the latter, the series editor specified that these were not to be "mere authors or so-called learned men."[47]

What did it mean to live a "public life" as a man of science in the early nineteenth century? Who could claim to stand in "a visibly historical relationship" to his time? When the early nineteenth-century learned world looked back on its recent dead, the line between a "master of *Wissenschaft*" and a mere "author" or "learned man" appeared particularly difficult to draw, as the *Zeitgenossen*'s first biography of a natural philosopher revealed. In the series' second volume, J. J. H. Bücking presented the life of his former teacher, the Helmstedt medical professor Gottfried Christoph Beireis. Bücking's biographical sketch neatly captured the ways in which the rapid expansion of print culture had devalued earlier marks of learned status. Writing in the 1810s, Bücking felt the need to justify his mentor's modest number of publications, and he offered his explanation in a defensive tone. The current age's immoderate fascination with print, Bücking suggested, provided a distorted measure of a man's real importance. As a university professor, Beireis had trained "many praiseworthy and treasured learned men." In the final analysis, he had done as much as, if not more than, "many learned men who attempted to teach through their writings, as the Zeitgeist demands."[48]

The reasons for Beireis's moderate publishing record were complicated. He never published much of his chemical work for economic reasons; he had invented several commercially useful chemical processes that were more valuable to him as trade secrets. He also had a passion for old alchemical works, and along with this taste came a love of secrecy that had been part of early modern alchemical practice. But his sparse publishing record could also be explained, according to his biographer, by the fact that he had poured such effort into his teaching, collecting, and personal correspondence. One should be ashamed to chide the old man for his lack of publications, Bücking admonished his reader, given the time that Beireis's other intellectual activities demanded of him. His extensive scientific correspondence, his lectures, and his collections—all of these pursuits had taken his energy and pulled his attention away from the printed word.[49]

In defending Beireis, Bücking was fighting a losing battle. "Publish or perish" was to become the order of the day in German intellectual life. But Bücking's defense of his quixotic professor is nonetheless interesting, and, as Brockhaus required, in some meaningful way "of its time." The gradual devaluation of early modern marks of learned authority was not a painless process, and in the 1810s the world was not yet so changed that other forms of intellectual distinction could be discarded out of hand. Publications had been only one vehicle for eighteenth-century men to make their intellectual reputations; fame in the learned world also came through teaching (if they were professors), handwritten correspondence, and collecting. Bücking's eulogy for his teacher makes it clear that the reduction of a learned man to his published record could produce discomfort and resistance.

Criticism in a Learned Crowd

The world of printed exchange was not just densely populated; it was also thoroughly critical. Review journals first appeared in German-speaking Europe in the late seventeenth century, and as the German book market grew, these periodicals also multiplied, vying with one another for the privilege of guiding readers through the vast new field of published writings. These journals offered evaluations of more substantial published works, and already in the late eighteenth century, the practice of critical review had become an omnipresent accompaniment to all kinds of writing.[50] The eighteenth century's most important general review journal, the *Allgemeine Deutsche Bibliothek*, stopped publication in 1806, but it had a number of nineteenth-century replacements. The most important general review journals of the period were the *Allgemeine Literaturzeitung* (published initially in Jena and then in Halle), the *Jenaische Allgemeine Literaturzeitung* (begun in 1804 when the preceding journal moved to Halle), the *Leipziger Literaturzeitung*, and the *Heidelbergischen Jahrbücher der Literatur*. (The word "literature" in all of these tiles refers to published works in general, and not just to literary works in the narrow sense.) More short-lived review organs also appeared in Vienna, Munich, Erlangen, Salzburg, and Kiel. The period's more specialized journals also often included reviews.[51]

This dense network of critical print forums spawned anxieties not unlike those expressed by the novelist Karl Immermann. *Naturforscher* also worried that the public for which they wrote seemed full of unsympathetic readers who formed their judgments too quickly. One of the most important German scientific journals of the early nineteenth century, Lorenz Oken's *Isis*, began its life with reflections on just this issue. The *Isis* was initially planned as a general review journal (its focus narrowed only later). In the journal's first issue, Oken explained that the *Isis* would be devoted to the "spread of all human discoveries, and the general and thorough judgment of all the intellectual products of science, art, industry and crafts," a universal purview that placed it in a long line of generalist Enlightenment review organs.[52]

Yet Oken's journal broke with Enlightenment critical practice in several ways. Anonymity had been at the heart of the German Enlightenment's ideals for proper critical practice. Supposedly, it guaranteed that the reviewer remained free of undue personal particularity in his judgments, and also ensured that his statements would be taken on the basis of their intrinsic merits alone.[53] In direct opposition to this earlier tradition, Oken hoped to reintroduce the force of personality into critical practice. The *Isis*, according to its editor, was to be the forum for a new "learned state," one free from the shortcomings of the current learned world. Central to this new learned state was a reformed kind of *Kritik*.

At the heart of Oken's desired reforms was an attempt to reintroduce the force of individual personality into critical practice. The right kind of personal sympathy, he argued, was essential for the fair assessment of an author's work. The identity of the reviewer mattered, too, and should not be hidden from readers. In moving into the world of print, a piece of writing should not become an isolated and detached object, divorced from the moment of its creation. The author remained a part of his writings, constitutive of their status and meaning, even after they left his pen. Furthermore, individual fragments of an author's work should not be evaluated in isolation. In criticizing a man's work, his fellows should show respect for the totality of his efforts and not dwell unduly on isolated flaws.[54]

For example, as part of his own critique of a Berlin-based botanical periodical, the *Jahrbücher der Gewächskunde*, Oken took issue with one of the Berlin journal's reviews. The *Jahrbücher* had printed an anonymous review of Alexander von Humboldt's work on plant geography, and Oken censored the reviewer for not identifying himself. The *Isis* editor claimed to have guessed the review's author, and since the author was "the only man besides Humboldt in a position to criticize his work," Oken argued that the review would have been much more valuable if this man had attached his name to it. He also chastised the review for being "strictly polemical," bringing out only the ways in which Humboldt was wrong. "Even if it were swarming with mistakes," he continued, Humboldt's book "was still superb, if only because it proposes ideas that had previously hardly occurred to anyone."[55] One should pay attention to the spirit that animated a work, not dwell on its isolated errors. Not only works of theoretical synthesis or particular intellectual daring should be treated in this way; encyclopedic works of compilation deserved similar respect. When a Dr. Steudel from Esslingen criticized small mistakes in F. H. Dietrich's *Lexikon der Gärterey und Botanik* in a letter to *Isis*, Oken added an irritated footnote. "In reviewing a work, one must look at the whole, not solitary mistakes (in cases where they really are solitary mistakes)—and one must also pay much attention to the goodwill and zeal [of the author]."[56]

In emphasizing the role of individual personality within the apparatus of criticism, Oken broke with earlier critical practice in significant ways. In the late eighteenth century, critics had typically not signed their names to reviews; anonymous Reason could speak for itself. In contrast, Oken claimed that he would publish reviews by anyone who had published a book, and said he would publish multiple reviews of the same book by different people. In other words, he refused to take on the authority that an editor like Friedrich Nicolai had exercised during the Enlightenment. He refused to pick the people who published in his journal; he would not just solicit reviews from his correspondents, he would accept reviews from anyone who sent them in, as long as they themselves had published something beyond a dissertation. Oken explicitly described his new policy as a challenge to the "tyranny" of earlier review practices, and in this new open forum, anonymity also had to fall by the wayside. The disappearance of these two things together—the discretionary power of the editor to choose reviewers and the anonymous book review—adds support to the idea that the anonymous reviews of the eighteenth century had in part received their authoritative status from the selectivity of the forums in which they appeared. Now that Oken was no longer going to vet his reviewers, people needed to identify themselves.

In some ways, Oken's complaints represented nothing new. Criticism of petty contention and dissent had been a standard part of German intellectual politics throughout the early modern period. Indeed, the Enlightenment's public literary men had criticized Germany's old-fashioned *"Gelehrten"* for their pedantry and their love of dispute. To the mid-eighteenth century, polite literary culture presented itself as rising above the niggling and narrow passions that fueled traditional learned arguments.[57] By the 1810s and 1820s, however, the institutions of public literary life, with their cycles of judgment and review, seemed themselves easy conduits for inappropriate emotional excess and bad blood.

From this perspective, in fact, the learned world, though crowded, was not particularly anonymous, at least not at the top. In some ways, Germany's leading *Naturforscher* were already too well acquainted with one another, whether or not they had ever met in person. Particularly within a given specialty, researchers knew one another, at least at a distance. Reading an anonymous review of Humboldt's botanical work, Oken was fairly sure that he could guess its author. Germany was, in Georg August Goldfuß's words, "large, wide and divided"; it was also "full of universities and professors, full of priority disputes and anticipations, full, from top to bottom, with honorable and dishonorable conflict." These men had insulted one another in mean-spirited reviews; they had struggled against one another for rank and credit. Should they eventually end up in the same place, Goldfuß joked, someone would have to call out the gendarmes to restore public order.[58] In learned conflict, the distinction between a literary and personal assault on a man's honor was often anything but clear. "Literary conflicts," Oken felt it necessary to warn his readers, must never be treated as if they were civil (*"bürgerliche"*) affairs, the kinds of disputes that could be brought up in the form of libel charges in front of a judge.[59]

Oken, Goldfuß, and their peers faced a dual dilemma. The republic of letters was more densely populated than ever before, and older markers of learned status no longer offered reliable clues to its precise boundaries. The ideology of *Wissenschaft*, as the preceding chapter explored, was one solution to this problem; in the name of a nonutilitarian and unified *Wissenschaft*, it was possible to ignore contributions that had once commanded attention in the Enlightenment public sphere. Within the world of elite *Wissenschaft*, however, *Naturforscher* now competed with one another for national and cosmopolitan renown with ever-greater vigor. Protecting one's personal authority in this crowded and contentious cultural world, as Karl Immermann intimated, was a matter of marshaling receptive, discerning, and attentive listeners, willing to give credit where credit was due.

Local learned societies provided this sort of audience.

Civility and Unity, Face-to-Face

In his 1807 proposal for a new university in Berlin, the philosopher Johann Gottlieb Fichte observed that when the first universities were founded, books were few in number and difficult to acquire, and most knowledge was passed on orally. In his own day, books were anything but scare, and the universities, Fichte complained, had not taken sufficient note of this fact. Professors still gave lectures (in German, *Vorlesungen*); they still read out loud to their students even though the necessity of doing so had long since disappeared. Despite the expansion of the book trade, and despite the fact that there was no longer any branch of *Wissenschaft* that did not have "a superfluity of books," professors continued to stand before their students and recite information that was readily available in print. In his proposal, Fichte offered a plan that would make the higher schools more than just "the simple repetition of the already existing stock of books," suggesting changes intended to give the face-to-face education provided at universities its own independent logic.[60]

What was the point of oral communication in the sciences now that the world was so flooded with print? Fichte's question might also be asked about another nineteenth-century institution, the local learned society. In particular, the proliferation of *general* societies, devoted to all the natural sciences together, seems in need of further explanation, especially given the fact that progressive specialization has usually been seen as the dominant trend in German scientific life after 1800. Why, in a period when specialized scientific journals were rapidly growing in importance, did leading German researchers still spend so much time lecturing to one another in general local forums? As Georg August Goldfuß, professor for zoology and mineralogy at the University of Bonn, noted acerbically in 1821, Germany possessed so many academies and learned societies that all a *Naturforscher* had to do was "put on his boots" and "he can go be lectured to [*sich vorlesen lassen*] and give lectures himself [*selbst lesen*] to his heart's content."[61] All of this seemingly atavistic talking was not the business of out-of-touch provincials; many of the period's new societies were in important cultural centers. By 1830, the majority of German university towns had a natural scientific society or medical-scientific society, most of fairly recent origin.

In fact, the local learned associations of the early nineteenth century are best understood as strategic responses to both the opportunities and the challenges offered by an increasingly specialized print culture. In discussing how one might go about restoring sympathy and civility to German science, a number of commentators hit on physically localized sociability as a way to counteract the indignities of the cosmopolitan learned world. Oken, for example, had a social complement for the *Isis*'s critical mission, his plan for a yearly meeting of German natural researchers. There was, after all, no more direct or immediate way to reattach the personality of the author to his printed work than through physical presence. Regular, physically localized sociability, Oken hoped, would make the world of printed exchange more civil. It might be impossible to remove ugly passions from learned discussion entirely, but perhaps their virulence could be lessened. In the late 1810s and early 1820s, Oken tried to convince his fellow researchers to join together for a yearly national meeting, and in 1822 he succeeded in bringing twenty of his colleagues to Leipzig for the founding meeting of what became a successful national institution.

In his first published announcement for the meeting, Oken claimed that the main benefit of the gathering would be that it would temper the antagonisms that emerged in printed debate. In addition to providing a forum for scholarly discussion, Oken argued that these yearly meetings would serve "to create a more mild tone in the literature. . . . People who have met and spoken to one another face-to-face tend to have, even at a distance, if not a certain respect for one another, then at least a kind of shyness that hinders them from judging literary works with bitterness."[62] Oken had many motivations for proposing a national meeting of German natural researchers (not least of which was the cause of German national unity), but in his published appeals he emphasized improving civility as a major advantage of his idea.[63] "The main goal of the meeting," he wrote, "is: to see one another, to get to know one another, and to learn to value one another."[64]

Oken was not alone in hoping that face-to-face sociability could provide a corrective to the incivility of contemporary intellectual exchange, and his plan for a national society succeeded in part because Germany had a growing number of local scientific societies willing to host and organize the GDNA's meetings, which changed location every year.[65] Like Oken's national society, these limited local publics were often described as much-needed intellectual havens. In the 1810s, the mineralogist and *Naturphilosoph* Henrik Steffens offered a particularly interesting account of the dialectical relationship between sympathy and conflict in intellectual sociability. The human social drive, he argued, existed between two extreme moments. Cosmopolitanism, a devotion to humanity as a whole, was its most completely "real" instantiation. At its most extreme "ideal" pole, the social instinct was turned entirely inward in self-contemplation. For Steffens, neither extreme alone should determine the horizon of intellectual life. Creative individuality emerged from the interpenetration of these two polar extremes. "There is therefore no doubt," Steffens asserted, "that every person who wants to be human in the fullest sense must be both hermit and cosmopolitan."[66]

For Steffens, learned sociability existed in this fruitful middle ground of interpenetrating individuality and humanity, but unlike many of his eighteenth-century predecessors, Steffens did not move seamlessly between discussions of local intellectual communities and communities of readers and writers. On the contrary, he was interested in explaining the benefits of regular face-to-face discussions. Furthermore, while in the eighteenth century the category of "friendship" stretched easily to accommodate many forms of learned exchange, Steffens distinguished productive intellectual interaction from "friendship in the narrower sense." Intellectual camaraderie took the form of *Zutrauen* (trust or confidence), but not necessarily friendship. "Who among us does not know people," he asked, "to whom we would never reveal our full being . . . , who might even create a strong sense of opposition within us, but who nonetheless are dear and important to us?" As long as "stupid vanity" did not intervene, this productive coexistence of mutual trust and genuine opposition was the greatest guarantor of intellectual productivity. This creative space, according to Steffens, only emerged in a protected, limited group in which the participants "know one another, where no one is judged on the basis of a single statement"—a group in which there were no "traitors" who would surrender their fellows to the unkind judgment of "the broad crowd."[67]

Sociability not only protected the integrity of the self; it could also, supposedly, preserve the unity of knowledge. Words spoken face-to-face carried a power that printed texts or letters could not, and physically localized conversations were essential to maintain connections among the diverse branches of *Wissenschaft*. In an 1807 essay entitled "On Learned Societies, Their Spirit and Purpose," the philosopher Friedrich Heinrich Jacobi argued that "through the gathering of the members of a society in a single place," things could be accomplished that "dispersed, single efforts . . . would never be able to produce." Social concentration effected not quantitative, but qualitative change. Through regular gatherings and discussions, "Sciences that seemed foreign to one another realize their close and increasing relation; narrowness of perspective disappears."[68] Another famous reflection on intellectual community from the period, Schleiermacher's essay on university reform (written one year after Jacobi's), articulated a similar faith that scholarly sociability would ensure the "inner unity of all *Wissenschaft*."[69]

In addition, physically localized communities were assigned an essential role in bringing self and world—both the natural and scholarly world—into proper alignment. Enlightenment societies had described communication across great distances as relatively straightforward and unproblematic, and the geographically dispersed nature of their memberships had not been a source of intellectual concern. Early nineteenth-century *Naturforscher*, by contrast, doubted the power and transparency of printed exchanges. In their theoretical reflections on intellectual sociability, commentators like Oken and Steffens claimed that personal sympathy had a unique potential to build bridges over epistemological fissures. Such sympathy, they argued, emerged most easily from proximity. While Berlin's eighteenth-century Society of

Nature-Researching Friends praised the advantages of its far-flung membership, early nineteenth-century commentators were more likely to emphasize the benefits of geographic concentration. Intellectual action at a distance could not achieve the same results as face-to-face discussion; physical presence heightened intellectual receptivity. Words spoken in person carried different qualities than those that were written, printed, and circulated.

If, in their more prosaic realities, local learned societies were often less than the dynamic, generative intellectual spaces Henrik Steffens and his contemporaries dreamed they could be, there is still much evidence to suggest that these commentators captured an important aspect of the emotional logic behind local learned sociability's continued appeal. In crafting their bylaws, early nineteenth-century societies showed heightened concern for creating polite, orderly social spaces where the corrosive potential for dissent was carefully contained. Eighteenth-century societies had been more delicate on the subject of potential disagreements among their members. There would have been something uncouth in the suggestion that enlightened "friends of nature" were likely to fall into excessive conflict. Many early nineteenth-century societies, by contrast, went to considerable pains to legislate trust and respect. Groups founded in the 1810s and 1820s showed a new concern for managing disagreement, developing rules of conduct to minimize disruptions.

Civility protocols and etiquette had played a central role in scientific societies from the seventeenth century forward.[70] New to the early nineteenth century, at least in the German case, was an increased formalization of etiquette demands, which were now written at great length into the founding documents of new groups. The statutes of early nineteenth-century societies contained detailed lists of the personal qualifications and forms of behavior they required of their members. Eighteenth-century societies had generally been brief on this topic, content to state that the personal honor of society members should be unblemished. A mineralogy society founded in Dresden in 1816, in contrast, felt the need to say much more. It warned its members that "too forceful arguments, which might lead to misunderstandings and tensions among the members, are to be minimized," and anyone who was guilty of "creating useless quarrels" would lose his membership.[71] Dresden's Society for Natural and Medical Knowledge stipulated that a man who was "sensitive to contradiction" or "stubborn in his preconceived opinions" was not welcome in its ranks. In addition to requiring that members be men of *"wissenschaftlicher Bildung,"* the society also noted its members should possess "amicability, unselfishness, and love of truth." They should avoid doing anything that could bring "a lack of peace among the membership." They should be sympathetic and measured in their criticism. "A member must support the work of others willingly, with a friendly cast of mind; he must share with them the mistakes he discovers and seek to lead his fellows to better ways of investigation, should he know of them." If a member brought "dissention and division" into the society, he should recognize that he did not possess the qualities necessary for membership and leave the group of his own accord. If he did not leave on his own, he would be voted out. While in eighteenth-century societies a man might lose his membership if he lost his public honor, argumentativeness was now grounds for exclusion. While the final version of the Society for Natural and Medical Knowledge's statutes made lengthy observations of this sort, the first draft had gone on even longer. Several entire paragraphs elaborating on these themes were edited down for the final draft.[72]

The statutes of early nineteenth-century societies also went to unusual pains to formalize and order the kind of speaking that occurred during their meetings. Dresden's Mineralogical Society listed twelve different kinds of contributions that might be offered in the society and specified the order in which each should occur. The meetings were to start with (1) longer presentations on mineralogical subjects, continuing on through (5) "new chemical discoveries and observations," until the members arrived at (12) "free mineralogical discussion about fossils and other similar things."[73] The statutes of Dresden's Society for Natural and Medical Knowledge included a similarly detailed list of the order in which different sorts of presentations would be made.[74] The Leipzig Nature-Researching Society opened its meetings with the reading of the previous month's protocol; written contributions were read next, followed by oral presentations. Next came reports from the society's officers, suggestions from the membership to improve the society, the reception of gifts for the library and collection, votes on new members, and "free oral discussion."[75]

From their earliest instantiations, learned societies had been about the production of audiences. Seventeenth-century societies provided certifying witnesses whose testimony made the strange new empirical particulars of the Scientific Revolution travelworthy; they turned local events and observations into cosmopolitan evidence.[76] Some nineteenth-century figures also thought that local societies could serve this particular end. J. S. C. Schweigger, for example, declared that the most important goal of learned societies should be the "demonstra-

CHAPTER 4

tion and testing of new observations and experiments" because, he argued, autopsy was "the most important thing in natural science."[77] At the time he made this statement, Schweigger was studying electrical phenomena, an area of experimental philosophy in which the reproduction of experiments was both delicate and difficult (he was best remembered at his death for inventing a device to magnify the effect of weak electric current to make it more experimentally tractable).[78] Schweigger's view was relatively atypical, however. For most of Schweigger's contemporaries, physical presence was valued primarily as a way to ensure that people listened to one another with careful attention and sympathy. If seventeenth-century societies had been communities of witnesses, early nineteenth-century societies were first and foremost communities of critical but respectful listeners.

Learned societies in more provincial places were much less concerned with such complexities, and this fact lends credence to the suggestion that men in important cultural centers, as regular participants in print culture, demanded different things from local sociability than did their less prestigious contemporaries. A society founded in 1831 in the Saxon province of the Voigtland, for example, took its name from Dresden's Society for Natural and Medical Knowledge, but little else. The Voigtland society's statutes also addressed the question of critical discussion, but it had different concerns about criticism and chose different strategies to manage it. The group was primarily a decentralized regional reading society that provided its members with access to new medical books. It did have some ambitions to produce new knowledge, however; the regional society planned to hold two meetings a year at which its members would present original work. In order to ensure that its discussions would not suffer from "limitations on the freedom of judgment" or "factionalism," the Voigtland society chose the Enlightenment's favorite tool—anonymity.

In these biannual meetings, the society's primary concern was to limit the effect an author's identity had on the evaluation of his work. Its bylaws specified that members' scientific papers should be composed succinctly and in "clear handwriting." Handwriting mattered because the society had decided it would be "more appropriate to the goals of *Wissenschaft* and impartial criticism," if papers were read aloud by someone other than the author. The author would then remain anonymous until the members of the society had given their judgment.[79] In setting out these rules, the society hoped to prevent an author's identity from unduly influencing the reception of his work. To ensure that its members could speak their judgments freely, the individual author had to be removed from consideration.

Cultural centers like Berlin, Leipzig, and Dresden developed more complicated forms of self-discipline, self-abnegation, and self-assertion to guarantee safe dissent. In addition to the requirements mentioned above, the Dresden Society for Natural and Medical Knowledge stipulated that comments on a presentation were not to be made to the speaker directly; they should be directed at the presiding director. In offering criticism, members were only to speak "on topic," carefully avoiding "all remarks of a personal nature."[80] Redirecting attention and altering patterns of address short-circuited potentially disruptive disagreement. In commenting on a presentation, members were to make formal statements on a narrow topic, and they were to direct their contribution to the society as a whole (through the person of the director) rather than offer criticism of a specific individual. This device preserved a sphere of personal assertion (the author presented his own work) and also allowed dissent, but managed the latter in such a way as to make it less likely to register, at least in theory, as a personal affront.

Learned societies also created formal strategies for promoting the unity of natural knowledge. J. S. C. Schweigger pointed out that Germany's many *Lokalakademien* (local academies) could not usually boast "multiple men from exactly the same discipline."[81] With specialization the rule of the day, a unified natural science could only be a collective product. As the Society for Natural and Medical Knowledge's statutes explained, every branch of natural knowledge reached so deeply into the others that maintaining an overview was essential, though impossible for a single individual to achieve on his own.[82] Upon joining the society, members had to announce the subject in which they primarily wished to work. The Society of Nature-Researching Friends in Berlin had similar requirements.[83] At one of its first meetings, the Dresden group compiled a list of twenty-eight different specialized fields of medicine or natural knowledge that it hoped to divide among its members; each member would agree to give the society periodic updates on important developments within his chosen field. Nineteen of the areas received subscribers immediately; the director asked for volunteers for the remaining nine.[84] In practice, the members were not always conscientious about performing this task, but the idea of mutual education remained central to the society's self-understanding. In a speech in honor of the group's tenth anniversary, B. W. Seiler celebrated the success of the group's central mission. Each member brought "his own,

CHAPTER 4

that which belongs to him, and [made] the others richer through educating them."[85]

Speeches like Seiler's presented a rosy view of local societies as places of lively intellectual ferment. But how successful, really, were these groups at weaving together different strands of scientific research? Society meetings certainly sometimes did host debates on issues of central importance to the sciences as a whole, and they also fostered exchange among different specialties. Ilse Jahn has argued that Matthias Jacob Schleiden's contact with mineralogy through the meetings of the Society of Nature-Researching Friends in Berlin played an important role in the development of his cell theory, and she cites several other examples of the society's importance as a site of interdisciplinary exchange.[86]

Nonetheless, epiphanies like Schleiden's did not happen on a monthly basis. When one got down to details, furthermore, *Naturforscher* in this period hardly agreed on a single unified vision for the pursuit of natural knowledge. The period's many specific proposals for a new unified science of nature created as much conflict as they did consensus. But learned societies continued to meet, as regularly as clockwork, year after year, and much of what society members heard at meetings would not have borne directly on their own most cherished intellectual interests. Why did boredom not set in? To raise again the question with which this section started, what did local meetings offer that printed discussion could not? General considerations of dissent, civility, and solidarity provide only part of the answer. To understand why local societies proved satisfying institutions, we need to examine in greater depth the specific kinds of exchanges that took place in society meetings, as well as the changing role of learned societies as producers of publications.

The Creation of Learned Events

Local societies' own role as publishing organs changed markedly in the early nineteenth century. The transactions of scientific societies and academies had been some of the most important scientific periodicals of the eighteenth century.[87] With the rise of more specialized journals and the continued multiplication of societies, the proceedings of these groups became less central to learned communication. Berlin's Society of Nature-Researching Friends, which had steadily published its papers between 1779 and 1803, struggled to continue doing so over the first half of the nineteenth century. When it took up publishing again in 1807, it altered the format of its journal, changing its name to the more general *Magazin für die neuesten Entdeckungen in der gesammten Naturkunde* (Magazine for the newest discoveries in all of natural knowledge). Unlike the earlier version of the society's journal, the *Magazin* was published four times a year in an effort to match the increased demand for timely publication. This journal was a reasonable success, and it lasted until 1818, but the collective publication efforts of the Berlin society floundered through the 1820s and 1830s (it published one volume of proceedings in 1829 and another three volumes in 1836–39). The Nature-Researching Society in Halle also tried to adapt its proceedings to a quicker pace of publication, releasing issues instead of yearly volumes, with different issues devoted to different specialties. The last volume of Halle's *Neue Schriften* appeared in 1819, and the society did not publish regular proceedings again until the 1850s. Financial difficulties were often behind these sporadic publishing schedules. The society in Marburg had trouble scraping up money for its first volume, and it had to forgo illustrations to lower the cost; seven years passed between the publication of the first and second volumes of the society's proceedings.[88] Such erratic publishing patterns were typical of the period.[89]

Nineteenth-century societies were less successful as publishing organs in part because of a transformation in the structure of their memberships. In the eighteenth century, societies had often had quite extensive relationships with their nonlocal members, many of whom provided regular contributions for their journals. In the nineteenth century, nonlocal memberships became primarily symbolic, and the growing number of societies diluted the meaning of any one bond. The Heidelberg botany professor Johann Heinrich Dierbach belonged to twelve scientific and medical societies throughout Germany and Europe. At his death in 1843, the Dresden anatomist Burkhard Wilhelm Seiler belonged to twenty-three different medical and scientific societies.[90] Such high numbers of memberships were typical for university professors or other prominent researchers. As a consequence, scientific societies became more meaningfully local than their eighteenth-century predecessors had been. Society journals had once been the product of wide-ranging exchange networks; they now recorded scholarly activity in a particular place.

Given the range of other possible publishing venues, however, society members' loyalty to a local journal was not always strong. Particularly in university towns or major cultural centers, a society's most prominent members usually already had connections to other kinds of journals. Halle's J. S. C. Schweigger was the editor of the *Journal für Chemie und Physik*; Leipzig's Ludwig Wilhelm Gilbert edited the *Annalen der Physik*. Society members were generally active contributors to other kinds of periodical publications. The publishing patterns of the president of Dresden's Society for Natural and Medical Knowledge, the anatomist and doctor Burkhard Wilhelm Seiler, can serve as an example. In the early 1800s and 1810s, Seiler had published in Johann Christian Reil's *Archiv für die Physiologie*, the *Archiv für medicinische Erfahrung*, and the *Kritische Jahrbücher der Staatsarzneikunde*. When the Dresden medical academy began publishing a journal, he contributed a number of articles, but when the journal folded in the late 1820s, he found other places for his work. Over this entire period, he also contributed reviews to the Halle, Jena, and Leipzig review journals.[91] Berlin's Society of Nature-Researching Friends mentioned the problem of its members publishing elsewhere when it explained its decision to publish a journal that came out more often. Publishing four yearly issues would hopefully allow the society to "keep some portion of the large number of papers presented in society meetings which otherwise would find a home elsewhere in order to be published more quickly."[92]

Other research journals, it is worth mentioning, retained marks of their overlapping genealogy with scientific societies. Journals in the nineteenth century dealt with their authors differently than their twentieth-century counterparts. Contributors did not always submit discrete pieces to be evaluated on an individual basis. For example, once a scholar had had one article published in the *Zeitschrift für wissenschaftliche Zoologie*, he was considered a colleague (*Mitarbeiter*) and could expect that future submissions would also be accepted.[93] Some early nineteenth-century periodicals boasted of their numerous "*Mitarbeiter*" on their title page. The 1821 volume of the *Journal für Chemie und Physik*, for example, announced that it was published "in connection with J. Berzelius, G. Bischof, R. Brandes," and eighteen other researchers.

As society proceedings became less prominent, another sort of printed report became more common—published protocols of society meetings. Collections of these protocols appeared sometimes as independent publications; more frequently, they were sent to other journals and newspapers. Descriptions of Halle's Nature-Researching Society appeared in the *Journal für Chemie und Physik* in the 1820s. Dresden's Society for Natural and Medical Knowledge specified in its statutes that reports of its meetings were to be sent to "several newspaper editors so that they could be taken up in the learned papers."[94] Oken's *Isis* and other general review journals published reports from a number of different societies.

In comparison to other kinds of society publications, these reports provided readers with a much fuller picture of the minutiae of society meetings. It is hard to find any narrow intellectual rationale for sharing this level of detail. The topics covered in society protocols were often too diverse and too briefly described to be of much use to a reader in any robust way. Many of the events recorded did not represent the kinds of contributions that filled other journals, although some of the society presentations mentioned did represent extensive original research. In the early years of the Leipzig society, for example, anatomy professor Ernst Heinrich Weber presented several papers describing his ongoing work on physiological acoustics; he also gave a talk on "life" and "organization" as philosophical categories. August Schweigger offered his thoughts on the quest for a natural system of botanical classification, currently the dominant project in his field.[95] Many presentations, however, were less ambitious. Members spoke about interesting recent publications they had seen, showed new specimens from their personal collections, or described other miscellaneous observations.

Society protocols, in other words, recorded many transactions that failed to register in other print genres. They also described members' scholarly reading and collecting. These lesser activities had been central to the identity of early modern learned men, and previous exchange networks had recorded them with care. Under the changed circumstances of the early nineteenth century, local societies provided an organized, attentive audience for forms of behavior whose power as markers of learned status had been fading. With their narrower, more socially exclusive memberships, these groups separated out the mundane intellectual activities of a certain group of men, removing these acts from a broader cultural continuum and certifying them as genuine learned events. This function also helps to explain the increased formality of early nineteenth-century societies. The rituals of the meetings served not only to ensure harmony but also to create a space that was clearly distinct from other kinds of less formal intellectual sociability. The organized silence and directed attention that marked society meetings were important as marks of respect; the members of local scientific societies provided one another with mutual recognition that *their* collecting, observations, experiments, and reading were different in kind from the activities of those outside the circle of self-appointed local learned men.

From month to month, this tacit cultural project—the collective certification of learned events—provided the most persistent unifying logic behind local scientific meetings. A variety of observations and activities were considered worthy of memorialization in a society's protocols. Original research that would find a wide cosmopolitan audience ranked highest on the scale of contributions, but it was not the only thing that counted. The Leipzig Nature-Researching Society's statutes, for example, listed several kinds of presentations it would accept at its meetings. It would hear longer "written treatises" but also "notes on presented *naturalia* and books" and "reviews."[96] More modest, discrete observations made up a substantial portion of the society's presentations from month to month, and many of these observations were taken from the daily experiences of society members. Interesting notes about the behavior of household pets, strange symptoms encountered in patients, an unusual local thunderstorm, the animals on display at Leipzig's commercial fair—the Leipzig society's protocols recorded brief presentations on all of these topics.[97]

Society meetings also allowed members to display their cosmopolitan connections for a local audience, and protocols implicitly assigned honor to the role of conduit and intermediary. The physics professor Ludwig Wilhelm Gilbert informed the Leipzig society about significant experiments published in his *Annalen der Physik*, such as Joseph Fraunhofer's work on the solar spectrum and Hans Christian Ørsted's work on magnetism.[98] In 1818, Christian Friedrich Schwägerichen showed the society a gift he had received, a suite of fossils from Greenland given to him by "the famous Gieseke, Prof. in Dublin." Two years later, when the Leipzig group received a few volumes of society transactions from the United States, Schwägerichen took the opportunity to comment on the general state of American science.[99]

Society protocols, whether published or not, produced a kind of credit that was something less than full authorship.[100] Meetings instead formalized and preserved events. In the early modern learned world, learned conversation and reading had also been "events" within the republic of letters to be carefully collected and passed on.[101] Early nineteenth-century scientific societies continued this older appreciation for such intermediary intellectual acts. The crucial empirical discovery or successful synthetic theory that would gain one recognition in a pan-European community was, of course, the highest achievement, but from week to week, being a *Naturforscher* entailed more mundane things as well—keeping up with the literature, acquiring individual natural specimens, receiving news and gifts from distant colleagues.

Society protocols preserved a record of these activities, the less remarkable aspects of the modern intellectual's habits.

In other words, early nineteenth-century scientific societies were complex switching stations that transferred credit between local and cosmopolitan forums. On the one hand, members' interactions with the world of printed exchange were translated into cultural acts performed before a local audience. Fellow specialists elsewhere might ultimately judge the value of a man's research, but generalist local societies provided a broader kind of cultural credit, giving achievements won in specialized forums local purchase. On the other hand, society meetings ensured that a range of more mundane intellectual activities received formal certification as learned events. Upon publication of a society's protocols, these transactions once again registered in the cosmopolitan world of printed exchange.

Societies also provided members with enhanced opportunities for collective local self-representation. Learned societies brought together disparate individual activities and forged them together to make coherent, noticeable spaces. Published protocols brought a society's collective activity to the attention of a wider community of readers, and learned societies made their members' research activities more visible in local culture as well. City guidebooks typically included descriptions of local learned associations, and these groups formed part of the broader educated public's sense of the cultural landscape of a town. For example, an 1829 guide to Dresden provided descriptions of each of the city's scientific societies. Such local guides were not just for visitors; this particular book billed itself as a "Pocket Book for Foreigners and Natives."[102]

Unlike later associations like gymnastic societies that counted their success in their numbers, the power of learned societies did not depend on the breadth of their recruiting. The exclusivity of such groups made them valuable to their members. Like other "closed societies" such as the Freemasons' lodges and dancing clubs, learned societies measured their status less by size than by selectivity. Members of the Society for Natural and Medical Knowledge also participated in other "closed societies" in Dresden. Several of them were prominent members in the Harmonie, a social club that catered to the elevated strata of the city's middle classes. A number of local *Naturforscher* were also Freemasons. When writing about Dresden associational life in the 1820s and 1830s, a local historian of the Harmonie, in fact, mentioned several of the city's scientific and economic societies as potential competitors to his own club.[103] Whether or not the Harmonie's historian

CHAPTER 4

was correct, the fact that this perception could persist in institutional memory suggests the degree to which learned societies carried broader local cultural significance. Unlike the Harmonie and the Freemasons' lodges, however, scientific societies provided their members with a local guarantee of their *learned* identity. The Harmonie thought of itself as a *bürgerliche Gesellschaft*, a bourgeois society. The Society for Natural and Medical Knowledge, as a learned society, cordoned off a narrower group within Dresden public culture, and its members enjoyed a particular kind of status, the distinction of being locally recognized learned *Naturforscher*.

Conclusion

Because of the polycentric world in which they lived, Germany's university-educated *Naturforscher* faced a complicated task when they attempted to shore up their cultural authority in the early nineteenth century. This was true in two ways: there was no single place where questions of hierarchy among leading researchers could be settled, nor a single place from which one could adjudicate the question of where the learned world stopped and the general public began. Like their German colleagues, leading French and British natural philosophers also complained that science's popularity had created false authorities and compromised the process of knowledge making. But in crafting their responses, elite researchers in these other places had resources that their German-speaking colleagues lacked. In countries with a powerful national center, the political and cultural capital served as a prominent stage from which questions of legitimacy and authority could be answered. After the 1790s in France, the world of French science became more nationalized and centralized around Paris and the Ministry of Public Instruction, at least for while. In Great Britain, this process of centralization was not so clean, but by the first third of the nineteenth century, the clubs and scientific societies of London had significant power to organize scientific discussion and dispense intellectual recognition. Of course, Paris and London did not always succeed in making the provinces dance to their tune. Even within the confines of London, the "gentlemen of science" did not always have the last word. A low scientific culture with a strong radical political bent flourished right in their own backyard. In Britain and France, the metropolis was not the whole story, but there was a metropolis.[104]

No such thing existed within the complex political geography of

Central Europe; Germany was a land of many provinces, each with its own center, and each situated in a complex relationship of cooperation and competition with every other. As a result, the boundaries of "learned Germany," with its multiple possible print and local audiences, were extraordinarily hard to lock down. Educated Germans participated in a national public that stretched across a diverse array of political and social contexts and had no single focal point. By 1815, the upheavals of the French wars had simplified German geography substantially, but Germany was still a world of many provinces—a few dozen now, instead of over three hundred. As a result, most of the local learned societies discussed in this chapter were not *provincial* in the way that a Midlands Lit and Phil Society was in the British context. They were not self-conscious attempts to replicate the cultural life of the metropolis in a culturally less powerful setting. The general natural scientific societies founded in German university towns and important political capitals were the organs of competing centers.

A general science of nature made an attractive common cause because it fit well with the complex set of cultural needs that local learned societies served. As we have seen in this chapter and the last, these groups did multiple things: they distinguished learned *Naturforscher* from their erstwhile compatriots in the old enlightened cause of useful natural knowledge; they stabilized the learned identity of elite researchers, separating them out from a wider sea of less distinguished authors. They offered elite *Naturforscher* a refuge from the slights of specialized debate, and allowed them to broadcast their cosmopolitan accomplishments to an appreciative local audience while simultaneously advertising the collective intellectual clout of their cities to the wider world.

In explaining the spread of scientific societies in this period, social historians have often gestured toward some deeper, protean growth in the middle classes' "need" for knowledge in the nineteenth century.[105] If intellectual curiosity was a necessary condition for the proliferation of nineteenth-century learned societies, a generalized desire for knowledge explains neither the form nor the popularity of these groups. These spaces of discussion were not what Nicholas Jardine termed "scenes of inquiry." The contents of their meetings were too disparate and disjointed.[106] At yearly banquets, society directors might celebrate the mutual education that members provided one another, and such statements were not disingenuous, but local meetings were pleasurable and productive for many other reasons as well.

Early nineteenth-century societies differed from their eighteenth-century predecessors in the range of the subjects they covered, but they also had a different structure and a different cultural logic. More self-consciously local than their predecessors, learned societies ensured that the credit acquired through research still had cultural valence in the everyday lives of educated men, giving specialized accomplishments a presence in local spaces. These associations also honored more localized, embodied forms of intellectual authority, keeping records of a variety of activities that flew below the radar screen of print culture but remained constitutive of learned identity. Through the publication of their protocols and networks of honorary members, local societies then projected local learned events back out into the wider world.

These local groups were central to the progressive consolidation of "the natural sciences" as a category in this period. Just as they provided the ideal of "pure *Wissenschaft*" with a concrete social referent, these societies were also the most important forum in which the unity of natural knowledge was not just preached but also practiced. Occasional philosophical pronouncements to the contrary, the "unity" that natural scientific societies provided was a messy, composite one, produced more through repetitive juxtaposition than through formal intellectual synthesis. Applied mathematics, physics, chemistry, zoology, botany, physiology, mineralogy, and the other subdivisions of natural knowledge recognized by early nineteenth-century natural researchers— these were all discussed in the same local forums. Even more than the grand synthetic systems of the *Naturphilosophen*, this mundane reality helped make the natural sciences a recognizable unity within educated culture.

FIVE

Wooing the Polite Public

In the past two chapters, we have watched elite *Naturforscher* come together under the banner of a general science of nature. Breaking faith with the previous generation's cause of useful natural knowledge, they set themselves the collective task of studying nature together as men of *"wissenschaftlicher Bildung."* According to Carl Gustav Carus, their job was to explore nature's "law-like unity in the multiplicity of appearances." Many other *Naturforscher* would have argued with Carus's specific formulation of this project; they would have balked at the idea that a *Naturforscher*'s goal was to explore "the relationship between the appearances of nature and the laws of reason."[1] But Carus and his colleagues could have agreed on several basic points: all of nature was interconnected somehow; as a result, the natural sciences needed to be interconnected, too. They should work together to create a composite account of nature as a whole. The first characteristic of a *Wissenschaft* was that it was general; it did not get bogged down in practical necessities or in disparate, isolated details.

This chapter explores how the new, improved learned men of the early nineteenth century pitched their cause to the members of "the educated estates," the wider literary and sociable public. At first glance, one might imagine that they did not need to work very hard to do this, given the connections their new general science of nature had to the concept of *Bildung*. As we saw in the last chapter, a polite lexicon like the Brockhaus could summarize the new ideal of a general science of nature quite succinctly.

CHAPTER 5

Natural science was the image of nature as created in the mind of the cultivated person. What more needed to be said? We might describe the situation as follows: a new general science of nature appeared when *Naturforscher* appropriated a central value of a new, emerging bourgeois public culture and cast their science in its image.

German elite interest in a unified science of nature was clearly closely related to the more general ideal of *Bildung*. In its standard nineteenth-century formulation, *Bildung* involved a form of moral self-cultivation through which the human personality could supposedly become well rounded, harmonious, and internally integrated. For this process to work, the object contemplated in the process of *Bildung* also needed to be harmonious and internally unified, too. Nature studied in fragments could not serve this end; it needed to be studied as a whole.

Just to stop there, however, would make the story too pat, for the kind of public interest that elite *Naturforscher* hoped to gain for their science often diverged significantly from the kind of public interest they actually attracted. Moving too quickly into a discussion of "bourgeois culture" also erases key features of the early nineteenth-century cultural landscape. "The educated estates" were not just bourgeois, and the upper reaches of the period's polite public had an important aristocratic component. Indeed, as we already know, the circles around Germany's leading courts continued to play a crucial role in cultural politics throughout the nineteenth century.[2] Finally, assuming that a gesture to *Bildung* can answer all our questions about the relationship between science and the broader public sphere would keep us from being able to examine one important issue left on the table: as elite *Naturforscher* reformed the learned world in the first third of the nineteenth century, what effect did this have on their relationship with the educated public, given that the organs of these two bodies were so closely intertwined?

In defending their learned dignity, elite *Naturforscher* were not just retreating behind the old corporate barriers of the learned estate.[3] To cement a reputation as a learned man of any consequence, it was not enough just to talk with one's peers, the other members of the learned world. One also needed to be visible to the broader public. When the history of the German learned persona is written primarily as an exchange between rationalizing bureaucracies and the holders of university chairs, the contours of this broader cultural world can be hard to see.[4] But these wider horizons still mattered intensely to early nineteenth-century elite *Naturforscher*, as I hope this chapter makes clear.

In order to explore the issues outlined above, this chapter moves away from general groups and patterns to look closely at two individuals, Christian Nees von Esenbeck and Ludwig Reichenbach, both men who did their primary scientific work in botany. In spending so much time with these two individuals, I am not trying to set them up as lost giants of early nineteenth-century German science, men who ought to be counted alongside Goethe and Humboldt. These men are interesting precisely as successful researchers who were nonetheless *not* cultural superstars.

Both Nees von Esenbeck and Reichenbach presented their science to the broader public using the language of *Bildung*; both men were steeped in the values of Weimar classicism and looked to Goethe as a model when crafting their public personas.[5] As Reichenbach put it, in Goethe's life, the "unity of *Wissenschaft* and art . . . [had borne] the most beautiful fruit."[6] As the preceding quote suggests, these were men who shared the Romantic era's interest in fusing scientific and aesthetic forms of understanding, a trend for which Goethe stood as a primary example.[7] Both wanted to win the attention of the fashionable upper echelons of German literary life, a mixed noble and middle-class group that contemporaries designated as *"das gebildete Publikum,"* the educated or cultivated public.[8] Both saw their promotional efforts as attempts to train the public to appreciate *Wissenschaft*, the proper kind of natural knowledge.[9]

For all of these reasons, they are useful people to examine if we want to see how German *Naturforscher* attempted to make their presence felt within the period's broader reading and sociable public. Nees von Esenbeck's and Reichenbach's activities illustrate the ways in which the reformed learned world of the 1810s and 1820s still remained intricately interconnected with broader public culture. The late eighteenth-century learned world, as we saw in chapter 1, had been embedded within the wider public sphere, and early nineteenth-century *Naturforscher* were loath to sever these ties completely, even as they worked to create new, more exclusive kinds of learned institutions. For an ambitious young man like Reichenbach, just showing up for meetings at the Dresden Society for Natural and Medical Knowledge was not enough; he also wanted to win the attention of the court city's cultural and political luminaries. Nees von Esenbeck worked to revive the old Leopoldina academy, but a life publishing specialized books and austere academic proceedings with Latin prefaces only partially fulfilled his ambitions. In fact, he had initially built his public reputation primarily through his essays in the *Jenaische Allgemeine Literaturzeitung*, a widely read general review journal. Goethe had hoped to bring him to Jena in

1805, but as long as Nees von Esenbeck's country estate was bringing in a decent income, the botanist saw no need to take on the burdens of a professorial position. After the agricultural crash of 1816–17, however, he agreed to take a chair in Bonn (the Prussian government made him this offer in 1818).[10]

As Reichenbach and Nees von Esenbeck worked to heighten natural science's position in public literary life, the unity of natural knowledge served as an important rhetorical resource. Weimar classicism placed great importance on the unity of the work of art, and according to many *Naturforscher*, works of science could possess this quality in analogous ways. The skill with which one crafted scientific representations mattered, and works of natural science could themselves be worthy of aesthetic admiration. Nees von Esenbeck and Reichenbach both hoped to teach their readers an appreciation for the beauty, not just of nature but of natural science—and not just its content but its form. Just as one might marvel at the elegant skill evidenced in a poem or a musical score, a properly trained reader could also admire a scientific text. In this and other respects, Nees von Esenbeck and Reichenbach imagined a more complex relationship with their audience than did later scientific popularizers. When they wrote for a wider audience, their goal was often to transfer scientific skills, not just information, to their lay readers. They continued to write as if the scientific world were open to autodidacts, at least to a certain extent.

Their efforts also allow us to see how devotion to a general science of nature dovetailed with, rather than working against, the promotion of individual scientific disciplines. Both Nees von Esenbeck and Reichenbach offered the public their version of scientific botany; in describing what this scientific botany *was*, they both talked a lot about the need to understand plants within the broader context of nature as a whole. In Reichenbach's case, we also have another chance to see how appeals to general *Wissenschaft* were linked with attempts to bring practical discussions more firmly under scientific tutelage.

Most important, these two men help us see that elite *Naturforscher* were not always able to translate the idealized learned personae they imagined for themselves into a concrete social form. Indeed, the *learned* aspects of their identities were both a source of prestige and a potential liability in the broader public sphere. This was particularly true given that the fashionable literary public they wanted to attract included both men and women. For Reichenbach especially, elite women were an appealing audience for botany; convincing fashionable ladies to take more interest in botany seemed an excellent way to help the field shed its more lowly associations with lower-ranking practical men like gardeners and apothecaries. According to the reigning gender norms of the period, however, women were not supposed to show too much interest in "learned" matters. Practically speaking, it could actually be quite challenging for the natural sciences to be learned and cultivated (*gebildet*) all at the same time.

Both of these two men had successful scientific careers, but in the two instances examined here, neither of them got exactly what he wanted. What Nees von Esenbeck wanted was a natural scientific supplement in the *Morgenblatt für gebildete Stände*, a platform from which he and his intellectual allies could display their polished and sophisticated science to a fashionable audience. To that end, he sent in a treatise explaining how the new scientific botany was part of an emerging new understanding of nature as a whole. His editor, Therese Huber, sent it back. The public did not want to read paeans to the unity of Nature; they wanted anecdotes, animal stories, and novelties. Could he please send in some more of those? Not, Nees von Esenbeck answered, and still preserve his dignity as a learned man. What Reichenbach wanted was a botanical society that drew in the leading figures of Dresden's courtly and literary life. In the end, what he got was a society full of gardeners. In the fluid cultural landscape of early nineteenth-century Germany, crafting the public persona one wanted was not a straightforward matter. Before moving on to Nees von Esenbeck's and Reichenbach's dilemmas, however, we should pause to look briefly at women's relationship to the sciences in this period, since gender emerged as a key sticking point in both of these cases.

Gender and Science in the Early Nineteenth Century

In Caroline Rudolphi's 1807 *Gemälde weiblicher Erziehung* (Portrait of a feminine education), natural history and natural philosophy appeared casually interwoven in the daily lives of young girls in a variety of guises. Girls botanized on their walks and in the garden. A thunderstorm offered their guardian an opportunity to discuss electricity. They also spent several hours a week formally studying these fields. But Rudolphi set clear limits on female children's introduction to learned pursuits. The father was to be the child's "representative of the True"; the mother, "the essence of the Beautiful and Good." This gendered division of labor mandated that, while a mother could do much to shape a child's appreciation for nature, "all truly scientific [*wissenschaftliche*] les-

CHAPTER 5

sons" should be taught by men. The children in her story received their formal instruction in natural history and natural philosophy from a local minister. Girls learned natural history to train their aesthetic taste and heighten their religious sensitivity, but Rudolphi's authoritative narrator proclaimed that women "cannot do research themselves . . . for well-known reasons."[11]

Rudolphi's pronouncements fit well with our current picture of the decades around 1800 as a time of narrowing opportunities for women in the sciences, a period when the increasing professionalization of science and the privatization of the middle-class family closed off the avenues that had allowed women access to natural historical and natural philosophical discussion in the past.[12] In the German context, the late Enlightenment and early Romantic periods offer a brief respite from this general narrative of increasing exclusion, with the Jena circle around the Schlegels providing the best example of how novel ideals of friendship opened up avenues for gender experimentation, allowing a select group of men and women to play games at the boundaries between the masculine and feminine. This tolerant, playful period of intergender friendship supposedly came to an end with *Naturphilosophie*, which once again set up more rigid limits on the proper sphere of feminine self-assertion.[13]

There is much to recommend this general account, but one can also find intriguing counterexamples. In 1830, Johann Christian Friedrich Harleß, a medical professor at the University of Bonn, used many of the same gendered images that Caroline Rudolphi had evoked but turned them to completely different ends. In his *Verdienste der Frauen um Naturwissenschaft und Heilkunde*, Harleß presented a historical overview of women's past contributions to medicine and the natural sciences, both as patrons and as scientific authors. In the preface, he wrote that he hoped to attract "intellectual *women* as readers [emphasis in original]." Harleß's book was dedicated to an aristocratic woman, the Electress Auguste von Hessen. He praised Hessen for her past support of "teachers and literati in the sciences as well as the fine arts."[14]

Harleß encouraged women to pay greater attention to the natural sciences and medicine, and not just as passive readers. He opened the book with the assertion that "it has not only been ordained for the man to research nature; it is not only in *his* spirit that the holy fire of *Wissenschaft* burns [emphasis in the original]."[15] According to Harleß, men and women gained insights into nature differently, and he relied on commonly employed gender dichotomies to explain these differences.[16] His opening passages extolled the piety, emotional sensitivity, and purity of spirit that characterized woman, "the more noble of the genders." He then went on to say that while the man, in responding to nature, looked for causal explanations and laws, the woman's soul "reveled, without demanding anything," in the beauty of nature, finding there the "highest thing that it sought"—love.[17]

Harleß's opposition between feminine emotion and masculine reason was a common one, but he employed it to unusual ends. This distinction was often used to argue *against* women's participation in the sciences.[18] Harleß used it as a starting point for his book-long argument that women were well suited to natural scientific pursuits. The woman's undemanding emotional intuition of nature was, he concluded, the "most natural and most vital understanding of the external world [*Aussenwelt*], at once both the most beautiful and the most elevated."[19] The book devoted long sections to royal and aristocratic women who acted as patrons to medicine and the natural sciences, but also listed women who were scientific authors. Among women's contributions to medicine and the natural sciences, he included cookbooks and other works on "*Haushaltungskunst*" (the art of keeping house) expanding the circle of the sciences wider than was standard in the early nineteenth century.[20] Many of the authors Harleß mentioned had published on topics that fit comfortably into prevalent ideals of feminine domesticity—they had written cookbooks, for example, or books on the education of young children. He did not, however, argue that these were the only proper areas for women's scientific efforts. In the introduction to his final section, he informed the reader that "here we will also meet some natural researchers in the fullest sense of the word—in botany Libert, Hutchens, de Bonay, Griffith, etc." If most of these authors had only published a few works, these had been "so valuable that they could only leave us with the wish that they would present more of such flourishing fruits to the public."[21]

Though women's literary activity was the frequent subject of satire in the early nineteenth century, it was not unusual for upper-middle-class and aristocratic women to put themselves forward in print. By one estimate, one-fourth to one-third of published German-language authors in the nineteenth century were women.[22] In his own survey of the contemporary literary market, Harleß noted that women had published far fewer works in the natural sciences than in belles lettres, but he asserted that when women *had* authored scientific works, they were of high quality. In medicine and the natural sciences, "the female authors of our time have shown such productivity and breadth that it can only inspire respect and admiration," while in other areas of literature, women's contributions "have often inspired misgivings and sharp criticism."[23]

Harleß turned contemporary stereotypes about female authorship to his own purpose. The problem was not women's authorship or intellectual activity per se; women had just focused their efforts in the wrong areas. Citing his fellow bibliographer Carl Wilhelm von Schindel, he noted that at least 550 German women had written works for publication since 1800. Most of these literary productions fell under the category of belletristic, and in Harleß's view this was a mistake. The natural and medical sciences, along with the household arts, were the fields to which women ought to direct their attention.[24]

Harleß's work brings our attention to an obvious and important fact—the German reading (and writing) public included both men and women. As a result, if we want to understand how the advocates of *Wissenschaft* envisioned their relationship to the wider reading public, looking at their discussion of women is a useful place to start.

Rudolphi's position, that the line around *Wissenschaft* was a gendered barrier women should not cross, was a widespread one, and this configuration had enormous structural weight behind it. State service, the universities, and the higher academies were exclusively male domains. If we look at broader patterns, Harleß's work pulls in the opposite direction from the dominant trends of the period. There can be little doubt that the progressive consolidation of a more elite form of institutionalized science after 1800 further eroded women's access to science. Harleß's work cuts against the grain in ways that are particularly illuminating, however. In the late eighteenth century, the presence of women had often been used as a mark that separated the learned from the larger educated public. Where we see early nineteenth-century elite *Naturforscher* addressing the issue of how their work might appeal to women, these are also good moments from which to explore how they saw the relationship between the learned world and other kinds of audiences and readers.

Nees von Esenbeck and the *Morgenblatt für gebildete Stände*

In 1816, the botanist Christian Gottfried Nees von Esenbeck began negotiating with the Tübingen publisher Johann Cotta to resurrect the *Acta* of the Leopoldina, a publication defunct since 1783. This austere periodical, which came back into print in 1818 with its Latin preface and Baroque pseudonyms intact, was perhaps the most vivid example of the persistence of early modern learned traditions into the first decades of the nineteenth century. The partial revival of older learned allegiances, however, was only one aspect of Nees von Esenbeck's cultural ambitions. Beyond the circle of his learned colleagues, there was the *"gebildete Publikum,"* the cultivated public, still to be courted. At the same time that Nees von Esenbeck was recruiting Cotta's help with the *Acta*, the two men were also considering collaborating on a different sort of periodical—a weekly natural scientific supplement to accompany another of Cotta's ventures, the *Morgenblatt für gebildete Stände*.[25] An influential literary paper founded in 1807, the *Morgenblatt* was at the height of its popularity in the late 1810s, when it offered about eighteen hundred readers across the Germanies daily doses of literary criticism, travel writing, and poetry.[26]

Both of Nees von Esenbeck's projects with Cotta, as it turns out, fell through. The *Nova Acta* eventually went to a publisher in Erlangen, and then later to another press in Bonn, and the natural scientific supplement to the *Morgenblatt* never appeared at all. The deal with Cotta for the *Nova Acta* died for contingent, practical reasons; the *Morgenblatt* supplement failed, in contrast, because Nees von Esenbeck and its editor disagreed over the kinds of scientific writing the journal's mixed-gender public would want to read. Over almost two years of negotiations, Nees von Esenbeck's interactions with the Cotta press produced an intriguing correspondence that sheds light on the challenges, opportunities, and limitations that members of his generation faced in crafting their public personas as *Naturforscher*.

In the early stages of their negotiations, Nees von Esenbeck and Cotta seemed in close accord over the form of their scientific supplement. Both agreed that it would provide a literature review that would be useful to all, both learned and *Liebhaber* (amateurs), who were interested in the study of nature. Writing to Goethe, Johann Cotta expressed his wish that the new natural scientific supplement would "become everything and give everything to the friend of nature" that his *Kunstblatt* provided for the friend of art; he hoped it would become "a *Communications-Blatt* that [would] bind together *all* the friends of nature" and "bring *Wissenschaft* into contact with real life."[27] In a letter to Goethe, Nees von Esenbeck described the venture in similar terms; the supplement would "bring natural philosophical views a little closer to the cozy life of the unlearned."[28] The articles would be accessible without losing their rigor; they would serve to make "those things that are generally understandable and instructive more widely known, without becoming unscientific [*unwissenschaftlich*]."[29] The collaboration seemed like a good fit in other ways, too. The *Morgenblatt* allied itself with the literary criticism of Weimar classicism (Schiller and Goethe both had close ties to Cotta),[30] and Nees von Esenbeck portrayed Goethe as a primary source of inspiration for his botany.

After his initial discussions with Cotta, Nees von Esenbeck began dealing directly with the editor of the *Morgenblatt*, Therese Huber, the widow of Georg Forster and Ludwig Ferdinand Huber, and daughter of the famous Göttingen professor Christian Gottlob Heyne. Huber had recently taken over editorship of the *Morgenblatt*, and would continue to run the paper for another six years. (She signed her letters to Nees von Esenbeck only as *"Redaktion"* (Editorial Staff), and the issue of her gender never entered directly into their correspondence; though Nees von Esenbeck doubtless knew to whom he was writing, he did not make that plain, continuing to use *"Wohlgebohrner Herr"* (Well-Born Sir) as his form of address through all their exchanges.) Nees von Esenbeck and Huber spent more than a year negotiating the *Morgenblatt* supplement, only to end in a stalemate.

Over the course of their debate, both participants delineated markedly different visions of what they thought the *"gebildeten Stände"* (the educated or cultivated estates) needed and wanted to know about the natural sciences. While Huber asked for discrete pieces of scientific "news," Nees von Esenbeck wanted to use the *Morgenblatt* supplement to build a public that could understand the broader historical significance of contemporary scientific work. He wanted to train readers who would respond to his scientific writing aesthetically as well as intellectually, and who were willing to appreciate the literary creations of natural researchers as works of art in their own right. In the service of this pedagogical program, Nees von Esenbeck sent in two kinds of articles—historical overviews of different branches of the sciences, and extended reports describing new discoveries or observations. Some of these he had authored himself; others were the work of colleagues with similar intellectual allegiances, such as the Erlangen professor Georg August Goldfuß (1782–1848) and the Würzburg professor Ignaz Christoph Döllinger (1770–1841).

Through the *Morgenblatt* supplement, Nees von Esenbeck wanted a publishing outlet designed not just to communicate the *results* of his (and his colleagues') research to a broader audience but also to educate this audience to appreciate their research in all of its historical sublimity. From the combination of research reports and historical overviews, the readers of the *Morgenblatt* were supposed to gain not only a sophisticated appreciation of new discoveries but also a proper sense of their historical meaning. For example, an essay on the history of botany that Nees von Esenbeck composed for the supplement, eventually published later in Oken's *Isis* with the title "On the Metamorphosis of Botany," described the development of botanical science from Theophrastus to Goethe; Nees von Esenbeck dubbed the latter the "Father of Botany." Goethe had earned this place of honor, in Nees von Esenbeck's opinion, because of his discovery that all plants emerged out of modifications to the form of an original archetypal plant; this insight provided botanists with the key to discerning the natural relationships among different vegetable forms.[31]

For Nees von Esenbeck and his fellow *Naturphilosophen*, the contemporary state of a science could only be understood in historical perspective. The dialectical character of history was such that progress came about not in a simple linear forward motion but always also as a process of return to the point of origin. The march of philosophy consisted in a complex circling back and moving forward, and the history of a discipline uncovered the progressive, historical unfolding of the ideal structures that were visible in, though not coterminous with, the real. It was the quality of a true *Naturforscher* to be able to grasp this fundamental ideal unity behind both nature and history.[32]

Nees von Esenbeck's editor was unimpressed with the grand historical narratives he penned for her publication. Huber considered the articles completely out of step with the requirements of the paper. "I really don't understand," she wrote to Cotta after receiving the first batch of manuscripts, "how a man who had even just glanced through the *Morgenblatt* for a month or so" could think that essays like the ones the botanist had sent would be "at all suitable for [the paper]."[33] Huber found both the lengthy historical essays and the research articles inappropriate. The historical overviews were too long and too learned, and the narrower articles were full-scale research reports, not the sort of short "notes" that would be published in a literary paper. She wrote to Cotta in frustration that Nees von Esenbeck had already sent her "over four hundred pages (lots of them folio size) . . . and in these he has yet to make it up to the present time."[34]

The historical overviews were the most serious source of disagreement. Huber considered these lengthy treatises anathema to a publication designed to report the novel and the new. Huber did find a place for some small portion of the work that Nees von Esenbeck sent her—a few short animal stories that had some "*new* worth." She decided to go ahead and publish them in the *Morgenblatt* itself, "since in fourteen days or four weeks they'll be old."[35] After the negotiations over the supplement had finally failed, she would note acerbically that Nees von Esenbeck and his colleagues had sent in so many manuscripts that "the agreed-upon space [in the supplement] would have been filled for an immeasurable time before they'd finally managed, according to their standards, to prepare the public sufficiently to be au courant of emerging scientific news."[36]

In Huber's opinion, the style of the submitted essays was also all wrong. The historical articles on magnetism and on botany, for example, were "too emphatic, too mystical, poetic, and lengthy to fit our space or our goal: impartial [*unparteiische*] description and instruction." Nees von Esenbeck and his colleagues' writings, far from being "impartial," were filled with invective; what's more, Huber asserted, the facts reported in many of the essays were already widely know.[37] The essays' extravagant language was also objectionable; the author could have easily made his point without all the bombast about "'angelic clarity' and 'a return to the ancestral state of purity' and 'a turning away from the life of the senses' and such." She found the essay on botany no better. It was "unclear, peevish, and, in the final analysis, copied out from Goethe."[38] In her letters directly to Nees von Esenbeck, Huber couched her criticism in more diplomatic terms. After receiving the first group of manuscripts, she wrote, "The material you have sent all seems to be too lengthy and treated in too scientific a manner." She reminded him and his colleagues that "our public is really more mixed; we are aiming at laypeople, and they want only short lessons—more facts and fragments than theories."[39]

Like his lengthy histories, Nees von Esenbeck's heavily metaphorical and emotional language also had a higher purpose, and the botanist was reluctant to alter his style. Many men in this generation of German scientists (*Naturphilosophen* like Nees von Esenbeck and Goldfuß as well as other figures like Alexander von Humboldt) were committed to the idea that aesthetic and scientific perception were closely related; consequently, a reader's aesthetic response to a scientific text or image could be important epistemologically as well.[40] For the *Naturphilosophen* in particular, deeply indebted to the debates sparked by Kant's philosophical work, the relationship between "things-in-themselves" and their representations had been a central issue in their own intellectual development, and the form that their writing took showed their awareness that their representations of nature *were* representations; use of metaphor and analogy could gesture toward fundamental, ideal connections that could not be represented directly.[41] What's more, as Nees von Esenbeck argued when he spoke of Goethe's botany, the products of science could *themselves* be works of art, worthy of aesthetic admiration.⁴²

In other words, the elements that Huber found inappropriate in Nees von Esenbeck's writing were exactly those things he considered most important. Consenting to her suggested changes would have meant abandoning his effort to initiate his readers into proper forms of scientific sensibility. In this respect, it is important to point out that Nees von Esenbeck still wrote under the assumption that it was not desirable to create too stark a separation between the *"wissenschaftliche"* and the *"gebildete"* public. He argued that the *Morgenblatt*'s editor had too low an opinion of her readership; he considered it reasonable to expect that the supplement would be useful to the learned as well as *Gebildeten*.⁴³ It should publish articles that were "popular without being common, because the German public is mature enough for scientific stimuli."⁴⁴ In choosing material, "the intelligent reader, if not also the learned one, must always be kept in view." It was necessary to include extensive scientific reports because even "among the so-called dilettantes one finds many friends of, and even many experts [*Kenner*] in, natural history and chemistry."⁴⁵

The final break between the editor and her recalcitrant contributor occurred when Huber returned Nees von Esenbeck's botanical essay with a terse request for revision, noting that "our publication must work toward the greatest simplicity and matter-of-factness [*Sachlichkeit*]."⁴⁶ From the botanist's perspective, the changes Huber desired were impossible. The essay was more than a history; it was a scientific manifesto that positioned Nees von Esenbeck's own work as part of a grand historical drama. He responded to Huber's criticism with the assertion that she did not know her own readership. "For among [the readers of the *Morgenblatt*] that I know, I am sure that the majority would certainly not read animal stories and book titles, but would be heartily thankful to me for my history of botany and for what I say therein about Göthe's *Metamorphosis of Plants*—and one would even find women in this group."⁴⁷ Huber's desire for "newsworthy" scientific articles, her demands for brevity, novelty, and "matter-of-factness," would make it impossible for Nees von Esenbeck to appear before the public in the guise he most desired to adopt. "The fact that you had sought out learned men," he upbraided Huber, suggested that "you were interested in a more sophisticated intellectual treatment of the material than just a newspaper article."⁴⁸ His collaborator Goldfuß echoed the sentiment in his own angry letter to Cotta. If the publisher of the *Morgenblatt* was unhappy with their work, then it was only because "you were thinking of something under the category 'natural historical treatises' that was different from what a *Naturforscher* would or could provide."⁴⁹ In her final note to Cotta on the whole affair, an exasperated Huber asked rhetorically, "If one were to ask an impartial third party . . . are these contributions written to conform with the character of the *Morgenblatt*, are they suited for a mixed public, which for the most part has no scientific [*wissenschaftliche*] training? He would answer, No! They are too learned."⁵⁰

As the references to women and to the "mixed public" in these final exchanges illustrates, the issue of gender emerged as a sticking point in Huber and Nees von Esenbeck's struggle to find a mutually agreeable form of accessible scientific writing. In her admonitions for clarity, novelty, and brevity, Huber made repeated reference to the "mixed public" of her publication. Nees von Esenbeck protested that "even women" would find his history of botany interesting. These opposing claims point to a potential problem that natural researchers encountered when writing for the broader public. Contemporary satire warned women not to be too "learned" or pedantic; at the same time, women were an important part of the cultivated public that read influential literary papers like the *Morgenblatt*. The next section looks at another natural researcher, Ludwig Reichenbach, who undertook even more concerted efforts to make the natural sciences the objects of feminine attention.

Ludwig Reichenbach, the *Abendzeitung*, and the Flora Society

The *Morgenblatt* affair was not the last time that Nees von Esenbeck wrote for an introductory audience. Like many other German academic botanists in the period, he published an introductory *Handbuch der Botanik* (Nuremberg, 1820–21).⁵¹ Among the most prolific authors of introductory textbooks was Nees von Esenbeck's Dresden colleague Ludwig Reichenbach, another acolyte of Goethean botany. The son of a *Gymnasium* teacher from Leipzig, Reichenbach came to Dresden in 1820 to serve as the director of a newly founded botanical garden and as professor of natural history at the medical academy. Soon thereafter, he also took charge of the royal natural history collections.⁵² Over the same period, he developed his own natural system of plant classification.

Reichenbach had spent the 1820s in a flurry of publishing activity, most of it, in one way or another, in the interest of popularizing his natural system. He had first presented an outline of his system at the 1823 meeting of the Society of German Natural Researchers and Doctors and promoted it industriously thereafter. He edited a revised second edition of Johannes Mössler's *Handbuch der Gewächskunde*, a three-volume work that served both as an introductory textbook and as a flora of common German plants. In a footnote in his preface to this work, Reichenbach referred readers to his previously published *Catechismus der Botanik*.[53] Later on in the work, he admonished readers that "the most important thing for a botanist these days is to have a *natural* way of viewing the plant kingdom," and suggested they consult his other published work on this topic.[54] The third volume of the handbook advertised herbariums being produced under his editorship, a project that combined the efforts of sixty-three botanists from all parts of the Germanies.[55] Reichenbach also produced volumes of serially published botanical illustrations, aimed both at serious botanists and at interested *Liebhaber*. One series publicized newly discovered plants from around the globe; another depicted hard-to-identify native German plants.[56]

For a taxonomist like Reichenbach, publicity was a useful ally. The "artificial" system developed by Carl Linnaeus, which classified plants solely according to similarities in their reproductive organs, still remained popular in early nineteenth-century introductory works. The system was simple to learn; to identify an unfamiliar plant, the collector only needed to consider one aspect of the plant's appearance—its reproductive organs, the pistils and stamens. Despite this system's popularity, however, many of Linnaeus's contemporaries (and indeed Linnaeus himself) had not believed that a single set of characteristics could capture the natural affinities among different plant species. In the eighteenth century, several botanists created competing "natural systems" that took multiple aspects of a plant's appearance into consideration. By the early nineteenth century, a number of German botanists had offered systems along similar lines.[57] The competition to become the next Linnaeus was fierce; over twenty different systems were vying for acceptance in the first half of the nineteenth century.[58] Systems of botanical classification, like other standardization projects, only realized their potential power if large groups of people employed them. For Reichenbach, then, success was in part a problem of publicity and persuasion.

Even after publishing two different introductory botanical handbooks in the 1820s, Reichenbach still felt the need to write yet another introductory work directed specifically at "women, artists, and friends of natural science"; his *Botanik für Damen, Künstler, und Freunde der*

CHAPTER 5

Naturwissenschaft appeared in 1828. A parallel set of local social efforts accompanied this published appeal, and through juxtaposing these interlocking efforts, one can see that Dresden's polite literary culture made women and artists a particularly attractive audience for an ambitious young naturalist like Reichenbach.[59] Before going on to look more closely at Dresden in the 1820s, however, it would be useful to spell out in more detail exactly what Reichenbach thought he had to offer both artists and women.

Reichenbach opened his *Botanik für Damen* with a statement that echoed Nees von Esenbeck's assertions about the "maturity" of the German public. "The *Bildung* of the public is now at a level that gives rise to high expectations, and *Wissenschaft* itself is different in its internal configuration than it has been in the past."[60] Reichenbach continued, "If I have dedicated this little book to cultivated, thinking women, I do not fear any reproach as a result." In what followed, he touched only lightly on how women's special "nature" might affect their relationship to botany. His greatest concern was to argue for the appropriateness of botany as an object of feminine attention. Botany had long been accepted as an appropriate activity for women, he argued; the book's dedication page, which listed eleven female botanists (most from the preceding century), provided evidence of the large number of women who had contributed to this branch of learning. We should not be surprised, he continued, "when they, the cultivated, the thinking, concern themselves with the loveliest, most mysterious of the kingdoms of nature in preference over other, less spiritually inspiring [*geistvollen*] pleasures, since striving toward the solution of hidden truths is so integral" and "at once so natural to them, that there is no lack of examples of them cultivating the most abstract sciences," like Greek philology and mathematics.[61]

The final section of Reichenbach's book gave directions for how one moved toward a scientific (*wissenschaftliche*) understanding of nature, beginning with the mundane details of creating a herbarium and ending with the heady philosophical heights of perceiving the Divine in each natural object. Reichenbach gave instructions for collecting, drying, and preserving plant specimens; he also directed readers to more detailed introductory works, guides for creating herbariums, and useful sets of botanical engravings (conveniently, Reichenbach had produced works in all of these genres himself). He also recommended a Leipzig instrument maker as a reliable provider of microscopes, and he emphasized the need to learn the Latin names of plants, as well as the technical terms for different plant parts.

Merely collecting and identifying plants was not enough, however, to qualify a person as a scientific botanist. The *Naturforscher*, Reichenbach argued, also paid attention to a plant's relation to its entire natural landscape. "A botanical excursion," he admonished, "is not simply going out to get plants." If that was all the collector did, then he or she was little better than a "haymaker." Scientific botanists paid attention to all of the natural relations in the landscapes through which they moved—the conditions of the soil, the general character of an area's flora, the groupings of different plant species.[62]

Reichenbach also described the historical progression that took one from a lower to a higher perception of nature—as the dismissive remark about a "haymaker" suggests, this was a movement from relationships of material necessity to relationships of spiritual understanding. After dispatching with the lowest stage of purely material interests, he described the different "Natures" that the "educated agriculturist," the "forester," or the "learned man" created around himself. "Only pines make the forester happy, and only oaks."[63] The learned man, with "his intellect, ever busy with thinking, eternally combining and abstracting," wanted his surroundings "to be arranged for thinking, combining, abstracting, and systematizing."[64] Here, we see the typical satirical figure of the *Stubengelehrte*; Reichenbach mocked the botanical pedant for his artificial conception of nature, his narrow focus on details, and his love of forced and altered cultivated plants.

At the end of this process of development stood the "scientific botanist," who builds for herself a microcosm of nature that reflects the true order of the natural world. Unlike "dilettantes," "real *Naturforscher*" were characterized by their "generality, their constant attention to the whole." Reichenbach cautioned against an overweening concern for superficial aesthetic virtues. *All* plants and *all* specimens were interesting to the *Naturforscher*. Exclusive interest in the flora of a single local region or a single plant family was the mark of a dilettante; the *Naturforscher* studied nature's particulars with an eye always fixed on a general pattern.[65] Reichenbach's discussion would have had two obvious referents in the scientific discussions of the period—the search for a natural system of botanical classification[66] and the project of Humboldtian plant geography, which attempted to capture the totality of natural relationships in a given landscape.[67]

It is important to note that Reichenbach's distinction between "scientific botanists" and "dilettantes" did not correspond to the later nineteenth-century distinction between "professional" and "amateur." Reichenbach drew an epistemological distinction between "dilettantes" and "*Naturforscher*," and he wrote as if any member of his targeted public, with training and self-discipline, could become "*wissenschaftlich*" in his or her approach to the plant world. He addressed his readers as future participants in botanical science. Similarly, in his preface to Mössler's handbook, he had written that he had decided to revise the handbook because he knew "that many German botanists have used it alone for their self-education, and through it have gained a not insignificant knowledge of plants."[68] He hoped that the new edition would "ease beginners into the study of botany," and that through his work he would "see the number of the eager admirers of this science multiply."[69]

Reichenbach's contrast between the learned and the scientific botanist also bears further examination. His insistence on distinguishing scientific from learned botanists does not have a precedent in his other publications. In the recent past, he had repeatedly referred to his academic colleagues as "learned botanists," and he obviously meant no offense in using the label.[70] Why, in this particular context, was he so eager to poke fun at pedants? In the conventional wisdom of the early nineteenth century, pursuits that were too "learned" were not fit topics for feminine attention. The "*gelehrte Frauenzimmer*" (learned woman) was a common figure in contemporary gendered satire, and in their private reading and study, women were admonished not to become too "pedantic" or "learned."[71] In his *Botanik für Damen*, Reichenbach was careful to argue that the "true *Naturforscher*" had nothing in common with the learned pedant. In presenting *Wissenschaft* as a cure and not a cause of pedantry, Reichenbach perhaps hoped to assuage any anxieties his readers (male or female) might have about botany's possible effect in shaping the character of its practitioners.

The advice literature on women's reading in this period encouraged reading as a form of conspicuous consumption rather than as a utilitarian pursuit. Elite women were supposed to read "*zwecklose*" (nonutilitarian) literature, though not purely for pleasure. In the service of spiritual self-edification, one should read works of high aesthetic quality.[72] Reichenbach's *Botanik für Damen* can also usefully be read against the backdrop of these rules for appropriate feminine reading. He presented scientific botany as a pursuit liberated from petty material interests. In his *Botanik für Damen*, the practical concerns so prominent in Enlightenment botany appeared as stages in an ascent toward the *Naturforscher*, who produced an image of nature for himself that corresponded, no longer to his own material exigencies, but to nature's own true order. This apprehension was at once aesthetic, moral, and spiritual.

What did Reichenbach believe he had to offer the second group mentioned in his title, artists? In an earlier serial publication, his *Magazin für aesthetische Botanik*, Reichenbach had argued that a correct understanding of natural order was essential to the proper aesthetic appreciation of nature. In the past, urban artistic taste had been warped by a distorted preference for the inappropriately artificial. Scientific botany could help correct this problem.[73] In his *Botanik für Damen*, Reichenbach continued to make his case. A truly scientific study of the plant world offered "charming, ever-renewing enjoyments, finally making us so at home in this pleasant world that we can only become convinced that judgments about the plant world remain one-sided without this true, inner knowledge."[74] Science was the only true interlocutor between humans and nature, the only true path to seeing nature's order, meaning, and beauty. Nature waited as a book to be read, "but the words of this book are written in hieroglyphics, and Nature appears before us as a great riddle if we attempt to understand her without order or preparation."[75]

In 1828, Reichenbach founded a local botanical and horticultural society designed to promote the jointly aesthetic and scientific forms of perception he described in the preceding passage. In part, the new botanical society, the Flora Society, was a familiar cross section of the same men who were involved in Dresden's earlier scientific society, the Society for Natural and Medical Knowledge. Professors from the medical and forestry academy and bureaucrats with technical responsibilities (in forestry, mining, or cartography) formed its initial core. But the society also recruited a respectable sampling of Dresden's most prominent cultural figures. The Flora Society included a number of the city's leading writers and artists, and also several high-ranking officials. To match the high cultural profile of its members, the group had a top-heavy organizational structure. The society initially had seventeen different sections, each with its own director, and within a year of its founding, it had grown to eighty-eight members. The board of directors was supposed to function as a research collective, covering all possible human relationships to the plant world. In addition to more common branches of botany and horticulture (systematics, plant geography, viniculture, pomology), the society also had sections for "poetic botany," "aesthetic botany" (Reichenbach's favored term for ornamental gardening), "classical botany" (the study of ancient botany), and painting.[76]

As the society's leading theorist, Reichenbach argued that all of human beings' engagement with the plant world—economic, aesthetic, and intellectual—needed to be grounded in a scientific understanding of nature. The society would transform Saxony's gardens, orchards, and vineyards while at the same time "researching nature in all its parts" in a truly scientific way.[77] In this project, Reichenbach had a strong supporter in his colleague Carl Gustav Carus, himself both a landscape painter and a *Naturforscher*. On Goethe's eightieth birthday, Carus spoke to the Flora Society on the relationship between painting and natural science (Carus's later book on this topic established his reputation as the leading theorist of Romantic landscape painting).[78] Goethe's scientific theories also came under discussion in the society. Johann Paul von Falkenstein offered a critique of Goethe's theory of the metamorphosis of plants; he presented his essay to Reichenbach with the obsequious disclaimer, "I feel my weakness in the field of the higher science of plant life only too well, and I certainly appreciate how presumptuous it was to set up another idea in opposition to the theory of someone like Goethe—after all, who does not recognize this giant's complete spiritual superiority?"[79]

Otherwise, the members' common interests revolved around issues that were very much in line with the traditions of enlightened patriotic-economic societies. Lectures given in the society's first year covered such topics as forest management, agricultural improvement, and useful exotic plants.[80] In the service of agricultural and horticultural improvement, the society also sponsored yearly public exhibitions of flowers and useful plants that culled the most impressive specimens from the region's private, royal, and commercial gardens, orchards, and vineyards.[81] The exemplary specimens of these exhibitions, it was hoped, would eventually spread throughout the Saxon countryside.

Reichenbach had not mentioned many of the "branches" of botany included in the society (poetic botany, for example, or classical botany) in a similar list of botany's subdivisions in his *Catechismus der Botanik* just a few years earlier. Tacking on these new topics allowed Reichenbach to cast a wide net among Dresden's cultural notables when recruiting members for his society. Including practical topics had a similar effect; he was able to attract the attention of state officials interested in agricultural improvement.

In the outlines of the Flora Society, we see another good example of a point made in chapter 3. The advocates of pure *Wissenschaft* did not want to sever ties completely with practical concerns. They wanted to reform the relationship between science and practice, and reaffirm the hierarchy between learned and practical men. The new scientific botanists of the day, Reichenbach explained, always kept one eye on nature as a whole. The nobility and generality of the insight that they won through their science would then enable them to refashion all of the other ways that humans interacted with the natural world.

True to the patterns we saw in chapter 3, Reichenbach showed remarkably little interest, at least initially, in incorporating lower-ranking men with practical knowledge of plants into his society. Indeed, in seeking allies for his new botanical society, he passed over a number of the city's citizens who had considerable knowledge of plants. Particularly striking in the society's initial membership list is the absence of gardeners. By the late 1820s, Dresden was becoming an important commercial gardening center.[82] Two royal gardeners were admitted to the group, but they were assigned to a subsidiary membership category and neither was given a directorship in the society's top-heavy organizational structure. "Garden botany" was assigned to the doctor and court councilor, Friedrich Ludwig Kreyssig, "in consultation with the court gardeners Seidel and Terscheck."[83] The absence of apothecaries in the society is also noteworthy.

Dresden's local literary and artistic scene, in contrast, was well represented. Many of the men who held directorships in the society also had ties to the Liederkreis, an influential local literary society, and Reichenbach had arranged for the *Abendzeitung*, a literary journal with close ties to the Liederkreis, to publish a new supplement entitled *Mittheilungen aus dem Gebiete der Flora und Pomona*. The journal, whose usual fare included poetry and theatrical reviews, claimed the new botanical supplement would help "to entertain its highly cultivated public in the most multifaceted ways possible."[84] The *Abendzeitung* habitually defended Dresden's literary, musical, and theatrical merit against slights from other German periodicals (its rivalry with critics in Berlin was particularly strong).[85] The paper's publisher brandished the new supplement as evidence that the citizens of Dresden also deserved renown for their exceptional "sensibility to the world of plants."[86] The new supplement promised to keep Dresden's plant enthusiasts comfortably up to date through reviews of books on botany and horticulture, news from the Flora Society, and notices of interesting plants in bloom in different city gardens.[87]

The Liederkreis and the *Abendzeitung* were generally seen as bastions of Romantic sensibility (or "pseudoromantic" sensibility, as one local historian disparagingly observed).[88] In the 1820s, these twin entities represented Dresden's standards of taste to the broader reading public

CHAPTER 5

in German-speaking Europe. When an unsympathetic Heinrich Heine satirized "Dresdner Poesie" around 1830, the Liederkreis and *Abendzeitung* were the butt of his joke. His poem ended with the lines

Arnold sorgt fürs Geld und die Verbreitung,
Zuletzt kommt Böttiger und macht Spektakel,
Die *Abendzeitung* sei das Weltorakel.[89]

Arnold takes care of the money and the distribution,
For the finale along comes Böttiger and makes a spectacle,
As if the *Abendzeitung* were the World Oracle.

Arnold was Dresden's most important publisher; after the late 1820s, he also printed most of Reichenbach's books.[90] The aging literary figure Carl August Böttiger, famous from his days in Goethe's Weimar, was the Flora Society's director for "classical botany" (among other things, Böttiger was a classical philologist). The Liederkreis author Friedrich Kuhn was the society's director for "poetic botany." The Romantic painter Carl Christian Vogel von Vogelstein joined his fellow Academy of Fine Arts professor Johann Friedrich Matthäi as director for painting. The director for "aesthetic botany," Karl Borromäus von Miltitz, had similar artistic allegiances.[91]

That a naturalist who hoped to win over the polite literary public of the *Abendzeitung* would decide to make a special appeal to "ladies" is, when placed in context, not entirely surprising. As recent research has shown, mixed-gender socializing, centered around open houses and female hostesses, was common in late eighteenth- and early nineteenth-century German cities. This form of sociability was far more usual than the literature on the supposedly exceptional "salons" of Berlin previously assumed.[92] Early nineteenth-century Dresden also had its share of celebrated female hostesses, from the painter and musician Therese aus dem Winckell, who welcomed guests to a tea circle in her modest abode in Dresden's Italian village, to the more affluent Elise von der Recke, described by a contemporary as the "high priestess" of the "true pantheon of friendship" that met in her home.[93] In all of Dresden's leading houses of the period, from the Koerners to the Caruses to the Tiecks, women were an integral part of informal urban socializing.[94] Women also took part in more formal literary gatherings. Elise von der Recke, for example, had ties to the Liederkreis, the literary society whose membership overlapped so heavily with the Flora Society.[95]

Women's presence in urban intellectual life can also be seen in the negative prism of much gendered contemporary satire. Not all men viewed women's presence in polite culture as favorably as Reichenbach, nor were all as sanguine about female intellectual abilities. Feminine authorship was controversial throughout the nineteenth century, and parodies of "learned women" were widespread.[96] In addition to drawing fire as authors, women could evoke ire as audiences. Wilhelm von Lüdemann, a critic of the Dresden Liederkreis, blamed the gathering's supposedly low artistic level on its members' lack of critical acumen—a problem, he suggested, that was caused primarily by the presence of so many women in its ranks. Lack of critical standards, he wrote, "is the unfortunate obstacle that causes the results of similar associations to be such failures—especially when women take part in them." Flattered by the audience's positive response, a lackluster author believes his work has merit; he dismisses the doubts he might have harbored in private and "recognizes his child, his mastery—because the praise of women, as many have said, is the only reliable kind. How false, how misguided! I have never written anything that wouldn't have received the praise of women!"[97] Women's influence, Lüdemann argued, was pernicious for male authors, sapping their self-critical energies and keeping them from a true assessment of their own artistic merits.

In contrast, Reichenbach hoped that the salutary influence of women might help win botany a more prominent place in "general *Bildung*." Despite these two men's different valuations of feminine judgment, both Reichenbach's encouragement of botanizing ladies and Lüdemann's satire shared a common assumption—women were an important part of polite urban intellectual culture. For this reason, it would be inappropriate to understand Reichenbach's appeal for women's participation in botany as a plea for broader feminine emancipation either by contemporary or by early nineteenth-century standards. In his writings, women's place in the educated public was not a future goal but an accepted fact. He merely wanted to redirect feminine attention to his own favored pursuits. The prevalence of mixed-gender socializing did not erase gender differences or establish equality, as historians of salon sociability have sometimes claimed.[98] But within her role as an appropriately feminine hostess, a woman could shape the course of conversation. Recognizing this power, Reichenbach hoped to recruit elite urban women to his own cause.

Taste and sensibility were the currency with which one traded in educated culture, and Reichenbach hoped to make natural history in the *naturphilosophischen* style an essential component of general *Bildung*.[99] In his *Magazin für aesthetische Botanik*, he had hinted that women were

CHAPTER 5

crucial to the success of his cause. Taking England as his idealized example, he claimed that

> the sensibilities of the inhabitants [of England] have always shown a general tendency to appreciate . . . the ownership and knowledge of the plant world as an essential part of their *Bildung*. A lady of social standing would be ashamed of herself if she did not own Curtis's *Botanical Magazine*, or did not know what Logdes [sic] illustrated in the last *Botanical Cabinet*.

Reichenbach expressed the hope that the situation in Germany would soon be so favorable, and he also noted signs of progress elsewhere on the Continent. "In France, a female botanical author has discovered a new genus of moss and written a flora for her region; in Italy, even female royalty botanizes."[100] Far from consigning botany to the private sphere in such statements, Reichenbach made women's knowledge the best marker of its public acceptance.

A Postscript from the 1840s

In his published work, Reichenbach was not particularly concerned with marking out a proper territory for feminine participation. In the Flora Society itself, however, the position of women was more complicated. The Flora, though it borrowed heavily from the city's literary circles, also owed a clear debt to the tradition of another kind of society—the Economic Societies (Oekonomische Gesellschaften)—in which women had not generally been included, at least not in large numbers. The Flora Society's ties to the Saxon bureaucracy meant that many of its concerns stemmed from the decidedly masculine context of official administration—initiatives to improve state vineyards or import new crops. Though some early records of the group suggest that women could be "honorary" members, no women were recorded in the society's official membership registry until the early 1840s. The society's first small private exhibition was given for a mixed-gender public and dubbed a "ladies' meeting," which suggests the "ladies" had not been present at earlier events.[101] In 1828, Reichenbach also began to give regular "botanical evening entertainments" in the city's botanical garden. A contemporary guidebook noted that these gatherings were attended by an elevated public, "*Personen der ersten Stände*," a public that would likely have included both men and women.[102] At the level of published appeal, Reichenbach drew no clear boundary for what might constitute appropriate female participation in the sciences. Within local associational life, women were initially invited to participate only as an appreciative audience for male efforts.

When a handful of women finally did appear as members of the Flora Society in the 1840s, it was under intriguing circumstances. Reichenbach's Flora Society had a long and successful career, but not in the form that he intended. The top-heavy gathering of literati did not prove sustainable; the *Abendzeitung*'s supplement lasted only a year.[103] Over the course of the 1830s and 1840s, the lifeblood of the society became those Dresden residents who had the greatest interest in practical knowledge about plants—the city's expanding circles of commercial gardeners. Unable to keep the society on what he considered to be the proper "scientific" course, Reichenbach resigned from the group in 1843.[104] By then, a new natural history society, examined in the next chapter, had become the main focus of his local efforts.

Several suggestive events occurred in Reichenbach's final months in the Flora Society, however. At the same time that he was trying to defend the "scientific" (*wissenschaftlich*) tone of the society against an influx of practical gardeners, someone (the protocols do not record who) raised the possibility of accepting women as members of the society. The fact that the proposal was accepted without dissent suggests that it had the society director's support, and it is not unreasonable to assume that Reichenbach himself was behind the suggestion. In the following months, several local noblewomen were put forward as members. But when Reichenbach finally threw up his hands and left the group, the noblewomen did, too; one by one, they sent letters politely explaining that the press of other obligations made their continued participation in the group impossible. The timing of noble female members' entry into and retreat from the society suggests that Reichenbach was making one final gambit with the basic strategy he had used in the late 1820s—seeking alliances with elite women to raise the status of botany, protecting it, in this case, from an excessive association with the practical activities of lower-ranking men.

Conclusion

"All art should become science and all science art; poetry and philosophy should be made one."[105] It was all well and good to admire Friedrich Schlegel's programmatic declaration, but how exactly did one go about producing such a science? And (more important for the questions addressed in this chapter) what sort of audience could one find to read it? Nees von Esenbeck's and Reichenbach's descriptions of natural science drew heavily on the vocabulary of Weimar classicism—they denigrated "*Einseitigkeit*" (one-sidedness) and praised harmonious unity as a quality in both the individual personality and scientific work. The links between scientific and artistic activity were not just conceptual, however; these two *Naturforscher* wanted to create a relationship between author and audience analogous to the one that pertained in the fine arts. They wanted an audience with educated ears, readers who knew how to appreciate a properly scientific work, even if they could not reproduce it themselves. They also assumed that the line between author and audience was fluid. Much like a member of a literary club, a "friend of nature," after listening for a while, might later decide to try her hand at composition herself, on however modest a scale.[106] Reichenbach and Nees von Esenbeck kept at least the narrative fiction that the individual picking up their books might learn to do science without formal training. Beyond more general concerns about their public stature, leading naturalists had an additional reason to cultivate relationships with a wider public. In the competition to replace Linnaean taxonomy with a natural system, publicity was an invaluable ally. And this larger public, by necessity, included women.

In this respect, prevalent limitations on women's intellectual activity had ambiguous implications for the wider cultural authority of the natural sciences. If elite women's reading was supposed to be both nonutilitarian and not too learned, then the natural sciences, in either their practical or their learned guises, were not obvious choices. In public literary forums and local social settings where the mechanisms of cultural authority ran in part through appreciative female readers, listeners, and conversational partners, pressing the case that both doing and reading science were appropriate for women became one strategy for keeping the natural sciences before the eyes of the cultivated public. Women's participation in formal scientific voluntary associations, however, was more problematic. As was often the case, inclusive rhetoric existed next to more restricted social practice.

Though the Romantic-era project of a unified art and science was not without progeny in the later nineteenth century, it did not emerge in exactly the form that Nees von Esenbeck imagined—as part of a literary public that could appreciate the historical sublimity of the *Naturforscher*'s works alongside the latest plays, poems, and novels. Alexander von Humboldt's *Kosmos* probably came closest to achieving this ideal. Like Nees von Esenbeck's essay on the history of botany, this was also a work that combined the literary style of Weimar with an account of the scientific state of the art. Humboldt's work sold a great many copies, but as Andreas Daum has suggested, it may well have been a book that was bought much more often than it was read. *Kosmos* was a challenging text; it was supposed to be a contribution to high scientific discussion and an offering to the educated public at once.[107] This was the divide that Nees von Esenbeck and Reichenbach also wanted to straddle, and their failures show that there were definite limits to how far the cultivated public was willing to play along. Reichenbach's vision of an expanded botanical science, a community in which leading philologists and poets would become his partners in the study of plants, proved unsustainable; Nees von Esenbeck's supplement never made it off the ground. One would not want to take this argument too far, of course. Members of the cultivated public were certainly interested in the natural sciences, as Cotta's initial interest in Nees von Esenbeck's proposed supplement aptly illustrates. Reichenbach's steady stream of introductory botanical publications could be used to make a similar point. But when elite *Naturforscher* tried to direct the channels through which public taste flowed, there were plenty of countervailing pressures that pushed back.

Nees von Esenbeck's and Reichenbach's efforts were not complete failures. Both men were respected researchers, and a number of their contemporaries built their public personas out of similar materials. Carl Gustav Carus, Reichenbach's fellow professor in Dresden, was admired as both a landscape painter and a morphologist, and his philosophical work attempted to bind these two spheres of practice into a single whole. Alexander von Humboldt's reputation was indebted to the aesthetic quality of his scientific writing; later in the century, he would be praised as the man who brought the clarity and beauty of Weimar classicism to the natural sciences.[108] More generally, the Romantic era's interest in moral and aesthetic concerns had lasting significance, both in German high science and in popular science writing.[109] For the most part, however, scientific associational life and print culture developed along lines other than those of the projects examined in this chapter.

In the 1830s and 1840s, many members of the Romantic generation would come under heavy censure from up-and-coming younger men. The next generation's objections to *Naturphilosophie* and scientific "Romantics" were often loud and caustic (masking, often, more intellectual debts than they openly acknowledged). It was not just the content of Romantic science that came under fire; its literary style and public ambitions were also the subject of critique. M. J. Schleiden ob-

CHAPTER 5

jected to the "so-called cleverness" of many previous botanical authors, alluding to the puns and analogies that had been ubiquitous in the *Naturphilosophen*'s scientific writing. Nees von Esenbeck was one of the primary offenders, to his mind. For the romantic generation, self-consciously metaphorical writing had been a mark of sophistication. This style of science had also been a strategy, albeit not always a successful one, to craft an authoritative persona before the mixed-gender literary public of the Restoration period.

Schleiden, one of the leading voices of the new "scientific botany" (*wissenschaftliche Botanik*) of the 1840s, wanted nothing to do with such extravagances, the "fashionable phrases" of this "salon wit."[110] "I consider such things superficial, tasteless, and repugnant in the face of the seriousness of *Wissenschaft*," Schleiden wrote. Rhetorical flourishes were at best suitable for the "flirtations of the salon"; they characterized the trivial exchanges of mixed-gender sociability and lacked the "strict honesty and conscientiousness" he considered exemplary of the best scientific writing.[111] A critique of his predecessors' stylistic ambitions, in other words, was also a critique of the public they had sought to attract.

In many ways, the frustrations and limitations Nees von Esenbeck and Reichenbach encountered mirror those of their literary heroes; Goethe and Schiller had also bemoaned the fact that the public was not able to distinguish between cheap entertainment and true genius. Such complaints reflected the realities of the book market, which stubbornly refused to submit itself completely to higher guidance on matters of taste.[112] Therese Huber's advice to Nees von Esenbeck—that her readers preferred to hear "facts and fragments" from a *Naturforscher* rather than grand theories—accurately diagnosed a persistent public preference. Furthermore, as the disciples of high *Wissenschaft* designed grand systems and founded more exclusive learned societies, more modest and eclectic networks of natural researchers continued to grow apace. Among the members of these networks, elite opinion carried weight, but as we will see in later chapters, it did not always have the last word.

SIX

The Nature of the Fatherland

In chapter 2, we followed the young drawing teacher Carl August Friedrich Harzer into the mountains of Saxony for a few days' vacation. Already in the 1810s, Harzer had some familiarity with natural history. In the decades that followed, this casual interest became the defining passion of his life. He provided engravings for works by better-known naturalists, and published works on butterflies and mushrooms under his own name. He also used his skill at rendering plants to create embroidery patterns for commercial sale. All the while, his modest lodgings slowly filled to bursting with specimens he collected on his ramblings through the Saxon countryside. In 1833, along with other naturalists from similarly unassuming backgrounds, he founded a new scientific association, the Isis Society, in his hometown of Dresden.[1]

In Harzer, we might imagine that we see a later version of Andreas Ficinus, the eighteenth-century Dresden apothecary who ended up in Hamburger and Meusel's learned lexicon. Here, we also have a man who started off with modest resources and who gradually earned a place in the learned world. Harzer, it might seem, was even more successful than Ficinus. Ficinus only *joined* a learned society. Harzer founded one. Ficinus had ties to noble and learned men; Harzer did as well. One of his correspondents, the Austrian nobleman and lepidopterist Joseph Emanuel Fischer Edler von Röslerstamm, described Harzer as a "most praiseworthy and well-known entomologist." Harzer also

CHAPTER 6

had a learned sponsor in Ludwig Reichenbach, who wrote a foreword for one of his books.[2]

To understand Harzer's life in this way, however, would be to miss the substantial differences that distinguished the late eighteenth from the mid-nineteenth century. The natural history society that Harzer founded now had only tenuous claims to learned status in the eyes of the German scientific elite; that was true for Harzer personally as well. Gone were the days when the Prussian Academy of Sciences appointed apothecary-chemists as academicians. By the 1830s and 1840s, even a career like that of Jacob Sturm, the engraver-naturalist discussed in the introduction, was becoming harder to imagine. Sturm, who began building up his scientific contacts in the 1790s, entered a different learned world from the one Harzer encountered in the 1820s. In Harzer's life and the lives of similar men, we see the learned world expanding, but also beginning to break apart. Something novel was happening: with the proliferation of groups like Harzer's Isis Society, a new, more organized low scientific culture was emerging, a culture whose social and geographic profile set it apart from the elite learned world. Heading into the second half of the nineteenth century, this low scientific culture would develop along largely separate tracks from German high science. Points of connection between these two levels of scientific life remained, of course. In the 1850s, Helmholtz would be active in the local scientific and economic societies of Königsberg as a young professor; Rudolf Virchow's career in Berlin was full of civic scientific activity. In the 1830s and 1840s, however, the divide that historians of the later nineteenth century have described—a divide between civic scientific societies and the academic establishment—began to take clearer shape.[3] This new low scientific culture was built using a familiar learned vehicle, the private scientific society. The new groups of the 1830s and 1840s, however, would no longer be fully absorbed into the older, integrative structures of a general republic of letters.

As we saw in chapter 2, the networks of natural knowledge had always been socially diverse. New to the second third of the nineteenth century, however, was the organization of nonelite researchers into societies of their own. The odd professor or elite researcher sometimes also participated in these groups. Particularly in smaller places, prominent local figures—the mayor, important regional bureaucrats—joined in, too. Nonetheless, the new scientific societies of the 1830s and 1840s began to mark out a second, lower stratum of scientific life. They still had some prominent scientists as members and could be prestigious local cultural forums. But the multiplication of these groups changed the overall complexion of German scientific associational life quite dramatically.

To characterize this transformation, we can start by saying the following: older elite scientific societies like Halle's Nature-Researching Society or Berlin's Society of Nature-Researching Friends retained their elite status into the later nineteenth century. The complexion of scientific associational life as a whole, however, looked significantly different in 1850 than it had in 1830. The new societies of the 1830s and 1840s were much more socially diverse, dipping deeper into the growing lower fringes of the educated middle classes for their membership. The local medical profession was generally well represented in these groups, but many other members were men whose lives had been made possible by the growth of German educational systems and state bureaucracies, by the expansion of consumer culture, or by the early stages of industrialization. They were schoolteachers, practical chemists, or early "engineers," the graduates of new normal schools and technical academies. Some of them made a living from the natural sciences' popularity as a form of consumable leisure—engravers like Harzer or the sellers of natural history specimens.[4]

The new societies of the 1830s and 1840s were also more meaningfully provincial than their predecessors, and this was true in two senses. More minor capitals or regional centers now had their own natural scientific society, and most of these groups also defined their scientific aims in provincial terms. Though the individual members of these groups often had varied intellectual interests, regional natural history was the signature scientific project of the emerging low scientific milieu. These new associations were "local" in a different way from their predecessors. Although early nineteenth-century societies had sometimes also collected information about regional natural history, they had aspired to be microcosms for the study of nature as a whole, bringing together the far-flung efforts of global science in one place. Many of the new societies of the 1830s and 1840s, in contrast, made studying local landscapes their primary goal. Earlier learned societies had seen themselves as responsible for nature in its entirety. The newer groups had a more limited aim: cataloging the natural riches of their regions. When the Munich professor Karl Fraas sneered at the local efforts of "the scientific proletariat," these were the people he was talking about.[5]

Most learned scientific societies founded in the 1810s and 1820s had been in major political capitals or university towns. In the 1830s and 1840s, associations appeared in a range of minor capitals and provincial centers. Many of these places, of course, would have been loath to describe themselves as "the provinces." Indeed, given the still complex political geography of German-speaking Europe, any number of towns could call themselves central places, however small their hinterlands might have been. Of the thirty-seven natural scientific societies founded in German-speaking Europe over these years, many were in the capitals of smaller dynastic states. A Natural Scientific Association appeared in Detmold, the capital of Lippe, in 1835; another was founded in Darmstadt, the capital of the Grand Duchy of Hesse-Darmstadt, in 1844. If the many independent political entities of the German Confederation did not offer geographic diversity enough, other groups dedicated themselves to regions that stood in unclear relation to the formal divisions of a political map—the Palatinate or the Wetterau.[6]

In order to understand these societies' appeal, however, we cannot be too hasty to remove them from the learned world, because this was very much where they wanted to belong. They exchanged publications with established learned societies in other German cities and abroad (often quite prestigious ones); their members presented papers and published reports. Labeling this activity "amateur science" does not provide us with much analytical traction, and it is in some ways misleading, for natural history was much more than a pleasant hobby to men like Harzer. As he cobbled together a livelihood on the fringes of the educated middle class, the study of nature provided him with both supplemental income and a minor public reputation. Many of the other, similarly modest men who founded the Dresden Isis were also published authors, and their individual work could still win them scientific praise. A colleague could still call Harzer "a well-known entomologist."

As the nineteenth century progressed, natural history continued to lose academic prestige, due in no small part to its association with Fraas's "scientific proletariat." The gap between academic science and regional natural history would continue to widen (although there were also points at which these arenas reconnected, as Lynn Nyhart has shown).[7] "Natural history" came to be a term that was used dismissively, even as many of its central concerns remained part of elite research.[8] "In the opinion of many from academia," Peter Stevens has written, "natural history in the late nineteenth century was that part of late eighteenth-century natural history that was not of academic interest."[9]

Wolf Lepenies once cast natural history as a pursuit that had to be left behind before natural science could appear in its modern form. In fact, *Vormärz* natural history societies made "*Naturwissenschaft*" their cause, too, though not in exactly the same way as their elite counterparts. The first German society to call itself a "Natural Scientific Society [*Naturwissenschaftliche Gesellschaft*]" was a regional natural history society in Blankenburg; several other regional natural history societies also took this name. In the 1830s and 1840s, the consolidation of the category "*Naturwissenschaft*" was as much the product of this mixed lower scientific milieu as it was the work of elite researchers, and for many educated Germans at midcentury, the canonical site of science was still the natural history collection rather than the laboratory. Andreas Daum has shown that natural history remained central to scientific popularization later in the nineteenth century; when Daum's popularizers promoted natural scientific *Bildung*, natural history was an important part of what they wanted to teach.[10]

Once again, we see the study of nature appearing as a more clearly delineated social cause in German-speaking Europe than in Britain or in France. Provincial societies that studied natural history also proliferated in the British Isles in the 1830s and 1840s; most of these, however, also studied antiquities. French provincial societies, too, usually combined the study of natural history with the study of civil history.[11] In Germany, history and natural history usually (though not always) were pursued in separate societies. A look at Dresden's Society for Saxon Antiquities, founded in 1824, makes it fairly easy to explain why this group did not later decide to join hands with the members of the Isis. The Saxon history society, to put it mildly, was a bit more prestigious than the regional natural history group. Its members were the highest-ranking officials in the Saxon government, joined by the city's most eminent artists, writers, ministers, and military officers. In the mid-1830s, Friedrich August II and his brother Prince Johann were the titular heads of the group.[12] Saxony's ruling family, the Wettins, drew heavily on Saxon history in crafting their dynastic image, and in using history in this way, they were entirely typical of German dynastic rulers in this period. Early history societies in Bavaria and Württemberg were linked with similar efforts.[13] Despite his personal interest in botany, Friedrich August II and his leading officials did not make the same use of the symbolic possibilities of natural history, at least not with equal eagerness and intensity. One of the founders of the history society, the Count von Einseidel, was also the head of the Economic Society during the 1820s; he and his fellow court councillors cared about the natural world when it took the shape of a productive farm, but they seem to have been comparatively less interested in supporting the study of Saxon natural history as such.

In other kinds of places, local elites did see natural history as an important vehicle for political representation, but another difference between the Isis and the Society for Saxon Antiquities suggests why history and natural history did not usually join hands even then. History and natural history attracted different branches of the learned professions. Many members of the Society for Saxon Antiquities have titles that suggest they studied law or theology. None of them have titles that suggest they studied medicine. Doctors of theology could be found in both groups but appeared in greater numbers among the historians. Law had long been a more prestigious course of study than medicine or cameralism, and the many prominent officials who belonged to the Saxon history society provide a good illustration of why this was true.[14] History and natural history recruited divergent groups of admirers. Collectively speaking, the cultivators of Saxon history would have been lowering themselves quite a bit to rub shoulders with the students of Saxon nature.

The dynastic agendas discussed above bring out another important dynamic that fueled the spread of new scientific societies in the 1830s and 1840s. Regional natural scientific societies were part of another important *Vormärz* process, the renegotiation of political allegiances within post-Napoleonic Germany. Some local natural history societies offered up their efforts under the banner of dynastic loyalty; others presented their work, in only barely disguised ways, as acts of resistance to a centralizing dynastic state.[15] Many German liberals considered the study of German nature as one of the best ways to cultivate a love of the German nation and hoped that the decentralized, grassroots project of regional natural history would further the cause of German national unity.[16]

In this respect, regional natural history societies were important precursors of the later nineteenth-century *Heimat* (homeland) movement.[17] They help us see how emerging German notions of *Heimat* owed a heavy debt not just to the pressures of German political unification but also to older, more cosmopolitan projects.[18] Drawing on earlier natural historical traditions, regional natural scientific societies marked out all that was distinctive and particular about a local region, but they did so with reference to more widely shared standards, and with an eye to the evaluative gaze of a cosmopolitan audience. In the process, they linked the personal passions of men like Harzer, with their lonely country excursions and their burgeoning collections, to the great public questions of the age.

The Porous Edges of the Learned World

In comparison to earlier learned societies, midcentury natural history associations were remarkably socially diverse. University-educated men provided only part of the cultural energy that fueled their creation. Generally, the university educated did not make up more than about a third of the membership, although they often occupied the highest leadership positions. Most of these men were either doctors of medicine or doctors of philosophy, the latter employed as teachers. They were joined in large numbers by men from the edges of the educated middle classes, as well as a smaller number of men in trade or manufacturing. For example, a decade into its existence, the Isis Society had two academy professors as its director and vice-director, Ludwig Reichenbach and Hermann Richter, respectively. The society's original founders, however, had come from the margins of the educated middle classes. Several were the orphaned sons of small-town ministers or teachers, men whose careers had been disrupted by a parent's early death. A few had graduated from normal school, but none from university.[19] Only one of the society's original 1833 members, a clerk in the royal natural history collections named Johann Heinrich Gössel, was also a member of the city's older learned society, the Society for Natural and Medical Knowledge.[20] In the mid-1840s, button manufacturer Moritz Bartsch was the curator for the society's zoological collection, and the library clerk Carl Nagel served as secretary.[21]

The Isis Society offers a useful snapshot of the range of people who filled the ranks of regional natural history societies, and its history also hints at the broader forces that helped such societies succeed. Natural history's long process of diffusion into middle class and aristocratic culture provided the first condition of possibility for the mid-nineteenth century's expanded scientific associational life. Once established as a form of entertainment and self-cultivation, its casual presence within German culture could seduce people into more serious involvement.

Harzer's life illustrates the feedback loops that emerged under these conditions. As chapter 2 explored, Harzer's interest in natural history grew out of his enthusiasm for hiking and regional travel, activities also linked to his training at the Dresden art academy, where professors had taken their students out into the countryside on sketching expeditions. In the second half of his life, the knowledge he had acquired provided him with irregular work doing scientific drawings, as well as contact with more eminent naturalists. One of his earliest efforts in scientific illustration was for Ludwig Reichenbach's *Magazin der aesthetischen Botanik*.

Other developments also increased the pool of people for whom natural history could be an appealing activity. The bureaucracies of the various German states grew significantly in this period, as did state-run school systems.[22] Schoolteachers provided a large percentage of the Isis's members; they were joined in smaller numbers by men from the technical branches of state bureaucracy trained in fields like forestry. The *Vormärz*'s fluid mélange of commerce, artisanal production, and manufacturing accounted for another large segment of society members. Most people from such backgrounds did not come from traditional trades, although the society did have one baker in its ranks. The Isis had two vinegar manufacturers, for example, who were part of a sector in the Dresden economy, the production of processed foodstuffs, which spearheaded the city's transition to large-scale production for a wider market.[23] Several Isis members came from the emerging technical professions, men who listed their occupations as *Ingenieur* (engineer) or *Mechanicus* (machinist). Alongside men engaged in various kinds of commerce, a handful of members described themselves as salesmen of natural history specimens (*Naturalienhändler*). Others earned their living through print or visual culture as booksellers, painters, and engravers.[24]

The basic social profile of the Isis, and the range of occupational groups on which it drew, was fairly standard for regional natural history societies in the period. Apothecaries and schoolteachers dominated the Natural History Association in Bonn.[25] The Saxonia of Gross- und Neuschönau, just like the Dresden Isis, had an engraver with entomological interests as its founder, C. G. Voigt. In the early years of the Saxonia's existence, roughly a fourth of the members were merchants or manufacturers, while 20 percent made their living in the book trade, either as engravers or as booksellers. The society also had several teachers, apothecaries, and estate managers in its ranks (approximately 13 percent of the membership in each case).[26] The natural history society in Brünn, to which the now famous monk Gregor Mendel belonged, had a similar composition.[27]

Normal school graduates, machinists, vinegar manufacturers, apothecaries, book merchants—none of these people fit easily into the scholarly world imagined by the preceding generation. The products of a transitional period, men like Harzer and his colleagues occupied an ambiguous place in the social order, their lives played out in an expanding middle ground between the learned and the productive estates. Many of these men could rely on neither the prestige of a university degree nor the honor of a traditionally recognized trade. They might possess some advanced education but could not put the title of "Doctor" before their names. If they were graduates of normal schools or other specialized academies, they were the products of educational institutions that were often recently founded and of insecure status.[28]

Regional natural history societies were themselves similarly liminal, and their intellectual standing shifted depending on the perspective from which they were viewed. The various labels assigned to the Isis Society illustrate the different ways these groups might be seen. The professor Hermann Richter, vice-director of the Isis in the late 1840s, described the group as a popular society devoted more to the spread of knowledge than to its production. For him, the older Society for Natural and Medical Knowledge was where the real work of *Wissenschaft* was done in Dresden.[29] Other members of the Isis, however, considered themselves to be members of a *learned* society. Normal school graduate Julius Bescherer presented himself this way, for example, on the title page of his *Methodik des naturwissenschaftlichen Unterrichts*.[30] The schoolteacher Carl Sachse argued that although the Isis's founders were self-made men without the benefit of a "so-called learned education," their experience with nature had brought them far enough that they deserved an "honorable place" alongside their fellow *Naturforscher*. He referred to the Isis as a *"wissenschaftliche"* endeavor.[31]

If there were general patterns to regional natural history societies' memberships, their precise complexion varied with the economic, educational, and bureaucratic landscape of the towns or regions in which they were formed. Priests from the local lyceum played a prominent role in Bamberg's natural history society; local forestry officials were well represented in the Pollichia, a society in the Palatinate. The origin stories of different groups varied as well. The modest standing of the Dresden society's founders, who were mostly young men not yet fully established in life, was not particularly typical. The first members of the natural history society in Cassel, to give a contrasting example, were local doctors, high-ranking officials, and other notable local men; only later did the society's rosters grow to include other natural history enthusiasts from lesser backgrounds.[32]

The attraction of regional natural history for local elites in second-tier cultural centers will be explored in the following section. The oc-

CHAPTER 6

cupational profile of the Isis founders, however, provides a reliable introduction to the range of people who made up the rank and file of similar societies in other cities. Interestingly, the occupational groups found within regional natural history societies appeared together in another context in the *Vormärz*—they were the kind of people often discussed collectively in debates about the educational value of the natural sciences. As the following two chapters will discuss in greater detail, educational reformers in this period often argued that occupations that dealt somehow with the natural world—from doctors and apothecaries to artisans, merchants, and cameralist-trained officials—needed a thorough grounding in the natural sciences. The rosters of *Vormärz* natural history societies drew heavily from these very groups; such men apparently agreed that natural science was a means to personal advancement.

Yet regional natural history societies were not (or at least not primarily) designed to bring science into conversation with economic practice or professional concerns. Regional natural history associations typically set self-consciously modest epistemological goals for themselves; they were often ambitious, however, in seeking ties to broader intellectual and political power structures. Their members were still, in recognizable ways, playing by the rules of the learned world; as this world grew larger and larger, however, all of the diverse activities of its members became harder to perceive as part of a single game.

Harzer's obituary, composed by fellow teacher Carl Sachse, illustrates this point well; it placed Harzer within the learned world but also acknowledged his circumspect status. In Harzer's name, Sachse offered up an ode to modesty as well as ambition. He noted that Harzer had passed away "on the same day that the learned world mourned the death of the greatest astronomer of the age," Friedrich Wilhelm Bessel, in Königsberg. By further coincidence, the drawing instructor was then buried on the anniversary of Newton's death. "The entire learned and cultivated world" memorialized Newton's passing, Sachse wrote, while Harzer's death was marked only in the "modest annals" of the society he had founded. Newton had explored the farthest reaches of the heavens; Harzer, the woods and mountains of his Saxon fatherland. Both the renowned astronomer and the local naturalist, Sachse implied, were part of a single enterprise. A ray of Newton's immortal glory refracted through Harzer's earnest scientific life. Capturing a small piece of cosmopolitan scientific glory and turning it to new ends was precisely the function of regional natural history societies like the Isis.

Regional Natural History and High Science

Not everyone agreed with Carl Sachse that a local naturalist like Harzer was a valuable member of the learned world. In the 1830s and 1840s, a new generation of university botanists and zoologists advanced novel standards for what counted as *"wissenschaftliche"* botany or zoology. The leading voices of the new "scientific botany" and "scientific zoology" championed plant and animal morphology, an approach they believed would uncover universal laws of organic development. They looked askance at the previous generation's fascination with systematics, *Naturphilosophie*, and Humboldtian natural history, and they also sharply criticized the (to their eyes) disorganized empiricism of many contemporary fellow naturalists. They had little patience, in other words, with the broader networks of German natural history. In the sharpness of their tone, one should see impatience not just with the preceding generation but also with Germany's emerging low scientific culture.

Carl Nägeli's opening essay in the *Zeitschrift für wissenschaftliche Botanik* scolded his fellow botanists for the "chaos" they created by failing to tie their observations to a search for general laws. "Making a truly good and useful natural historical observation is not nearly as easy as one often believes," Nägeli wrote. Much effort was wasted in activities that served no real purpose, and a large number of his fellow botanists, in Nägeli's view, were anachronisms, stuck in a previous stage of scientific development. As science progressed, "the sum of existing material content grows ever larger, and the circle becomes ever smaller in which one meets with the correct form."[33] The circle attracted to Nägeli's cause was small indeed, not wide enough, as it turned out, to sustain more than three years of his periodical. The journal's more successful zoological analogue, the *Zeitschrift für wissenschaftliche Zoologie*, founded four years later, included similar rhetoric. It excluded casual observations, "simple notes and natural history news"; everything submitted for publication should serve the cause of discovering causal relationships and uncovering universal natural laws.[34]

In contrast, "simple notes and natural history news" were the stock-in-trade of regional natural history societies. Their collective efforts were not, for the most part, concerned with uncovering universal laws, though regional naturalists did not necessarily cede epistemological supremacy to more global forms of science. The natural history society in Württemberg (the Gesellschaft für vaterländische Naturkunde) hinted

CHAPTER 6

that more ambitious scientific associations might simply suffer from hubris. "Other societies may set more extensive tasks for themselves," wrote society secretary Theodor Pleininger, but the Württemberg society would take as its main focus the territory of Württemberg. This limited terrain, Pleininger argued, "offers to research a goal that is perhaps all the more inviting" because it provides the possibility that with time a complete whole could be achieved. "This is less likely when research expends its power in infinite space, which science [*Wissenschaft*] in its generality perhaps vainly strives to encompass."[35] In striving for universality, Pleininger suggested, other researchers might lose themselves in the grandeur of their cause.

In place of global theories, regional natural history societies created detailed lists and descriptions. Catalogs of the plants, animals, or minerals found within a given area were the most common products of society members' efforts. In the mid-1840s, for example, the Isis's entomologists joined together to create "the most complete possible index of Saxony's butterflies."[36] In the Pfalz, indexes of local flora and fauna filled the pages of the Pollichia's yearly reports. The society also kept manuscript catalogs in its library, where a Herr Studienlehrer Frank deposited "an index of the insects that he found in the area around Annweiler," the village where he taught.[37] Similarly, the Verein für vaterländische Naturkunde in Stuttgart published a running series of specialized catalogs in its reports, as did Württemberg's natural history society. Members also published the results of their studies independently. Several members of Dresden's Isis Society authored local collecting guides. For example, Friedrich Reichel, who made a living alternately as a lending librarian and a vinegar manufacturer, compiled a botanical collecting guide to Dresden as well as two general travel guides to the area.[38]

In their enthusiasm for amassing new specimens, regional naturalists' collecting habits diverged from those of elite researchers. With the appearance of large, state-funded collections in the early nineteenth century, personal collections had declined in importance for leading naturalists.[39] In 1836, Ludwig Reichenbach repeated this widely shared view, writing that private collections were only useful if they focused on a narrow range of natural forms. There was no reason to create a more comprehensive collection of naturalia when one could use "a larger, already existing museum."[40] Already two decades earlier, the Count von Hoffmannsegg, drawing on his experience in Berlin, had advised the Dresden Society for Natural and Medical Knowledge that learned societies ought to get out of the business of collecting. Such labor-intensive efforts were better left to the state.[41]

Both individually and collectively, many regional naturalists ignored this sort of advice with impunity. In the dwelling of library clerk Carl Nagel, "algae was growing in glasses on the windowsills, mushrooms under glass stood on the tables." The walls were crowded with books, plants, minerals, and insects, "making the already small room narrower still."[42] Harzer's living space was equally crowded with natural objects; "he had small collections for almost all of the subjects of natural history." Assembling naturalia was Harzer's comfort and his passion. When a fire destroyed his collections, the "only trusted and quiet companion of his life," he was inconsolable.[43] Despite Harzer's violation of contemporary elite natural historical wisdom, his fellow Isis member Sachse described him as "the model of an honest, eager researcher for many of us."[44] If Harzer's collecting activity did not fit easily into the programs of new state-funded institutions and panregional networks of elite practitioners, it still served as a respected mark of skill and knowledge to his local peers.

Though regional natural history diverged in significant ways from elite science, it was not completely separate from it. In Württemberg, several professors from the University of Tübingen, including the botanist Hugo von Mohl and the zoologist Wilhelm Ludwig von Rapp, were on the governing board of the regional natural history society, which was centered in the political capital of Stuttgart.[45] In the 1840s, Gottlieb Wilhelm Bischoff, who held the chair of botany at the University of Heidelberg, regularly traveled to the annual meetings of the Palatinate's natural history society, the Pollichia, which met in his original hometown of Bad Dürkheim.[46] University professors' involvement with regional natural historical societies also persisted into the second half of the nineteenth century.[47] Men with academic appointments also still wrote regional or local floras as well. Ludwig Reichenbach spent almost twenty years compiling a *Flora Saxonica* that included extensive references to all geographical locations in which different plants had been found.[48] Not surprisingly, many of the men above (though not all) had serious interests in systematics, and their own continued focus on taxonomy and collecting gave them a certain degree of common intellectual ground with regional naturalists.

In addition, the specialization that characterized German high science can also be found in less elevated circles. We have already seen that Harzer was accorded the title of "entomologist" by a fellow lepidopterist. The *Dresdener Tageblatt* gave another Isis member, Johann Friedrich Hübler, a dealer in mineralogical specimens, the label of "geognost." "Is there a geognostic collection in all the Saxon fatherland—indeed, even abroad—that does not have, among its Saxon specimens, a few items gathered and finished by the hand of the tireless Hübler?" the paper asked. Isis Society members also declared a specialty when they joined—physics, chemistry, zoology, and so forth—which was then listed by their names in the membership roster.[49]

Non-university-educated men were well aware of the gulf that separated them from the highest echelons of German intellectual life, and they employed various strategies for addressing this gap. Men without official academic credentials often adopted a posture of protective deference, using self-effacing rhetoric to preempt the charge that they were trespassing on the territory of the "learned." They might explicitly forgo a claim to learned status, or claim it only in a qualified way. In his illustrated work on mushrooms, Harzer asked his learned reader to remember that "it could not have been my intention to want to put myself forth as a learned man. . . . As an unlearned man, it never would have occurred to me to write for learned men."[50] The Saxonia society adopted a modest stance in its first published report, also in the interest of deflecting criticism. Because they were not a group of *Fachmänner* (experts), the society wrote, they did not need to fear that "the severe weapon of criticism" would be used against them. They were "a small circle of diversely educated [*gebildeten*] friends, who are filled with love and enthusiasm for nature," and they trusted that their work would be received in that vein. The Nature-Researching Society in Danzig described itself as a learned society but also noted that its accomplishments, in its own estimation, were less extensive than those of other groups, because its members, "chained by their varied vocations," only practiced natural science as a *"Nebenstudium"* (a secondary study) in their leisure time.[51]

The original core of the Isis, predominantly schoolteachers and apothecaries, had recruited the city's leading naturalist, Ludwig Reichenbach, to their ranks with a similar show of self-conscious deference. The Isis's founders first approached Reichenbach with deliberate formality, clearly aware of the sizable gap in social and scientific prestige that existed between the academy professor and themselves. The society sent a formal committee to the professor to invite him to attend one of its meetings, despite the fact that several of the groups' founders already knew Reichenbach personally.[52] Harzer had provided engravings for some of Reichenbach's published works, and Gössel worked under him in the royal natural history collections. Once the academy professor agreed to join, the Isis memorialized his entry as the most significant event in the group's early history. In the members of this society, Reichenbach gained the only really enthusiastic converts that his proposed natural system of botanical and zoological classification would ever win. The members of the group's botanical and zoological sections met together to studiously organize their collections according to Reichenbach's dictates.[53]

Ties to professors and the directors of major collections were highly useful to men like Gössel and Harzer. Higher-status naturalists could certify the work of their less elevated colleagues, acting as patrons in their clients' interactions within broader natural history networks.[54] Reichenbach provided a foreword to Harzer's book on mushrooms, vouching for the author's industriousness and reliability.[55] In the absence of such direct personal endorsements, membership in a society also offered artisans, schoolteachers, and apothecaries a useful status marker to employ on the title pages of their published writings, as teacher Julius Bescherer's aforementioned description of himself as "a member of several learned societies" illustrates.

The early years of Halle's Natural Scientific Association for Saxony and Thüringen offers an excellent example of both the growing social and epistemological divergence of the regional natural history milieu and the lingering ambiguity about its relationship to high science. The Natural Scientific Association was founded in 1848 by a small group of young university-educated teachers and *Privatdozenten* (the latter were men who had the right to hold university lectures but did not have a professorial appointment). Whereas the older Nature-Researching Society, closely tied to the professoriat and the rhythms of the university calendar, aspired to a dignified formality, the new society was casual.[56] Its members met weekly for their discussions in a local pub. In social profile, the society's founders were similar to the men who founded the Physical Society in Berlin, which was another group founded by ambitious young men who had not yet become part of the city's academic elite. In a matter of years, however, the Halle Natural Scientific Association evolved into a more socially diverse group with regional natural historical interests at its core.[57]

One obvious aspect of the Halle society's initial purpose was to raise the visibility of its founders. Soon after forming, the group made up an ambitious international list of other natural scientific societies with which it hoped to establish relations, and also bestowed honorary memberships on numerous professors throughout Germany.[58] Closer to

CHAPTER 6

home, it also tried to forge ties with the older, more venerable Nature-Researching Society. In 1851, the Natural Scientific Association invited the members of the Nature-Researching Society to attend its meetings. The new group also asked for permission to participate in the older society's gatherings, being careful to indicate that it did not mean for this arrangement to threaten in any way the independent existence of either group. In a polite letter a few months later, the Nature-Researching Society agreed to the proposal. Despite their formal acquiescence, however, the professors evidently had little interest in joining their younger colleagues at the pub. The Natural Scientific Association repeated its invitation several times over the course of the next year, apparently without ever gaining a significant response.[59]

At least one contemporary observer lamented the division between Halle's two groups of *Naturforscher*. Christian Keferstein, a state official whose membership in the Nature-Researching Society dated from before the period of professorial dominance, found the increased formality that had come to characterize the older society regrettable (and he had also gotten into trouble for not paying his dues).[60] In his 1855 memoirs, he described the difference between the two societies in his hometown as a purely generational one. He thought a merger would be advantageous; the two groups were working, after all, toward the same goal, and the "integration of the young limb with the old body" would benefit everyone.[61]

Age and status were not all that separated the two groups, however. By the time Keferstein was writing in 1855, one could question whether they were working toward the "same goal" at all, despite the fact that both pledged their allegiance to the cultivation of natural science. By the mid-1850s, the Halle Natural Scientific Association's membership stretched far beyond its initial core of youthful university-educated men. Booksellers, merchants, schoolteachers, apothecaries, and engravers—the usual diverse cast of middle-class natural researchers filled the ranks of the new group, giving it an entirely different profile from its university cousin. Professors in other cities accepted honorary memberships in the group; a number of prominent foreign societies responded to the association's requests to exchange publications. Closer to home, however, the society's distance from the heights of academic science could be seen more clearly. Much as in the eighteenth century, when physical distance had helped ease relationships that bridged status or gender gaps, the formal integration of regional natural history societies like Halle's Natural Scientific Association into the networks of international science existed alongside status differences that were thrown into clearer relief in local contexts. Both of the Halle societies continued to exist separately for the rest of the nineteenth century, and the Nature-Researching Society remained the domain of university professors.

For the Love of the Fatherland

The world of elite science was not the only place where regional natural history societies sought approval. Most of the honorary members of the Isis, for example, were not university professors but Saxon officials, men of aristocratic birth at high levels of the bureaucracy. Over the first two decades of its existence, the society worked hard to capture the attention of the royal family and the state, with mixed results. Its efforts to gain official recognition as a "royal" society, for example, never came to much. Individual officials accepted honorary memberships, but the royal family itself remained distant.[62]

In other places, local and regional elites embraced natural history societies as a vehicle of civic self-promotion with more enthusiasm. This was certainly not an entirely new phenomenon in the 1830s and 1840s. The earlier groups discussed in chapters 3 and 4 were also, in important ways, invested in the honor and reputation of a particular place. Furthermore, when urban elites joined hands with science, the link between these two things did not always run through the study of regional natural history. The patrician commercial elites of Frankfurt and Hamburg supported natural scientific societies, and the museums these groups founded became important local cultural ornaments with decidedly global dimensions. More general natural history, itself the product of far-flung exchange networks, was a particularly good fit in such places, where the wealthy local merchant and banking class also made their living from wide-ranging relationships of trade.[63] In Ayako Sakurai's study of scientific associational life in Frankfurt, one sees little evidence of the intense connections to local landscapes that characterized groups like the Dresden Isis. The Senckenberg Society proudly tended a collection of *naturalia* as cosmopolitan as the ties of Frankfurt's commercial elite.[64]

For societies that did make it their project to study the nature of the "fatherland," the referent of this word was often far from obvious, and groups made strategic choices in crafting their allegiances. Seemingly simple sentiments—a pride of place, a sense of local distinctiveness—were often anything but straightforward given the complex political

CHAPTER 6

and cultural geography of the German Confederation. The ideal of the local *Heimat*, or homeland, that famous particularistic trope of later nineteenth-century German culture, was beginning to emerge in this earlier period, and it was crafted using a range of cosmopolitan tools, not least of which was natural history.

The act of cataloging the "nature of the fatherland" gained its meaning against the backdrop of *Vormärz* efforts in state building. After the territorial consolidations of the late eighteenth and early nineteenth centuries, recently expanded states such as Württemberg, Bavaria, and Prussia attempted to fashion cohesive political entities out of a patchwork of new territories, while the casualties of the new post-Napoleonic order struggled to adapt to their diminished political and cultural importance. Throughout the German Confederation, aspirations for a unified German nation added another layer of complexity to the question of political belonging.

The contrasting development of two societies in Bavaria illustrates the ways in which natural history societies could became linked with local and regional elites' evolving responses to the post-Napoleonic political order. In both the former Catholic bishopric of Bamberg and the predominantly Protestant Palatinate, the promotion of natural history dovetailed with much wider cultural, political, and religious negotiations with the Bavarian state. In the former case, natural history served a mediated form of dynastic loyalty; in the latter, the natural historical society cultivated a regional identity that minimized the Palatinate's connections with Bavaria.

The first decade of the nineteenth century had struck a number of blows to the Franconian town of Bamberg. Previously home to a university and center of a politically independent Catholic bishopric, the town became a part of Bavaria in 1802 and saw its university downgraded to a theological lyceum a year later. Tension with Munich ran high in the decades following Bamberg's annexation; the city's secular and religious leaders engaged in a number of battles with their new rulers. The centralizing efforts of Munich also struck a considerable blow to the city's cultural institutions, which lost their state support when their patron, the Bamberg bishop, lost his political powers. The creation of a Bavarian constitution in 1818 and Bamberg's elevation to an archbishopric in 1821 helped improve relations with Munich.[65] In a move that signaled a growing sense of Bavarian allegiance, Bamberg's lyceum held a festival in honor of the Bavarian constitution in 1822. King Ludwig I visited the city for the first time in 1830.[66]

The most visible symbol of Bambergers' growing accommodation to

Bavarian rule was the yearly Theresienfest, a festival in honor of the queen. First held in 1833, the festival included parades, competitions, and tournaments meant to recall the glories of an imagined medieval Franconian past. A tribute to the crown, the Theresienfest also celebrated Bamberg's cultural particularity. In the classic form of invented traditions, this sense of difference owed as much to the experience of Bavarian rule as it did to any organic tradition of Franconian distinctiveness.[67] Speaking to a smaller and more elite audience, the local historical society (founded, supposedly, at the suggestion of the Bavarian king) also celebrated this regional identity.[68]

Bamberg's Nature-Researching Society, formed a year after the first Theresienfest, had a similar relationship of mediated loyalty to the Bavarian state, designating the new local administrative district, the Obermainkreis, as the field for its efforts.[69] In addition to accepting Bavarian administrative geography, the group also courted personal patronage from members of the royal family and bureaucracy. At its founding, the society had solicited and received the special protection of Bavaria's crown prince, Maximilian, and the heir to the throne visited the society's collections in 1837.[70] A number of high-ranking Bavarian bureaucrats were also named honorary members.[71] In 1837, the Bamberg local paper reported that the highest regional administrator in Upper Franconia, the Freiherr von Andrian, came and "viewed the new, highly interesting acquisitions of the natural historical society and convinced himself of the active striving and tireless progress of the association toward the achievement of its chosen goal."[72] A few weeks later, the local paper announced that Andrian had agreed to become an honorary member of the society.[73] In 1840, the new regional governor, Melchior Ignatius Nicolaus von Stenglein, accepted his honorary membership with expressions of "well-meaning feeling and the most powerful support for the flowering of the association."[74] The society also opened its collections to the public as part of the Theresienfest.[75]

Most of the Bamberg lyceum's faculty joined the natural historical society, as did a number of other local clerics. In fact, 36 percent of the society's founding members were priests (fourteen out of the original thirty-nine members). The remaining members were doctors, apothecaries, or local secular officials.[76] A local doctor, Friedrich Kirchner, served as society chairman; the second chairman was Wilhelm von Lerchenfeld, a high-ranking local church official. The secretary, Adam Martinet, was a philosophy professor at the lyceum (and an ordained priest), while the treasurer, Peter von Hornthal, was the son of Bamberg's first preunification mayor.

CHAPTER 6

The natural scientific society's promotion of local natural history also coincided with a growing civic interest in the surrounding countryside as a source of leisure and of profit. For example, restorations had begun in 1818 on the Altenburg, a ruined castle on a hill outside the city. New walking paths had been put in for the use of Bamberg's citizens in the mid-1830s, and a founding member of the Nature-Researching Society, the Freiherr von Stengel, oversaw the construction of a small tavern at the top of the mountain.[77] The society secretary, Peter von Hornthal, was also important in early industrial efforts in the town. He was both president of the Main Steamship Society and one of the primary investors in a mechanized cotton mill built in nearby Gaustadt during this period. He owned the city's first political newspaper, the *Fränkische Merkur*, as well.[78]

Local priests' involvement in the Nature-Researching Society continued an older tradition of clerical scientific activity, disrupted but not destroyed by the cultural and material losses of the early nineteenth-century secularization of much church property.[79] The city of Bamberg's natural history collection, for example, had originally belonged to a nearby convent; it came to the town along with its priestly curator, Dionysus Linder, when the Banz convent closed in 1803.[80] The lyceum teachers in the Bamberg society filled the niche occupied by secular schoolteachers in other places, joining the Nature-Researching Society in comparable numbers.

For the clerics involved in the Nature-Researching Society, their participation was also a mark of their new, if limited, loyalty to the Bavarian state. The Bamberg society was founded against the backdrop of heated debate over church-state relations and papal supremacy, one year after the German publication of the ultramontane tract *Triumph des Heiligen Stuhls* (Triumph of the Holy See). Most of the priests in the society were younger men who had been ordained in the 1820s during a time of relative concord between the Bavarian state and the local church hierarchy. Several members of this younger generation were also open local opponents of ultramontanism. Michael Deinlein, a teacher at the Bamberg lyceum, spent the 1830s struggling to stem the pedagogical influence of his ultramontane fellow professor Leonhard Clemens Schmitt.[81] Another lyceum professor, Adam Gengler, regularly corresponded with two prominent critics of the doctrine of papal supremacy, the historians Ignaz Döllinger and Johann Adam Möhler.[82] The theological debates of the 1830s also touched on the church's stance toward contemporary philosophy, science, and history, and the Nature-Researching Society's clerical members were the heirs to the older, integrative intellectual traditions of the Catholic Enlightenment. One of the oldest society members, Friedrich von Brenner (ordained in 1807), had been a student of influential reformer Johann Michael Sailers.[83]

In the 1830s and 1840s, the Nature-Researching Society focused its attention on a local natural resource that easily attracted attention from the outside world—the impressive deposits of fossils in the nearby countryside. In the invitations they received, potential members were informed that the society had been created because of "the high worth of fundamental natural scientific research, the particular predilection and participation of many educated men of the city of Bamberg, and in particular, the richness of the area . . . in fossils and other products of organic nature."[84] Indeed, almost all of the recorded society activities from the 1830s and 1840s dealt with fossils in some way. The association called several special meetings to examine particularly spectacular finds. In 1838, the society met to examine the most complete ichthyosaur remains ever found in the region. A year earlier, Adam Martinet and several other men traveled to the Baunach Valley in order to research its "geognostic relationships" and fossil remains. The results of the trip were reported at the next society meeting. In 1841, the society met in the lyceum to hear a lecture on "the relation of geological strata and their fossil remains to biblical reports of the creation and the Flood."[85]

In Bamberg, regional natural history simultaneously defined the city's local particularities and courted the attention of a broader national and cosmopolitan community. In this respect, it fit neatly into the local elite's broader patterns of self-presentation. A concern for documenting Bamberg's excellence and uniqueness was a common mark of the city's public culture in the *Vormärz*. The Nature-Researching Society's activities were regularly reported in the new daily newspaper, the *Tagblatt der Stadt Bamberg*. Until the late 1840s, the news in the *Tagblatt* was almost exclusively local, presenting the town's literate citizens with announcements of government mandates, notification of club meetings and concerts, and lists of the out-of-town guests staying at the town's inns. When prestigious outsiders, either scientific or noble, viewed the natural historical cabinet, their favorable judgments were recounted in the local press. In this way, the paper provided literate Bambergers with images of their city through the eyes of notable visitors. Local readers were informed that "Professor Jäger from Stuttgart . . . an outstanding connoisseur of fossils, admired the richness of our primeval flora, which can be found in the Bamberg area."

Similarly, "Mr. Elie v. Beaumont from Paris and Mr. von Buch from Berlin visited the collection of our natural scientific society." "As we all know," the paper continued, "the founder of this endeavor is Dr. Kirchner . . . and what he and other similarly exceptional men have done for this branch of scholarship with their expenditure of time and trouble shines back on the city itself and its inhabitants." When the national Society of German Natural Researchers and Doctors met in nearby Erlangen in 1840, members of the geognostic section made a special trip to Bamberg to see the society's natural historical collection; this fact was also duly noted in the local paper.[86]

Both in the *Tagblatt* and in other forums, members of the local civic elite of Bamberg showed great concern for documenting their activities, both for the benefit of outsiders and among themselves. The lyceum's librarian, Heinrich Joachim Jäck, the founder of the *Tagblatt*, was also the most important chronicler of Bamberg's polite culture during this period. In addition to books such as his *Pantheon of Bamberg's Writers and Authors*, Jäck also composed a travel guide to Bamberg and the surrounding countryside.[87] In both works, Jäck argued for Bamberg's continued cultural importance. The descriptions of gifts to the natural historical collection reflected a similar concern for comparative status. Accounts of visitors to the collections emphasized both the eminence of the people involved and their favorable judgments on the collections themselves. To define a sense of local distinctiveness, Bambergers used standards that came from a cosmopolitan community of authors, travelers, and readers.

Cosmopolitan provincialism did not always coincide so neatly with growing allegiance to a centralizing political state.[88] The Bavarian state had a much less cordial relationship with the regional scientific society in one of its other new territories. The patronage strategies of the Pollichia, a natural history society in the Palatinate, stood in stark contrast to the Bambergers' overtures to the Bavarian crown. While avoiding ties to the royal family or central bureaucracy, the Pollichia was much more assiduous than its counterpart in Bamberg in fostering permanent connections with influential scientific figures in other nearby cultural centers. Beginning in 1842, the Pollichia's members published an annual report of its activities and initiated exchanges with other civic scientific societies. In addition, they sought out eminent researchers as honorary members. Initially focused on men in southwest Germany, these efforts to recruit scientific patronage quickly expanded to Europe as a whole.

Another recent acquisition of the Bavarian crown, the Palatinate was cut off from Bavaria by both geography and religion (Bavaria was predominantly Catholic, while most inhabitants of the Palatinate were Protestant). An important center of liberal dissent in the 1830s and 1840s, it was viewed with particular suspicion by the Bavarian bureaucracy.[89] Although the Pollichia was informally founded in 1840, the Bavarian state took two years to approve the society's statutes. The organization gained official recognition only with the condition that it limit its efforts to the Bavarian-controlled section of the Palatinate (this geographic designation had historically included land that now belonged to Baden). In addition, all of its members had to be Bavarian subjects. This stipulation was a significant blow—as a result, G. W. Bischoff, professor of botany at the University of Heidelberg, had to resign as the society's president. The Pollichia used the category of "honorary member" to get around these restrictions, however, and Bischoff continued to attend the group's annual meetings under this label.[90] Among the society's regular members, about a fifth were local officials (mostly forestry officials); the various branches of the medical profession (doctors and apothecaries), schoolteachers, and local landowners were all represented approximately equally. A handful of pastors and merchants completed the society's rolls.[91]

In the early years of its existence, the Pollichia sought out scientific allies in cultural centers in southwestern Germany—Karlsruhe, Mannheim, and Heidelberg—rather than in Munich. In 1843, the society had thirty-four honorary or corresponding members. Out of this group, the largest concentration of members was in Heidelberg (five). A number of these "honorary" members (some initially among the founding members of the society) regularly attended its annual meetings; the ties with others were less significant. Soon after its founding, the Pollichia had contacts as far afield as Algeria, the southern Tyrol, and Athens. Its only member in Munich, however, was a Russian diplomat.[92] The society finally named a faculty member of the University of Munich to its rolls in 1848 (the botanist Philipp von Martius); however, at this point it already possessed three times as many honorary members in Paris (six) as in the Bavarian capital (two).[93]

The Pollichia met yearly in the spa town of Bad Dürkheim, and the society enjoyed significant support from the city government. The town council provided the organization with a yearly financial allotment, and the mayor was also a society member. The society's yearly meetings were attended not only by its members but also by a number of curious onlookers. In addition, the society's natural historical collections, containing specimens from the region's flora, fauna, and mineral wealth, took up several rooms in the town hall. Year round, the collections provided both visitors and Dürkheim natives with a composite image of the "Nature" of the Palatinate, prominently displayed in a central civic space. Its specimens of Palantine plants and animals were joined by a collection of typical German birds. In the visual assemblage of the Pollichia collection, the Palatinate was a region within "Germany." "Bavaria" was nowhere to be seen.[94]

Here we can see an early example of the process described by Celia Applegate, the use of regional identity as a mediating term in understandings of the German nation. Along similar lines, several members of the Dresden Isis founded a journal in 1846, the *Allgemeine deutsche naturhistorische Zeitung,* devoted to the "promotion of knowledge and study of natural history in Germany for people of all walks of life." The publication hoped to find an audience among both academics and "friends of nature." If their journal's task was successful, the editors claimed, "German natural science would finally be accorded its due importance in the entire development of the German nation and in general *Bildung.*" The publication would provide an overview of current scientific work in natural history and related fields, not just for the purposes of disseminating information but also so that "men from different practical positions and occupations" would participate in natural history "through word and deed."[95] The *Allgemeine deutsche naturhistorische Zeitung* imagined a socially diverse community of researchers, shoring up natural science in the broader life of the *Volk*. All who were "passionate for the progress of science and for the promotion of its influence on the life of the *Volk*" were invited to contribute to the publication.[96] For the scaffolding of this national natural scientific community, the journal hoped to build on Germany's expanding network of regional natural history societies.

Conclusion

The men who filled the ranks of this period's new natural scientific societies found membership in such a group appealing for a variety of reasons. For men who, for reasons of either education or geography, were marginal to the leading centers of German intellectual life, belonging to a local scientific society put them on the map of the learned world, even if only in modest measure. As was the case with more elite societies, regional natural history societies also made their individual members more visible locally. Scientific or learned reputation was one element in the hierarchies of distinction that structured local urban life, and though fame throughout Europe was certainly a desirable goal, a more limited reputation could also carry local weight.

At the same time, the multiplication of these groups, along with their social diversity, made it harder to incorporate them completely into older conceptions of a unified learned world. Around 1800, the spread of *individual* marks of learnedness such as collecting and authorship had caused the value of these status markers to decline. The enlightened world of learned friends became the more self-consciously formal world of a new emerging scientific elite. In the 1830s and 1840s, one can see a similar process of devaluation going on as the field of scientific societies expanded; now this *collective* marker of status mattered less, too. By the mid-nineteenth century, most elite German researchers had stopped listing their memberships in learned societies on the title pages of their books.

SEVEN

The Wellspring of Modernity

At the 1838 meeting of the Society of German Natural Researchers and Doctors, Karl Josef Kreutzberg gave a speech entitled "On the Influence of Natural Science on Industry and Life." The natural sciences, he claimed, were now channeling their beneficent power "through all civilized people like a moral electricity." Their accomplishments were legion. Alongside steamboats and railroads, he counted new mechanized forms of production among natural science's marvels. He only regretted that he could not live forever to see all of the other wonders that the conjoined powers of industry and natural science would produce.[1] Sixteen years later, the historian Johann Droysen described the same situation, but with all the value signs reversed. According to Droysen, the natural sciences (which were fine when they stayed in their allotted channels) had flooded over their banks. They were now spreading out through all of German culture on a dangerous wave of materialist enthusiasm. In the process, they threatened to choke off everything noble and spiritual in national life.[2]

When one looks more closely at the early industrial press of the 1830s and 1840s, however, one finds that the position of *Naturwissenschaft* is somewhat more complicated than either of these two statements would lead one to believe. For example, though the natural sciences did receive praise in the early industrial press, the general ideal of *Wissenschaft* loomed far larger. In the 1837 volume of Cologne's *Allgemeines Organ für Handel und Gewerbe*, the words "natural science" or "the natural sciences" were used just three times, and in only minor and insignificant passages. "*Wissenschaft*," on the other hand, appeared forty-six times and was a concept at the heart of much of the journal's rhetoric.[3] Similarly, though the period's many manufacturing associations were very interested in building up a *Wissenschaft* of technology, how exactly this new *Wissenschaft* (itself called *Technologie*) would relate to the basic natural sciences was by no means always clear. By the late 1840s, the library of Dresden's Manufacturing Association had over a hundred volumes that it categorized under the heading *Technologie*, mostly the publications of manufacturing associations from other parts of German-speaking Europe. The natural science section of the library had five books.

Nonetheless, Josef Kreutzberg was not alone in arguing that natural science, for good or ill, had become a dominant cultural force, a force that was transforming both material life and the social order. J. F. Binder, the Nuremberg mayor and board member of the Ludwig Railroad Company, also thought natural science and mathematics had improved manufacturing and industry; this was a major point of his 1835 speech inaugurating the Nuremberg-Fürth railway line, the first steam-powered railway line built in the German states.[4] Economic historians and historians of technology looking back on this period have generally greeted such claims with skepticism. Most now agree that scientific discoveries were not the engine behind mid-nineteenth-century industrialization. It was only in the late nineteenth century, with the growth of the chemical and electrical industries, that basic scientific research became a regular source of economically relevant innovation.[5] In the 1830s and 1840s, too, people interested in improving manufacturing often greeted elite *Naturforscher*'s claims to tutelage with a certain skeptical distance. The general ideal of *Wissenschaft*, in contrast, had broader appeal. So did the communicative practices of learned sociability, which seemed to offer a powerful example of how craftsmen and manufacturers could build new forms of connection and consensus in a period of economic and political transition. But *Wissenschaft* also did not reign supreme; plain old practical experience had some strong defenders, too.

People who placed special emphasis on the importance of *natural* science were usually, though not always, active *Naturforscher* of various kinds. They might be university professors, mathematics teachers, or faculty at the new higher technical schools; they might also be from the period's thin but growing group of technical experts: chemists, machinists, or factory managers. The members of manufacturing associations and the editors of journals devoted to economic improvement had a broader agenda and a broader set of intellectual allegiances, and they typically used somewhat different rhetoric. They placed the general concept of *Wissenschaft*, not the more specific *Naturwissenschaft*, at center stage.

Of course, professors and technical experts also wrote for the emerging industrial press and joined manufacturing societies, and they also talked a good deal about the value of *Wissenschaft* in general. The people they were speaking to in these contexts, however, did not typically share their dual conceptual allegiance to both *Naturwissenschaft* and *Wissenschaft*. The craftsmen, manufacturers, state officials, and educated liberal reformers who filled the ranks of manufacturing associations were all highly interested in technical and economic improvement, but they did not always have a similarly strong attachment to the category "*Naturwissenschaft*." That is not the same thing as saying that they did not value natural knowledge. A provincial bureaucrat like Karl Preusker thought chemistry, natural history, and natural philosophy were all very valuable, but in his mind, it was the power of systematic thought and knowledge *in general*, not something peculiar *just* to the natural sciences, that would breathe new life into crafts and manufacturing.

In the 1830s and 1840s, the cause of economic improvement overlapped significantly with liberal interest in educating the *Volk*, or common people. This fact had significant implications for discussions of both *Wissenschaft* and *Naturwissenschaft*. *Vormärz* liberals wanted to create an educated, active society of citizens, and science in all its forms was an important part of those efforts.[6] When they wrote about spreading science to the people, the people were not supposed to just passively receive knowledge; they were to be active (albeit second-tier) participants in an ongoing conversation. Here, we can hear clear echoes of enlightened attempts to create philosophical farmers and learned artisans. The language used to describe this process of intellectual transmission had changed, however, and in significant ways. As we have already seen in the preceding chapter, learnedness was no longer an attribute that spread very easily; doing "learned" things like belonging to a scientific society or publishing scientific work was no longer enough to make someone a learned man. Now, artisans and manufacturers would not become philosophical or learned, but *"gebildet,"* educated.

Nonetheless, people in the 1830s and 1840s still continued to draw on a familiar learned repertoire of behaviors, and recognizing this fact helps us better understand how they thought about the power of science. Both *Naturwissenschaft* and *Wissenschaft* were seen not only as valuable tools for future reform but also as powerful forces that had already done much to transform the world in which Germans lived. As I have already mentioned, the conviction that natural science (or *Wissenschaft* more generally) had deeply transformed German material culture by the 1840s is not one that most contemporary historians of technology share. Many educated nineteenth-century Germans, in contrast, believed that they saw the hand of *Wissenschaft* at work everywhere around them. The evidence they offered in support of this contention was often as much social as it was intellectual or technological: that is, writers often talked about the spread of activities and communicative behaviors rather than pointing to the transformative power of key intellectual insights.

The conviction that science was changing society made a central contribution to the emotional charge carried by the term *"Naturwissenschaft."* In practice, elite researchers occupied a complicated position within broader public discussions of economic improvement. In celebratory speeches, however, it was possible to gloss over this fact. When Kreutzberg described science as a moral electricity, this force did not just flow through a narrow group of elite researchers; it came from the activities of a disparate and far-flung community. When Kreutzberg spoke of "the genius of researchers . . . transforming the world," he had a fairly large cast of characters in mind. *Naturwissenschaft*, according to him, was no longer the monopoly of the learned estate. It was "observing everywhere, making itself useful everywhere, offering the simplest artisan as well as the most profound learned man instructive new objects." This did not mean that "the profound learned man" and "the simplest artisan" were doing the same thing, but it did mean that science could be understood as a project in which "the most diverse civil [*bürgerliche*] stations and individual forces can participate."[7]

The participatory community that Kreutzberg described was not a "learned" one; the learned were an identifiable group within it. It was clearly a community, however, that was different from the kind of public for which later nineteenth-century scientific popularizers wrote. When Kreutzberg claimed that natural science was "observing everywhere, making itself useful everywhere," this way of conceptualizing what the spread of knowledge entailed owed a heavy debt to the older ideal of the republic of letters. Information was not all that traveled; forms of *activity* did, too. One can certainly see Andreas Daum's dialectic of the professional and the popular emerging in the 1830s and 1840s. The concept "*Naturwissenschaft*," however, gained significant emotional and rhetorical power from the fact that this dialectic was not yet fully established. Kreutzberg argued that unlike other fields, which were "bound to partly outdated disciplinary norms," natural science was "freer, less constrained, more accessible."[8] Only a few university professors and *Gymnasium* teachers cared about the study of antiquity, wrote a technical academy professor named E. Freese, but the natural sciences had "more connoisseurs [*Kenner*], friends, and admirers than one could ever count." This diverse new public, Freese thought, had emerged over the past few decades and had left intellectual life transformed.[9] A reviewer of Justus Liebig and J. E. Poggendorff's *Handwörterbuch der reinen und angewandten Chemie* made a similar observation. In the current age, chemistry had become "the common property of all educated people."[10] Rhetoric like this could make a very wide field of cultural activity—the early industrial press, the multiplying number of manufacturing societies—look like the fruits of natural science, at least if one was observing from a certain distance. Little wonder that the historian Johann Droysen thought the natural sciences were overflowing their banks.

Saxony, one of the most heavily urbanized states in mid-nineteenth-century Germany, provides most of the examples explored in the following pages. The center of a growing textile industry, Saxony was also one of the most economically modern regions of the *Vormärz*, and a particularly politically volatile one as well.[11] Educated Saxons were quite conscious of belonging to a region where controversial new forms of production were making significant inroads. In this sense, Saxony was far from typical. As a site of marked economic change and growing instability, it is, however, an interesting place to look at discussions about the relationship of knowledge to social and economic order.[12]

For this very reason, Saxony is also a good place from which to make the point that public enthusiasm for science and natural science did not just spring from enthusiasm for the technologies of large-scale industrial production.[13] Even in Saxony, promoting industry, "*Gewerbefleiß*," did not mean working toward a future dominated by concentrated, large-scale capital and huge mills. The advocates of natural science did not imagine this kind of future either. Their future was one that still had many artisans and small manufactories, and it was in this human-sized world that they wanted to plant the seeds of science.[14]

The Advocates for Natural Science

In the 1810s, 1820s, and 1830s, many German states founded new technical academies or upgraded previously existing trade schools. The Viennese academy, founded in 1815, became the model for many of the schools that appeared in other states over the next few decades. These institutions, the precursors to the late nineteenth-century technical universities, were the seedbeds for the modern profession of engineering in Germany; they were designed to train a new elite class of architects and machinists who would enable the Germanies to challenge the economic power of Britain and France.[15] Many German states also continued to put resources into agricultural education as well. Germany had eleven agricultural academies by 1860. Though lower in prestige than the universities, these new technical and agricultural schools were not part of a totally separate world. Many of their professors were university-educated men, and technical academy professors sometimes went on to posts at universities.[16]

Though their enrollments remained small until after the 1850s, these new institutions had enormous symbolic importance. In his famous essay lambasting the state of chemistry in the Prussian universities, Justus Liebig singled out the technical schools as a point of light in an otherwise grim landscape. His obscure provincial colleague August Beger agreed, claiming that these schools were on their way to making all the professions (*Berufsarten*) that dealt with the material world, from miners to merchants, into truly scientific pursuits.[17]

Dresden's technical academy, founded in 1826, was one of the more successful schools of the period, at least from some perspectives. Alongside the academies in Berlin and Vienna, it became one of the most important institutions of its kind. Its most famous professor, J. A. Schubert, had notable success in reverse engineering English machinery. Over the first several decades of its existence, it also forged close ties to Saxon industry, a development that furthered its strength and prestige.[18]

The Dresden Manufacturing Association, which Schubert helped to found, was also initially structured to give special status to the technical academy's faculty. Its statutes described a group in which a small number of natural scientific experts would serve as advisers to a much wider circle of practical men. Society meetings would include presentations "on topics in the natural sciences" as well as on manufacturing, *Gewerbekunde*. The natural science lectures were supposed to be "generally understandable," not the stuff of specialized science. The society's statutes made extensive use of the adjective "practical" in describing the epistemological content it considered most relevant to the discussions of the society members as a whole.[19] It also used divisions among scientific fields to break the association up into sections.[20] One section corresponded roughly to the emerging modern discipline of physics, along with its accompanying practical applications; it was devoted to "topics that are grounded in mathematics, natural philosophy, and mechanics." The second was devoted to "topics dealing with knowledge of materials and chemistry."[21]

Here, professors like Schubert would seem to have achieved with industry what an earlier generation of learned men had failed to do with agriculture: they had placed themselves squarely at the head of an emerging field of practical discussion, setting themselves up as the tutors of a broader practical public. In agriculture, too, one can find similar efforts at midcentury. Justus Liebig's efforts are probably best known, but there were others as well. The Jena professors M. J. Schleiden and E. E. Schmid, for example, authored an *Encyclopedia of All the Theoretical Natural Sciences as Applied to Agriculture*. They admitted that their book entered a crowded field of similar works but claimed to have corrected the many mistakes, both methodological and substantive, of their predecessors. Many past attempts to connect scientific theory to agricultural practice had been botched; most other handbooks had been either too theoretical or too practically oriented. Schleiden and Schmid promised to finally make theory "usable" ("*anwendbar*"), to show how "the stale, unconscious lazy habits of the peasant could be elevated to the conscious and sensible actions of the rational farmer."[22]

Such sunny optimism about the subordination of practice to theory masked a far more complicated and contested reality. State officials, military officers, schoolteachers, and technical academy professors made up a small percentage of the Dresden Manufacturing Association; but 74 percent of the local members were manufacturers, artisans, or merchants. These men were a mixed group. Some, like the chemical entrepreneur Gottfried Reichard, were the owners of small manufacturing concerns; others used the guild title "master" to describe themselves.[23] As it turns out, there is much to suggest that the rank and file of the Manufacturing Association was not really all that interested in sitting quietly at the feet of the city's elite *Naturforscher*.

By the mid-1840s, one can find the university-educated men in the group complaining that Dresden's businessmen lacked respect for the sciences.²⁴ For the academy professor and chemist Alexander Petzholdt, tensions escalated to an open break. Petzholdt was a well-respected young chemist who later went on to a university chair; the local businessmen of Dresden, however, were less impressed with his abilities. In the early 1840s, Petzholdt conducted a series of experiments on electroplating techniques at the behest of the Dresden Manufacturing Association. After giving Petzholdt the funds to carry out his experiments, the Manufacturing Association's members expressed dismay at the "impractical" nature of his research. Angry to have paid for experiments they did not find useful, they refused to pay to print Petzholdt's results as originally agreed. Petzholdt had to rely instead on a subvention from the Saxon state to publish a brief monograph of his work.²⁵

Even the celebrated Schubert was not immune to criticism. In the late 1830s, for example, he became embroiled in a public dispute with the Dresden Steamship Company. The company had bought some faulty steam engines, and many of the shareholders cast the blame for this mistake at the feet of Schubert, the company's machinery expert. In the 1838 shareholders meeting, Schubert vigorously defended himself. He was not at fault, he insisted. He had told the company's directors all along that they ought to buy high-pressure engines; they had ignored this advice, and once the engines they did order arrived, they had failed to call in a competent engineer to make sure that the machines were in good working order. "It is a mistake of great consequence," Schubert intoned, "when the directors labor under the delusion that they know something about technology and only half trust the engineer [*Techniker*], when they ought to leave everything to his discretion."²⁶

When one moved from blithe generalizations to brass tacks, the proper relationship between scientific expertise and economic practice was always a fraught and difficult issue, and this remained the case throughout the nineteenth century. Schmid and Schleiden, the authors of the previously mentioned agricultural handbook, may have believed that they had finally figured out how to make natural science useful to farmers; the subsequent history of agricultural education suggests otherwise.²⁷ In the 1830s and 1840s, as in later periods, even those who were convinced that *Wissenschaft* was useful to farmers, artisans, and manufacturers had varying ideas about the kinds of science such people needed to know. For example, when the Saxon educational reformer Karl Preusker spoke of the particular scientific information that would benefit artisans, many of his claims came straight from the repertoire of eighteenth-century cameralists. "The craftsman [*Gewerbetreibende*] must at the same time be an expert [*Naturkundiger*] in relation to his materials," he wrote. It was important for a carpenter to know the properties of different kinds of wood, their density and their hardness; the stonemason needed to know about the properties of different kinds of rock, and he needed to know the best places to quarry them.[28] Preusker saw traditional forms of natural history collecting as very useful to the educated artisan. In contrast, the chemistry professor Justus Liebig was skeptical that the old natural historical tradition had much to offer. Natural history collections might inspire "even the simplest farmer" to feelings of wonder for the works of his Creator, but they "would not make him more skilled in earning his keep, more capable of carrying out his obligations to the state, or more useful to his neighbors." That was a job for Liebig's own program of experimental chemistry.[29]

When we look at Dresden's civic scientific life, we can see that the cause of natural science overlapped with the cause of improving manufacturing only at certain points. In the Isis Society, one finds chemists and machinists, the skilled technical experts that academies like Dresden's had been created to train. In other natural scientific societies in nearby Meissen and Schneeberg, one also finds men who described themselves as "managers" of manufacturing concerns—one man ran a porcelain manufactory, and another oversaw a textile mill. The owners of these two enterprises, however, did not appear on the membership rolls.[30] The mixed group of artisans and small entrepreneurs that made up the rank and file of the Manufacturing Association were also not very well represented in the Isis. One would not want to build too much out of these still fragmentary patterns; as we know from work on industrial entrepreneurs in later periods, there is a great deal of regional variation in the kinds of cultural and social alliances they formed.[31] In general, though, it seems that participation in the natural sciences seemed most attractive to men whose livelihoods were built around their possession of novel forms of technical expertise.

Before moving on to look more closely at manufacturing societies and the early industrial press, we should examine one more way in which the Dresden technical academy's professors worked to make their scientific status more visible in the city's cultural landscape. In the early 1840s, several professors at the academy founded a new Natural Scientific Society, bringing the city's number of generalist natural scientific societies to three. This group looked quite different from the city's earlier learned associations. It had a small core of active researchers, and then a large group of "extraordinary members" whose participation in the society essentially amounted to a subscription to a lecture series. The active elite met together in closed sessions to discuss their own scientific work (weekly during the winter, biweekly during the summer) and then held public lectures for the rest of the group.[32]

The Natural Scientific Society's lecture subscribers, though labeled "members," were no longer subject to any vetting process, and this represented a break with the practices of the city's other learned societies, in which even extraordinary members were usually subject to a vote. This new permissiveness weakened the meaning of "extraordinary" membership, which lost even more of its already greatly diluted honorific function once it could be purchased. In the Natural Scientific Society, anyone "who needs knowledge of the natural sciences" could buy their entry tickets to attend lectures, and the society statutes specifically noted that women were welcome as members in this category.[33] The lectures were also frequent. The statutes designated that they should be held every fourteen days in the winter and once a month in the summer, and the group kept a regular lecturing schedule for the first several years of its existence.[34] The lecture subscribers had the right to pose questions to the regular members, but they were not allowed to attend their scientific meetings or participate in the governance of the group.[35]

The gender gap between the society's scientific core and its paying audience reinforced the structural divisions written into its statutes. Women made up about a third of the society's 154 extraordinary members, while all of the ordinary members were male. Many of the group's female members had male relatives who were either extraordinary or ordinary members, but not all did. Subscribing to the lecture series appears to have been something that female friends and relatives might do together. Out of the 51 female members, 9 married women joined with other female relatives; 8 more married female members had no obvious family ties to anyone in the society. Among the single women, there were two sets of sisters, along with 11 other women who had no other obvious relatives in the society.[36]

Here, the new structures typical of later nineteenth-century scientific popularization were coming more clearly into view. The Natural Scientific Society was devoted to "the cultivation and spread of the natural sciences for their own sake and in their relationship to life," but the group clearly separated the cultivation of knowledge from its spread. It was designed to share knowledge with an interested lay public without the guiding fiction that new researchers might be created in the process. The Natural Scientific Society's lecture audience was not passive in any absolute sense, of course; reception always involves an active

CHAPTER 7

process of appropriation. Yet the society's internal structure captured an important moment of transition in the evolution of Germany's intellectual public, a point at which the traditions of intellectual community inherited from the early modern republic of letters were being abandoned for a new model of popular science. The Natural Scientific Society did not survive past the 1840s, but the assumptions it embodied became increasingly common in the second half of the nineteenth century. Over the next fifty years, the ideal of an ever-expanding, diverse intellectual community would become increasingly implausible as high science became more securely professionalized and popular science became a more clearly delimited field of cultural production.[37]

The Natural Scientific Society was definitely different from its predecessors, but it would be a mistake to see it as the vehicle for an entirely new kind of intellectual activity—to see it as the moment, for example, when earlier forms of *Naturforschung* were finally replaced by modern *Naturwissenschaft*, rigorously mathematical and free from philosophical baggage. In intellectual content, it did not represent this kind of clear break, though there were of course new ideas on the table (Alexander Petzholdt, for example, was a great enthusiast for Liebig's theories). In the society's presentations, natural history lectures still appeared alongside lectures on the physical sciences. Philosophy was not gone either. A Dr. H. Geyer gave a talk entitled "On the Relationship of Philosophy to the Natural Sciences" that was very much in the vein of the kind of epistemological discussions that had been taking place since the early nineteenth century. Geyer did not argue that natural science could dispense with philosophy; he examined which components of natural knowledge were just empirical and which could properly be called "philosophical."[38]

All in all, the practical tasks of the Dresden Technical Academy also played a relatively minor role in the presentations of the Natural Scientific Society. In 1845, J. A. Schubert gave a talk on the principle of inertia; a steam engine provided one of his examples, but only one of them.[39] The society was primarily a forum in which the academy professors, joined by a handful of local schoolteachers, military officers, and medical doctors, demonstrated their scientific expertise to a well-heeled, mostly upper-middle-class public. Their subscribers drew primarily from the city's service elite: they were military officers or officials, or the wives thereof. These people were joined by a handful of merchants. The men on the governing board of the Manufacturing Association, in contrast, did not generally appear as members of this other audience.

The next section looks more closely at the role accorded *"Wissenschaft"* in discussions of economic improvement. Within this broader context, the status of university and academy professors was decidedly ambiguous. The learned elite could be celebrated as heroes, but sometimes cast as villains as well. Arleen Tuchman and Timothy Lenoir have shown how ambitious academics capitalized on broader liberal interest in science to secure institutional support for their own agendas. The laboratory-based field of experimental physiology was a particularly successful beneficiary of this alliance.[40] Liberal enthusiasm did not always line up neatly with high scientific innovation, however, and as often as not, *Vormärz* writers built the case for science using well-worn materials. Basic chemical analysis, natural historical nomenclature, and systems of measurement—these classic tools from the preceding century often received as much attention as groundbreaking new discoveries. Indeed, general tools and techniques, often those that had been a part of the sciences for a long time, provided many of the specific reference points that writers employed when praising the power of science. The high science of the 1830s and 1840s, although not absent, was hardly omnipresent in their rhetoric. Typically, authors spoke of microscopes, not of cell theory, and of chemical analysis in general, not the (still controversial) specific discoveries of Justus Liebig.[41] Indeed, to the eyes of many in the 1830s and 1840s, the primary evidence that *Wissenschaft* was extending its empire lay not with Germany's emerging cadre of professional experts but in the expanding intellectual public that had come into being over the preceding decades, and that continued to grow.

Technological Development and *Wissenschaft*

In 1837, the Brockhaus firm in Leipzig announced the first volume of a new publication, the *Bilder-Conversations-Lexikon*, a compact reference work that was Germany's first extensively illustrated encyclopedia. At a mere four volumes (comparatively brief for books of this kind), the work was intended for an audience who could not afford or did not desire more voluminous encyclopedias. "It was our intent," the editors wrote, "to offer all classes of the German people a work that avoids a strict scholarly tone and that speaks of everything that belongs to ordinary life in an understandable manner and with preferential concern for German and practical interests." The encyclopedia was to be, its subtitle proclaimed, a "handbook for the spread of useful knowledge

CHAPTER 7

[*gemeinnütziger Kenntnisse*]." "*Gemeinnützig*," that adjective on which the romantic era had heaped so much scorn, here again appeared as a form of praise.[42]

At first glance, the *Bilder-Conversations-Lexikon*, with its practical agenda and suspicions of scholarly abstractions, fits neatly into familiar narratives about German culture in the 1830s and 1840s. Hegel died in 1831, and his death is often used as a placeholder for the end of idealism's reign as the dominant force in German intellectual life. The 1830s and 1840s unquestionably represented a watershed within the world of German academic philosophy. The crisis of idealism, however, is generally portrayed as much more than an academic event. It is cast as only one aspect of a wider cultural transformation. In the second third of the nineteenth century, supposedly, a new appreciation for the power of natural science, sparked by industrialization and an advancing material culture, made the intellectual projects of the preceding generation seem both inadequate and excessively otherworldly. Cholera may have killed off Hegel, but the natural sciences have taken a good portion of the blame for idealism's demise. In this view, the new adulation of natural science represented a radical departure from the older culture of neohumanism and idealism. The former was practical, worldly, and empirical, where the latter had been antiutilitarian and abstract. From this perspective, the idealist tradition was much more than a group of philosophical systems. It was a form of life, one apparently ill suited to the emerging realities of an industrializing society.[43]

The *Bilder-Conversations-Lexikon*, for all its brevity and accessibility, tells a more sophisticated story. The publication's social agenda, to be a work for all classes, certainly stood in sharp contrast to certain strains within the Romantic tradition—to Steffen's brash assertion that being understood by the general public was none of his concern, or to Nees von Esenbeck's attempts to train a refined scientific readership that could appreciate the subtleties of high science. But on closer inspection of the encyclopedia's content, any simple opposition between early nineteenth-century aesthetic and philosophical concerns and the practical "realism" of the mid-nineteenth century falls apart. Among the useful and entertaining things the *Bilder-Conversations-Lexikon* chose to include were extensive treatments of Kant, Fichte, Hegel, and Schelling, and the picture these entries painted was not one of idealism's decline in the face of ascendant natural scientific and practical concerns. Hegel's entry emphasized his early studies in "philosophy, theology, mathematics, and natural science." Fichte was praised for his practical contributions to education. And while the encyclopedia criticized

Schelling's *Naturphilosophie*, it was not the *idea* of a philosophy of nature that came under fire but Schelling's particular branch of the same (which, according to the lexicon, was "formulated with more fantasy and wit than disciplined thought and careful knowledge"). The *Bilder-Conversations-Lexikon*'s entry for "*Naturwissenschaft*," while it celebrated humanity's practical control over the natural world, also relied heavily on the rhetoric of the preceding generation of romantic *Naturforscher*. *Naturwissenschaft* was defined in holistic terms (it was "a *Wissenschaft* of nature as a unified whole"), and this science supposedly achieved its highest form in a unifying *Naturphilosophie*, a project, according to the *Bilder-Conversations-Lexikon*, still under construction.[44]

The *Bilder-Conversations-Lexikon* did not overturn the legacy of the Romantic generation. It attempted to popularize it. Between its pages, neohumanist and idealist ideas flew under the banner of *Gemeinnützigkeit* (public utility). This combination of the practical and ideal, the material and moral, all paired with a commitment to social inclusiveness, neatly captures the evolving trajectory of "*Wissenschaft*" as public category in the 1830s and 1840s. The concept of *Wissenschaft*, once used to erect a dike against overweening practical demands and an expanding public culture, was now turned to new ends.

The intellectual resources of the idealist period, in particular its many pairs of binary, dialectically combining opposites, shaped a great deal of practical political and economic discussion in the 1830s and 1840s, and it is important to keep this creative and dynamic legacy of the Romantic period firmly in view.[45] For the Romantic generation, one did not choose either the Real or the Ideal, the subjective or the objective, Spirit or Nature. Binary oppositions were not mutually exclusive; they stood in dynamic interrelationship to each other, and this penchant for dialectical thinking continued to structure *Vormärz* debates about knowledge and nature. Terms that appear as static opposites in many secondary accounts of the period (realism and idealism, for example) were often framed in the mid-nineteenth century as mutually desirable poles between which one moved, not exclusive choices.

Just as in the Romantic period, science and art still appeared as close cousins in the early industrial press. Writing to his brother Werner about the family business in 1870, Carl Siemens complained that his skilled machinists were resisting his attempts to introduce more deskilled, subdivided forms of labor onto the shop floor. The "artists" and "gentleman mechanics," he noted derisively, considered it beneath their dignity to work in this way.[46] The label that Siemens used sardonically, "artist," was one many *Vormärz* liberals had employed in all earnestness. When they looked into the future, they saw not the triumph of material interests but the resources of science and art joined together in good Goethean harmony, shaping the course of productive life. Truth and Beauty would work hand in hand in the creation of even the most ordinary objects; they would guide manufactories and artisanal workshops alike.

In the early nineteenth century, the classical ideal of the unified artwork had first been formulated as a critique of popular taste, a category used to mark off "high" and worthy art from the junk enjoyed by the wider public. Real art was also defined by its lack of practical utility. Art existed to be beautiful.[47] In the 1830s and 1840s, many liberals tried with renewed vigor to elevate the taste of the masses, and in the process they also argued that ordinary, useful objects could aspire to be complete, internally unified works of art. The spread of *Wissenschaft*, in turn, was seen as a crucial precondition for the improvement of public taste. "Throughout history and in all nations where manufacturing and trade blossom, *Wissenschaft* and *Kunst* have also been powerful and strong," opined the *Gewerbe-Blatt für Sachsen*. "If the spirit of great inventiveness is to remain vibrant, the pure scientific research out of which this spirit emerges must also remain strong and pure."[48] The influential Christian Peter Wilhelm Beuth in Prussia, the chief architect of Prussian industrial policy, had a similar vision of the industrial future, one he attempted to realize in Prussia's early technical exhibitions.[49]

Bureaucrats like Beuth focused their attention on new forms of mechanized or larger-scale production, but *Wissenschaft* was also cast as a potential ally in the small-town artisan's struggle against the new mechanical factory. In a speech to the Thüringian Manufacturing Society, the Freiherr von Pfaffendorf advanced the idea that *Wissenschaft* was the traditional artisan's best hope in his fight against his new competition. "How should the artisan survive the fight with his great enemy, the factory and machine production, if it is not possible for him to use *Wissenschaft*, which helped the latter to such astounding accomplishments, to serve his own interests?"[50]

In addition to helping the artisan become more productive and inventive, *Wissenschaft* would supposedly also help his work become more beautiful. Karl Preusker advised his artisan readers to consult works of anatomy and natural history in order to train their senses to appreciate higher forms of beauty, and he suggested they read Johann Wolfgang von Goethe and Philipp Otto Runge's works on color theory.[51] The editors of the *Gewerbe-Blatt für Sachsen* were also fond of

Weimar classicism. References to Goethe's works were common in the journal's articles. The paper's editor opened its review of C. H. Dieckmann's *Versuch über das Schöne und Geschmackvolle für Handwerker und Künstler* with favorable comments on Goethe's *Farbenlehre*.[52] The *Gewerbe-Blatt* occasionally decorated its front page with quotes from Goethe, such as *"Warum willst Du weiter schweifen? Sieh', das Gute liegt so nah."*[53] The newspaper regularly offered prints of decorative patterns, which were supposed to be copied as ornaments for fabrics, wall coverings, and furniture. In explaining how his readers should use the patterns, the editor referenced Schiller and explained that "the creation of new ornaments from plants requires more than the mere capacity to copy patterns, plaster casts, and even nature." Artisans should have an active understanding of the natural world. They should be *"Gewerbekünstler,"* manufacturing artists, whose aim was "the complete artwork" of Weimar classicism, even when they were producing for "the narrower circle of practical use."[54]

In this regard, the early industrial press painted a relationship between science and production that differed dramatically from the images of mechanical, mindless reproduction found in much Romantic polemic. The artisan was not supposed to use science mechanically in his work; he was not merely the hand that uncomprehendingly copied the higher spiritual achievements of others. Instead, with the help of *Wissenschaft*, the educated artisan became an artist in the Goethean mold. Here, the Romantic era's concept of the generative and creative artist appeared in more democratic dress.

Many *Vormärz* authors celebrated their age as one that had given birth to a new union of theory and practice, *Wissenschaft* and life. Times were changing, and *Wissenschaft* pointed the way forward. "All objects of human effort at home as in the field must no longer be produced according to lazy, inherited tradition, but according to scientific [*wissenschaftlichen*] principles," intoned an anonymous contributor to the weekly provincial journal *Die Biene*. "Without denying practical experience its rightful and significant place," the schoolteacher August Beger wrote, "the conviction is almost universal that *Wissenschaft*, the ruler and queen of every art and form of knowledge, is the best guide."[55] *Wissenschaft* had supposedly earned the right to rule through her past accomplishments. Having already wrought astonishing changes to craft, trade, and industry in the past, science would continue to do the same in the future, assuming she was accorded her proper role.[56]

It was also possible to argue the reverse in the mid-nineteenth century, as Engels did in his notes for his uncompleted *Dialektik der Natur*.

CHAPTER 7

"Already from the beginning, the origin and development of sciences determined through production."⁵⁷ In defending the historical importance of the productive classes, *Vormärz* liberals occasionally made comments that might be marshaled in support of such a formulation. The director of the Craftsmen's Association (*Handwerkerverein*) in Chemnitz, a Herr Webermeister Rewitzer, thought artisans were central to the history of science. The "sensibility for the sciences [*Wissenschaften*]" had blossomed in preceding centuries along with "industriousness [*Fleiß*] of artisans."⁵⁸ The liberal bureaucrat Karl Preusker also credited urban artisans with the rebirth of the sciences after the Middle Ages, though in his opinion their salutary influence came from their strong moral character rather than from the demands of economic activity.⁵⁹ But for the most part, *Vormärz* commentators emphasized the ways in which *Wissenschaft* shaped economic practice instead of the reverse.

Historians have sometimes pointed to science's increasing professionalization as the primary reason for its cultural prominence at mid-century.⁶⁰ In contrast, many nineteenth-century writers argued precisely the opposite: it was the escape of science from the control of the learned elite, the social heterogeneity of its practitioners, and the accessibility of its forms of knowledge that made it powerful. In most cases, the transformative force purported to be at work was "*Wissenschaft*" in general, not just natural science. "It is with good reason," wrote schoolteacher August Beger, "that one sees it as a characteristic feature of our time that *Wissenschaft* has been transplanted out of the lonely, dark rooms of the scholar's study and into the open country of life."⁶¹ Karl Preusker agreed. He considered one of the defining qualities of his age the fact that "the *Wissenschaften* are no longer the sole property of only one class of the *Volk*, namely, the academically educated (*akademisch-gebildeten*), as once was the case."⁶²

Writers mustered various kinds of evidence to support these assertions; *Bildung*, in their minds, was spreading through many channels. Periodicals like the *Mittheilungen für das Erzgebirge* or the *Gewerbe-Blatt für Sachsen* claimed to bring the fruits of *Wissenschaft* to artisans, merchants, and manufacturers, to keep them up to date on the latest useful discoveries and innovations. The German book market grew 330 percent between 1815 and the mid-1840s, and periodicals and newspapers multiplied. With the spread of the steam press in the 1820s, the costs of books and periodicals also fell. Publishers like Brockhaus capitalized on the new lower rung of the print marketplace with products like the illustrated lexicon discussed earlier.⁶³ Every week, the back pages of journals like the *Gewerbe-Blatt für Sachsen* held countless advertisements for introductory handbooks and reference works promising scientific enlightenment, written by authors both famous and obscure.

In addition to the expansion of print culture, *Vormärz* authors also pointed to growing networks of manufacturing associations as evidence for the growing power of *Wissenschaft*. These kinds of civic societies were particularly dense on the ground in urban, industrializing Saxony. By the late 1830s, there were manufacturing societies in Dresden, Schneeberg, Großenhain, Leipzig, Chemnitz, Bautzen, and Annaberg, some of which were very large. The Chemnitz Craftsmen's Association had over 1,000 members in 1839.[64] The Manufacturing Society in Annaberg had 331 members in 1836–37, and its library had close to a thousand volumes, which had been taken out a collective 3,721 times by its members. The society's rolls included 1 minister, 3 lawyers, and 13 state officials; otherwise, all of its members came from various artisanal trades. Though *Gewerbevereine* typically drew their leadership from the state bureaucracy or the professors of technical academies, the bulk of their membership came from outside the educated middle classes.

With a healthy allowance for overly rosy self-reporting, the *Gewerbevereine* appeared to be reasonable instantiations of Enlightenment reformers' dreams of a philosophically informed productive estate (*Nahrungsstand*). These societies were communities of practical men who made the union of *Wissenschaft* and *Praxis* an important part of their collective self-image. In addition to amassing libraries and holding meetings, manufacturing societies also sometimes sponsored Sunday schools for artisans, providing another way to spread a taste for *Bildung* among the productive estates.[65] When Saxon liberals spoke in glowing terms of *Wissenschaft* as an omnipresent force, the state's growing network of manufacturing societies provided them with an important concrete referent.

Although this was not their only aim, these groups often saw the creation of a new *Wissenschaft* of technology as part of their mission.[66] Unlike in the eighteenth century, however, the educated middle classes no longer dominated the publicly organized community of people interested in improving economic productivity. In the late eighteenth century, patriotic-economic societies had included a respectable minority of actual estate managers, artisans, and the like. They were far outnumbered, however, by the nobility and university-educated commoners. With the spread of manufacturing societies in the 1820s, 1830s, and 1840s, the social base of societies for economic improvement widened considerably, following the patterns seen earlier with agricultural societies. What the eighteenth century had joined together, the mid-nineteenth century now divided up; different societies covered different spheres of economic practice. In Dresden, for example, the Economic Society (*Oekonomische Gesellschaft*) dealt almost exclusively with agriculture. The botanical-horticultural society, the Flora, became the province of the city's strong commercial gardening industry, while the manufacturing society dealt with manufacturing and trade.

When discussing how a *Wissenschaft* of technology might emerge, authors generally agreed on a few basic assumptions about what differentiated *Wissenschaft* from "lazy and blind" tradition. In the widest colloquial use of the term, "*wissenschaftliche*" knowledge was knowledge transmitted publicly, taught in formal schools rather than through apprenticeships, and published in books and journals. These were not, of course, exhaustive criteria for *Wissenschaftlichkeit* in any strict sense, but discussions of the usefulness of *Wissenschaft* to industry and manufacturing often shifted into more general debates over the usefulness of reading and writing about technical topics.

The term "*Wissenschaft*," in other words, acted as a shorthand for the multiple changes wrought by Germany's conjoined educational and reading revolutions, and defending *Wissenschaft* often meant defending the relevance of serious reading to people's everyday lives.[67] *Wissenschaft* was not just a body of knowledge; it was a form of communicative practice—and the power of organized, written, published, or protocolled communication was an important constituent of what authors meant when they spoke of the power of *Wissenschaft*.[68]

In the 1830s and 1840s, Saxony's numerous manufacturing societies had an ongoing debate over the ways in which they were or were not like learned societies, and this discussion offers a useful window into what an involvement with *Wissenschaft* meant in concrete terms to the men who made up these groups. Being "scientific" involved disciplined conversation, written records, and (at least potentially) publications— this was the clearest consensus answer. The *Gewerbe-Blatt für Sachsen* was a strong proponent of the idea that manufacturing societies ought to publish their protocols after the manner of learned societies. If technical science was ever to come into its own, the paper insisted, the active participation of practical men in printed technical discussion was essential. Technical science would mature only "when practical businessmen [*Gewerbsleute*], not only the academic theoreticians [*Theoretiker vom Fache*], are willing to read a lot and also eventually to write." The current problems with technical literature lay in this gap— learned men wrote, practical men did not. It was little wonder that artisans often came to the unfortunate conclusion that reading was a waste of time, thereby robbing themselves of the benefits it might have brought them.[69]

The Freiherr von Pfaffendorf, director of a manufacturing society in Thüringen, took a more skeptical view. Societies full of authors and learned men, he thought, were perhaps the least likely to be of use to manufacturing. A society full of practical men could sometimes accomplish a great deal without stepping forward in print. Even if a society did not publish proceedings, however, it at least needed to keep good protocols and maintain discipline and order in its conversations. The point of such societies was not to promote gratuitous writing (*Schriftstellerei*); but neither would these groups achieve their aims through "unsuccessful talking back and forth, where after the meeting ends no one thinks anymore of what was said." Words needed to be captured and made permanent if they were to have any force in the world.[70]

Inaugurating new people into the circle of *Wissenschaft*, then, meant more than teaching them a collection of facts or laws. It meant introducing them to a set of behaviors that brought them into contact with broader networks of intellectual exchange. It also meant teaching them concrete scientific skills. Karl Preusker, one of Saxony's most prolific authors on the topic of *Volksbildung*, laid out this ideal in particular detail in his extensive writings.[71] A provincial Saxon bureaucrat, Preusker published voluminously and also devoted a great deal of organizational energy (with mixed success) to bringing enlightenment to the artisans of his local post of Großenhain.

The model artisan that Preusker hoped to create was one who engaged in activities that one hundred years earlier would have qualified him for membership in the republic of letters, and this is a fact that is easy to overlook if we think of Preusker's efforts only as "educational" in the modern sense of the word.[72] In an 1838 speech given to the Großenhain Manufacturing Society, Preusker described the new form of life he hoped to inspire in the rank and file of his town's craftsmen. The modern artisan, according to Preusker, should regularly use the descriptive and communicative tools of the sciences—which ones exactly would depend on his trade, and Preusker went through the different branches of handiwork in lengthy detail, explaining the skills useful to each. A few common themes emerged. The modern artisan knew botanical, zoological, and mineralogical nomenclature. He quantified whenever possible, weighing and measuring rather than being guided by rules of thumb. He knew how to perform chemical analysis. He read on chemical topics, made experiments, and corresponded with other chemists. He had a number of models in his workshop to allow him to compare different designs.[73]

Preusker's imagined model artisan was also a critical reader. He would sometimes dispose of a book with the statement, "a corrupt pirated edition . . . translated without expert knowledge of the topic . . . a theoretical fantasy not based on practical experience." Like learned European men since at least the Renaissance, this artisan kept a commonplace book, in which he recorded "interesting things he had read or experiments he had tried." He owned a microscope; he kept a natural history collection.[74] Elsewhere, Preusker made it clear that he did not expect that most artisans would engage in "deeper research, which seems to be more appropriate to learned experts and the independently wealthy." Yet he also believed that a man of middling rank could gain sufficient knowledge of botany, zoology, chemistry, physics, or mineralogy to enter into conversation with those more learned than himself. After attending a Sunday school or studying a good handbook, the artisan entered a border zone of participation in the learned world. In his Sunday wanderings through the countryside, the educated artisan might well come across some rare plant or mineral deposit of interest to the more learned, which he could then share with them.[75]

Within the manufacturing society he founded, Preusker fell short of realizing these grand ideals. To his frustration, many of the society members often seemed more interested in socializing than in listening to his improving lectures.[76] Preusker's imagined model artisan is instructive nonetheless. He illustrates the fact that for *Vormärz* liberals, the growing power of *Wissenschaft* came not just from the spread of information but from the spread of scientific skills and habits. Enthusiastic heirs to an Enlightenment tradition that saw the spread of knowledge as the key to material and moral improvement, men like Preusker also championed the Enlightenment's porous, expansive vision of an intellectual public. Men of simple means and basic education, it was assumed, might not be able to scale the highest peaks of the learned world, but they could still contribute their modest bit to the progress of knowledge. In the mid-eighteenth century, Jacob Schäffer had imagined a world where his more humble townsmen might bring him interesting specimens they found; eighty years later, Karl Preusker advised his readers to do the same.

At midcentury, there was still a productive ambivalence about whether or not *Wissenschaft* was knowledge produced just by a narrow elite, or something of much broader provenance. When promoters of industrial improvement spoke of *Wissenschaft* becoming the common property of all, part of what they meant was that new segments of the population were becoming integrated into forums of printed exchange and organized discussion. If initiating new people into the world of *Wissenschaft* just meant teaching them a fairly basic set of skills, the reach of *Wissenschaft* could be wide indeed. It was in part this generous notion of what counted as "scientific" that made *Wissenschaft* seem such a potent thing. The persistently fuzzy line between the community that made knowledge and the one that merely received it allowed *Wissenschaft* to seem ubiquitous in the 1830s and 1840s, a force everywhere transforming German lives.

Wissenschaft as a Form of Alliance

In the 1830s and 1840s, the natural sciences also still retained their place within the pantheon of improving leisure pursuits. Karl Preusker offered a particularly clear discussion of the social benefits that natural science offered to an enterprising young man. An interest in natural history, Preusker thought, often introduced a young man into "successful relationships" of mutual exchange, gaining him friends who might prove useful in the future. Preusker put these words into the mouth of the local schoolteacher in a *Bildungsroman* he wrote for young artisans, *Der Sophien-Ducaten* (The tale of the Sophia-Ducats). Such connections, Preusker's fictional teacher went on to say, proved especially useful during a youth's time as a journeyman. Journeymen with scientific interests enjoyed numerous advantages through the contacts that they made; they found themselves introduced into the family circle of admirable local men and were the beneficiaries of preferential treatment and useful support.[77] The study of nature secured a young man's character as well, inspiring him to greater piety and protecting him from less morally salutary socializing. It was an alternate, morally beneficial form of leisure that occupied a man to better advantage than card games or time wasted in "empty conversation." These other alternate, dangerous forms of sociability were usually feminized in Preusker's novel, taking the form on one occasion of a profligate, excessively social wife who led her husband to financial ruin, and on another the more seductive shape of a maidservant of dubious morals.

In the mid-eighteenth century, the behaviors Preusker described would have been the raw material of a learned reputation. He advised young artisans to start collections, spend lots of time reading serious books, and establish contacts with more knowledgeable men. By the

CHAPTER 7

mid-nineteenth century, these behaviors no longer had nearly the same potential to alter a man's status, but the aspirations Preusker had for his artisanal audience were clear. An engagement with science would fundamentally alter the relationships in which an artisan was embedded, and Preusker and his fellow liberals imagined many ways in which these new kinds of relationships might serve the cause of recasting German society. *Wissenschaft* would produce new, more general allegiances out of the particularist world of the early modern economic order; it would provide a valuable model of consensus and community building.[78]

Against the background of fractious debates about the Customs Union, the future of guilds, and the causes of pauperism, *Wissenschaft* was sometimes evoked as an alternative model for organizing collective interests, a replacement for the integrative, regulatory functions of the guilds.[79] The industrialist Friedrich Georg Wieck argued that manufacturing associations, which made "*Wissenschaft* . . . the property of all," would bring new prosperity to artisans, forging "modern corporations [*Korporationen*] and guilds [*Innungen*]," which, unlike the old guilds, would be based "on the principle of the most complete freedom."[80] And in actual fact, the new manufacturing societies did indeed offer a sharp contrast to older models of corporate identity, which had been tied to a particular craft or group of crafts. In the place of guild ties, a manufacturing society offered a generalized community of manufacturers and craft producers joined in the free exchange of skills and information (that, at least, was what it purported to do). The Annaberg Manufacturing Society's members, for example, came from fifty-eight different trades.[81]

Models taken from the world of *Wissenschaft* could also serve as a solution to the problem of negotiating the relationship between center and periphery. Throughout much of the 1830s and 1840s, Saxony's provincial manufacturing societies were involved in an acrimonious dispute with the Dresden Manufacturing Association over a plan to centralize all of the kingdom's associations under one umbrella, and this debate prompted a number of reflections on what the Germanies' various manufacturing societies shared among themselves. The provincial *Gewerbe-Blatt für Sachsen* and many of the local societies it represented were opposed to any strong institutional connection between the different Saxon societies. Their collective, panregional interests, the paper argued, would be better served through an organization like the Society of German Natural Researchers and Doctors, a yearly centralized meeting for a community of knowledge makers. The paper suggested "meetings of artisanal and technical men, after the fashion of the meetings of German natural researchers." These meetings would promote "the exchange of opinions both in writing and orally" and allow the community to make collective decisions on important questions.[82]

Even joint-stock companies, another newly popular vehicle for collective action, could take on a scientific sheen in the minds of *Vormärz* social commentators. A contributor to the *Mittheilungen für das Erzgebirge und Voigtland* complained in 1836 that the Germans had been less successful than the French and the British in bringing scientific knowledge into contact with practical life. Though Germany had many "hardworking learned men, experts, voluminous theories, and thousands of recipe books," it still lacked "the powerfully unified societies and the lasting, influential joint-stock companies [*Aktiengesellschaften*]" that were the product of the successful unity of *Wissenschaft* with economic life.[83]

Dissenting Voices

Wissenschaft was often praised as the wellspring of all that was prosperous, creative, and beautiful, but not everyone sang to this tune. Experiential know-how could also hold its own in the public sphere. Alongside odes to the wonders of *Wissenschaft*, people also groused about empty theory and the ignorant bluster of impractical professors. Already in the 1830s and 1840s, Beuth's dream of shaping Prussian industry through *Wissenschaft* and classical aesthetics was foundering on the rocks of entrepreneurs' more pressing practical concerns, and for the rest of the nineteenth century, technical and agricultural academies would struggle to answer the charge that "scientific" study did little to prepare students for the real demands of running a profitable farm or building good bridges. Agricultural interest groups often lobbied to have men with practical farming experience take over professorial chairs in place of candidates with more impressive scholarly credentials.[84] In other areas, too, professors were sometimes ignored when they offered advice on practical issues, particularly if their advice ran counter to what their audience wanted to hear.[85]

Indeed, the educated advocates of *Wissenschaft* often lamented the suspicions their cause engendered among other groups within German society. Like a secondary theme in a minor key, such complaints regularly appeared alongside optimistic celebrations of scientific progress. Villagers in mining districts, for example, were said to distinguish between an engineer (*Ingenieur*) and a miner in the following way—the former might understand how things worked but could not actually do anything, while the latter, despite his lack of formal understanding, actually knew how to get things done.[86] Small-town artisans purportedly dismissed the idea that books could be of any use to them in their work, defending tradition with a brusque "Our ancestors were no fools."[87] Along similar lines, a contributor to the *Mittheilungen für das Erzgebirge*, identifying himself only as "a practitioner," wrote that as far as he was concerned, "technologists [*Technologen*]" were fine as long as they kept themselves to purely scientific matters ("*bei dem rein Wissenschaftlichen stehen bleiben*"). When they tried to move into the sphere of practice, such authors did more harm than good. The craftsman did not need a professor, however well intentioned, to tell him how to go about his business.[88] A science of technology was just fine, in other words, as long as it did not try to actually tell people what to *do* in their work.

Even in well-heeled journals like Leipzig's *Illustrirter Zeitung*, an illustrated paper that catered to a loftier clientele than provincial journals like the *Mittheilungen für das Erzgebirge*, one can find evidence that people often chose to take the suggestions of practical know-how over the advice of *Wissenschaft*. The texts of book ads suggest that practical experience sold well. C. G. Forster, ornamental gardener in Leipzig, offered a book on gardening that built on "many years of practical experience." Adolf Friedrich Magerstedt's *Der praktische Bienenvater* was "complete, clear, to the point, and based on many years of practical experience." What the book was *not*—the product of empty, learned theory—was clear by implication. Even in the case of books written by the university educated, ads often emphasized experience over *Wissenschaft* in touting the author's qualifications. According to its publisher, Dr. Friedrich August Günther's *Der homöopathische Thierarzt* was based "on *pure experience* [emphasis in original]."[89] By any definition, social, technical, or epistemological, *Wissenschaft* never achieved anything like the universal sway its *Vormärz* prognosticators claimed for it.

Conclusion

From one perspective, *Vormärz* claims about the influence of science on economic practice were clearly exaggerated. As historians of technology have shown, new theoretical insights did not, in fact, do much to transform the world of economic practice in this period; the sources of innovation lay elsewhere. Nonetheless, mid-nineteenth-century insistence on the power of *Wissenschaft* and *Naturwissenschaft* was not just false consciousness. The language of *Wissenschaft* served as shorthand for any number of very real changes in the organization of social and economic life. The industrial and agricultural associations are often treated as precursors to later nineteenth-century interest groups; they also sometimes appear under the rubric of "adult education."[90] To the eyes of observers in the 1830s and 1840s, the heavy debt these groups owed to their learned precursors was much more visible than it has been to subsequent historians. By the mid-nineteenth century, a variety of people were taking up habits that had once belonged to learned men. Seen in this light, *Wissenschaft* was indeed a transformative force, an important resource in the creation of new conceptions of general interest and new forms of collective action.

Over the last several chapters, we have encountered many different kinds of groups—agricultural societies, manufacturing societies, gardening societies, elite societies for natural research, and less prestigious regional natural history groups. On the ground and in local settings, the distinctions among these different kinds of groups were extremely important, and also clearly on display, in the varied composition of membership lists and in the differing ways in which these groups described their aims. These differences were part of what helped to mark out the outlines of Germans' idiosyncratic concept of "the natural sciences." Natural science was separate from practice, though still relevant to it; natural science was a social cause that was distinct from (though not hostile to) the more general pursuit of literature, philosophy, or other areas of *Wissenschaft*. "*Naturwissenschaft*" was what the professors of a technical academy had to offer the rank and file of a manufacturing association, and also the members of the broader polite public.

Yet these different societies also still shared many features, a fact that becomes more visible if we step further back and look at them from a distance. When viewed as part of cosmopolitan networks of exchange, their commonalities become much easier to see. In the 1865 report of the Smithsonian Institution, one can find a long list of German societies that donated their publications to this preeminent organization of American science, and the list ran the full gamut. Elite natural scientific societies were there, but so were regional natural history groups. The Nature-Researching Society in Halle had sent its yearly report; so had the less prestigious Natural Scientific Society for Saxony and Thüringen. The Smithsonian had also received gifts from a wide range of practical associations, from an industrial society in Graz to a gardening society in Gotha.[91]

EIGHT

The Particularity of Natural Science

In the later nineteenth century, German intellectuals discussed knowledge making using terms that diverged in significant ways from those of French or British scholars. German thinkers frequently argued about whether there might be not one scientific method but two. They also consistently divided the sciences into two branches, the *Naturwissenschaften* and the *Geisteswissenschaften* (the social sciences and humanities). These categories did not have exact cognates in French or in English. Even when nineteenth-century French or British scholars wrestled with similar philosophical issues, an analogous epistemological distinction never achieved the ubiquity or importance that it did in the German context.[1]

The division between the *Geisteswissenschaften* and the *Naturwissenschaften* has been fundamental to German intellectual life ever since. Philosophical disputes over the significance of these two categories continued well into the twentieth century, and these disputes often centered on a supposed distinction in method that separates the human and the natural sciences. Exact descriptions of this methodological difference have varied, and many thinkers have claimed that the distinction is specious. A recurring issue in these discussions, however, has been the difference between the kind of knowledge that can be produced from the study of texts and language, and the kind of knowledge that can be gained about the natural world.[2]

The split between the human and the natural sciences

emerged in the mid-nineteenth century out of a more complex set of earlier classificatory schemata. Alwin Diemer, who has traced the history of the *Naturwissenschaften/Geisteswissenschaften* distinction, noted that this binary pair first appeared in its modern form in an obscure pedagogical tract, Eduard Calinich's *Philosophische Propädeutik für Gymnasien*, published in Dresden in 1847.[3] Diemer glossed over Calinich's book in a single sentence, and though clearly puzzled to have found the first use of this foundational intellectual opposition in such a lowly place, did not regard the fact as worth further consideration. After all, Calinich's obscure work had no demonstrable influence on later debates.[4]

Nonetheless, Calinich's not-so-famous first is instructive. Though the life of this schoolteacher was far removed from late nineteenth-century high philosophical developments, the context in which his book appeared is suggestive. Calinich wrote his treatise against the backdrop of an acrimonious debate within Saxony over possible reforms to the *Gymnasien*, the classical high schools that prepared students for university. These Saxon discussions were part of a broader struggle within German secondary education, the ongoing battle between *Humanisten* and *Realisten* that pitted defenders of the classical curriculum against reformers who wanted to include more modern history, science, and languages.[5] Within Saxony, Calinich's native state, the most controversial issue of the *Vormärz* educational debates was the possible addition of natural history to the *Gymnasium* curriculum. Advocates of the natural sciences found natural history's exclusion from academic secondary education particularly troubling; without early training in the apprehension of nature, they argued, students would struggle to understand the natural sciences later in life. Classicists, however, found natural history, with its lowly and miscellaneous subject matter, an unacceptable tool for training young minds. Calinich's book was composed in the midst of this argument about whether the study of natural objects was compatible with an educational ideal (the neohumanist notion of classical *Bildung*) that was grounded in the study of classical texts. The terms of this argument paralleled the basic divisions of later, more famous epistemological debates. To better understand the fissure that runs through German intellectual life from the mid-nineteenth century forward, the school debates are precisely the place we need to look.

This chapter examines the emergence of the idea of a natural scientific method, a concept that had previously been absent from German public rhetoric. The distinctiveness of this German concept can be seen when it is compared with its British cousin, "the scientific method." As Richard Yeo has shown, the idea of a single "scientific method" took on heightened importance for British intellectuals in the mid-nineteenth century, when this concept was used to defend the educational value and public standing of the sciences, often (as in Germany) against supporters of the traditional classical curriculum.[6]

In Britain, however, advocates of the classical curriculum relied primarily on moral arguments to make their case. Studying the classics built character. German neohumanists argued this, too, but they also argued something else: that the classical curriculum was the best preparation for boys whose lives would be devoted to *Wissenschaft*. Studying classical languages trained the mind to think scientifically. German *Naturforscher*, as a result, wanted to show that the natural sciences had their own distinct epistemological contribution to make. Natural science provided intellectual skills different from those that could be gained studying texts. Claiming that the natural sciences had their own method was one succinct way to make that point. The British, in contrast, more easily equated the methods of natural philosophy with the methods of knowledge making per se.[7] The nineteenth-century English phrase "the scientific method" could be translated into German two different ways: *"die wissenschaftliche Methode"* or *"die naturwissenschaftliche Methode."* The distinction between these two things, which could be easily expressed in the vocabulary available to mid-nineteenth-century Germans, was much harder to draw in English.

The educational debates did not create a strong common *detailed* account of how the natural sciences were different from other kinds of knowledge, but they did transform the loose set of clichés and rhetorical strategies that natural researchers used to define their commonalities with one another. They gave greater prominence to the idea that the natural sciences shared a distinctive way of knowing about the world. This was by no means a universally accepted claim, but it was an often-repeated one. The educational debates also markedly altered the temper of public rhetoric about the natural sciences, mustering *Naturforscher* of diverse political, institutional, and philosophical commitments around common resentments. In response to their educational marginalization, *Naturforscher* defended the intellectual and moral value of the natural sciences with a cutting defensiveness and heightened intensity. Against the backdrop of the school debates, a collection of phrases—"the natural scientific method" or "natural scientific *Bildung*"—took on new resonance and importance; they became badges and rallying cries that would be used throughout the rest of the century.

The altered status of the adjective *"naturwissenschaftlich"* marks the change I am describing. Through the early 1830s, this word had little polemical force. Something *"naturwissenschaftlich"* was simply something that had to do with knowledge about nature. If you wanted to indicate that you approved of the content and form of a piece of research, you said it was *"wissenschaftlich."* By the middle of the nineteenth century, in contrast, some German intellectuals wanted to apply the "natural scientific method" to things that had little to do with nature at all. Legal scholars proposed a natural scientific school of Roman law, and the ethnologist Adolph Bastian, a natural scientific account of human culture.[8] These proposals would have been uninteresting or even meaningless in earlier periods, when the natural sciences were not thought to have any particular epistemological method to call their own. This subtle shift in adjective use marks a profound transformation in the basic scaffolding of German intellectual life. The Bavarian naturalist Mathias Flurl had called his 1799 speech a "Disquisition about the Influence of the Sciences [*Wissenschaften*], Especially Natural Knowledge [*Naturkunde*], on the Culture of a Nation," and it was not a lack of rhetorical skill that left him with such an unwieldy title. By 1850, things had changed.

Knowledge and *Bildung* in the *Gymnasium* Debates

Neohumanism, the leading force in educational reforms of the 1820s, 1830s, and 1840s, had its roots in the preceding century. This loose and diverse movement, woven out of many strands of late eighteenth-century intellectual culture, spread out from its north German origins to become a powerful force both in school policy and in educated culture more broadly.[9] Neohumanist pedagogues captured key bureaucratic posts in numerous German states. Holding up ancient Greece and Rome as models for the regeneration of contemporary culture, they defended the study of Greek and Latin as the centerpiece of preparatory schooling and breathed new life into a curriculum that had come under fire during the Enlightenment.[10] In offering a revised and updated version of an older curriculum, the neohumanists ensured that Latinity continued to be a defining hallmark of the well-educated man.

The Latin schools of the eighteenth century had fulfilled multiple functions, educating boys destined for the university alongside the sons of urban families bound for other walks of life. The neohumanists deplored this mishmash of educational ends; they wanted the classical high schools to be exclusively learned institutions, educating only pupils who would go on to university. This ambition was never realized in practice in the first half of the century, and local schools often proved recalcitrant to state efforts at standardization.[11] But once this narrower vision of the *Gymnasium* was articulated in theory (and in some places in law), the question of how to best serve those young *Bürger* not bound for the learned professions also took on new urgency. School reformers in the 1830s and 1840s directed urban families who wanted other kinds of intermediate schooling for their sons to the so-called *Bürgerschulen* or *Realschulen*, though in practice these institutions were still thin on the ground.[12] According to most neohumanist pedagogues, these *Realschulen* were the proper home for the natural sciences.[13]

The period's educational reforms had different chronologies and complexions from state to state, but neohumanism's wide influence gave these developments a meaningfully pan-German complexion, providing common themes that cut across regional differences in school policy. The new school plans followed two basic models, and both were heavily weighted toward the study of classical languages. Prussia and most of the middle German states adopted curricular guidelines that included a smattering of "modern" subjects (the natural sciences, for example, and the study of modern languages) alongside Greek and Latin. Bavaria and Austria, in contrast, approved plans based almost entirely on the classics. In both cases, the natural sciences played a minor role in the curriculum, and the natural historical fields in particular were especially weakly represented.[14]

According to defenders of the classical curriculum, it made pedagogical sense to limit the time devoted to natural science. Natural science, orthodox neohumanists claimed, did not possess the necessary qualities to fulfill the *Gymnasium*'s central purpose—the task of *Bildung*, or self-cultivation. Christian Grossmann, an educational official in Leipzig, argued that students would never develop the necessary moral and mental character if they spent their formative years wandering between classical texts and natural historical objects, going from "Solon's . . . laws to baobabs, quartzes and pebbles . . . from reason to unreason, from the ideals of humanity to the beasts, from the high and the eternal to the changeable, the ordinary, and the trivial." In fact, Grossmann claimed, only a man with a solid early training in the universals of classical language and culture would have the mental and moral capacity necessary for later greatness in the natural sciences. A philological education was the best preparation for future intellectual achievement, regardless of the specific field one hoped to enter. Grossmann cited Herder to the effect that it was "the intensity of the spiritual powers" fostered by their humanist education that had made Bacon, Kepler, Newton, Leibniz, Haller, Euler, Linné, and Buffon the great men—and the great natural researchers—that they were.[15]

The preceding passages capture many of the neohumanists' favorite arguments against the natural sciences. A philological education supposedly provided students with a set of universally applicable intellectual skills, skills that would allow them to master other subjects with ease later on in their lives. Indeed, a philological education was the best possible preparation for the future university student. Boys whose future lay in the world of *Wissenschaft* needed training in abstract, ordered thought. This kind of education, referred to as "formal *Bildung*," involved mastering "strict formal relationships" and "strict external lawfulness," as *Gymnasium* director Friedrich Lindemann explained. Grammar was the subject most suited to teach students these habits of disciplined, law-governed thought.[16] Since the neohumanists considered classical languages superior to modern tongues in the purity and logic of their grammar, learning Latin and Greek supposedly offered students the best initiation into a world governed by elegant and internally consistent rules.[17]

The classical curriculum was also supposed to mold the student's character, not just his intellect. The major sin that neohumanist pedagogy sought to avoid was the familiar evil of "one-sidedness" (*Einseitigkeit*), a lack of intellectual and moral harmony in the personality of the young student. To prevent this fault, students should spend their time studying subjects that were themselves internally consistent and harmonious, qualities that Greek and Roman culture supposedly embodied to perfection.[18] The subjects that young boys studied shaped the structure of their souls, and the natural sciences, according to neohumanist critics, were too utilitarian, too base in their subject matter, and too internally diverse to offer the right moral resources to young students. Even in cases where natural science might merit inclusion as a minor subject, it could never serve as a cornerstone of a *Gymnasium* education. It did not lend itself to formal *Bildung*, training in abstract, internally consistent forms of knowledge.[19]

The opposite of formal knowledge was knowledge that was merely "positive"—knowledge that dealt not with law-governed abstractions but with specific empirical content. Another often-invoked opposition was between the spiritual and the material, with the classical curriculum aligned with the former. The *Gymnasium* trained the *Geist*, the spirit, and subjects that dealt with the material world were supposedly poorly suited to educating the spirit. While material or positive knowledge might be valuable, it was out of step with the *Gymnasium*'s primary mission. For the university-bound student, the natural sciences, with all their messy material particularities, could wait until later. Teaching the specific content of the *Wissenschaften* was the university's job, not the school's, and students should begin studying natural science only after their minds had been properly forged through the study of the classics. An earlier introduction to these fields might warp their development. This was not a concern for boys destined for other spheres of life. The natural sciences could be taught in *Realschulen* or *Bürgerschulen* without fear of negative consequences, since these were schools intended for students whose future did not lie in the intellectually rigorous world of *Wissenschaft*.[20]

While the study of languages held pride of place, many neohumanists believed that there was another body of knowledge that met the criteria for generality and internal consistency they required in a *Gymnasium* subject—mathematics. Alongside the classical languages, mathematics was also a respectable medium for formal *Bildung*, and for many neohumanists, mathematics served as a sufficient proxy for the natural sciences as a whole. Math was "the grammar of natural phenomena," and like the grammar of languages, it provided "a system of laws."[21]

The question of what should be taught was closely linked with the question of who should be educated. Wilhelm von Humboldt's conception of the neohumanist *Gymnasium* had originally called for an *"allgemeine Nationalschule* [general school for the nation]," but actual state policies, both in Prussia and elsewhere, had much narrower aims—preparing students for the university, and from there for the state bureaucracy or the learned professions.[22] By the 1830s, this model of the *Gymnasium* was widely recognized throughout Germany, although actual educational reality was more complex. As we have seen, many Latin schools had started off the century serving multiple functions; by 1850, both the social composition of the Latin schools' student bodies and their pedagogical function had begun to change. Many German *Gymnasien* attracted students from more diverse familial origins than they had in earlier periods, but the students' postschool trajectories became more uniform. *Gymnasium* students became more exclusively oriented toward learned careers, continuing on to university at a higher rate than had preceding generations.[23]

In general, advocates of the classical curriculum saw a *Gymnasium* education as something that drew a firm line between the university-educated man and other sorts of people. Even philologists who were sympathetic to the minor inclusion of the natural sciences often expressed the fear that too much science would dilute the special character of the learned schools, turning them into mere *Bürgerschulen*, schools that catered to a broader cross section of the German middle classes. A *Gymnasium* principal in Zwickau, Franz Raschig, acknowledged that students needed some knowledge of the natural sciences but insisted that if the natural sciences were given equal status to classics, the *Gymnasium* would gain a second goal, imparting "material *Bildung*," which was alien to its mission. The job of the *Gymnasium* was to train future university students, and these students were better served if they received their training separately from other boys. Indeed, the introduction of natural science into the *Gymnasium* on an equal footing with the classical languages was equivalent to a call for "a new organization of our entire public school system"; it would erase the special character of a learned education completely.[24] Friedrich Lindemann bemoaned the "counterproductive fusion of popular and learned schools." The learned estate had a special role, and it should be educated separately. It should remain, as it had been in the past, "the teacher and educator of the *Volk*."[25]

Here, facets of natural science that in other contexts were cited as strengths—its social heterogeneity and connections to economic improvement—left *Naturforscher* vulnerable to criticism, lending support to the claim that natural science's strongest affinities were with the productive estates, not the learned one. Here philologists were driven by status anxieties of their own; they could point toward public enthusiasm for practical knowledge and to Germany's new technical academies to claim that humanist scholarship was under siege. Hermann Köchly, whose reform proposals I will discuss shortly, wrote of the need to remedy the "lack of respect, not to say the general derision, in which contemporary public opinion holds philology and philologists."[26]

Those who supported the inclusion of the natural sciences often (though not always) had a socially broader vision of the higher secondary schools. They also championed a definition of *Bildung* that included educated artisans and merchants. Lorenz Oken complained that the neohumanists were perpetuating a "shameful difference within society," a situation in which educational differences meant that "no estate fits with any other."[27] Freese argued for a combined school that joined together elements of the *Gymnasium* and the *Realschule*, pointing out that in practice the *Gymnasien* already served a wider clientele than neohumanist theorists wanted to admit. Besides, "influential learned men are never closed in on themselves, nor do they move only with those who are similar to them; rather, they are in contact with the educated [*Gebildeten*] and live among the people [*Volk*]." To that end, they needed a more general education than the classics provided.[28] For pedagogical writer Karl Mager, school reform was a vehicle for raising the influence of the *Bürgertum* relative to the "guild" of learned state servants.[29] The aforementioned Hermann Köchly, though a philologist himself, condemned the arrogance of neohumanists who ignored the great gains made by the natural sciences and who considered themselves above "educated [*gebildeten*] merchants and artisans."[30]

Many *Naturforscher* echoed this charge of cultural arrogance, and the neohumanist reforms provoked a sustained and often angry response. Friedrich Thiersch's 1829 Bavarian school plan represented an important watershed in this regard. The publication of this school plan happened to coincide with the annual meeting of the Society of German Natural Researchers and Doctors, and it caused an uproar at the meeting. Thiersch's plan, which contained no natural science at all, inspired universal outrage among the *Naturforscher* assembled in Heidelberg. Lorenz Oken reported that everywhere he went, people were talking about the new Bavarian curriculum.[31]

Men as intellectually distant from each other as Lorenz Oken and Justus Liebig, though they agreed on little else, saw eye to eye on this one issue and found, in the neohumanists' pedagogical agenda, a common enemy. Both of these men—Oken, the aging hero of Romantic science, and Liebig, the vocal critic of *Naturphilosophie*—complained about the overweening ambition of philologists, men who set up their own form of *Bildung* as the gold standard by which all other kinds of knowledge should be measured. Oken had been attacking the narrow horizons of the classically educated since the early nineteenth century, and in the 1830s and 1840s, he used his journal *Isis* to complain about the marginal role of natural science in the schools.[32] Similarly, Liebig's famous 1840 speech on chemistry in Prussia, long seen as the defining document for the "new" science that followed the Romantic era, was also written as an intervention in the school debates.[33] Liebig contrasted the characteristics of the *Naturforscher* with those of the classically educated, arguing that the natural sciences would train a "newer, more powerful generation" than the one raised on philological learning; he ended the speech with an extended attack on the neohumanist model of the *Gymnasium*.[34]

Science and Method

Proper method had been the rallying cry of neohumanists since the late eighteenth century, and when philologists argued for the universality of a classical education, they often did so by citing the universality of the "method" this education imparted. By this, they meant that the study of classics provided a set of basic intellectual capacities that could be used to advantage in any branch of *Wissenschaft*.[35] This claim gave *Naturforscher* increased incentive to emphasize the particularity of their own forms of knowledge. Before turning to these arguments, however, it is worth pausing to look more closely at what the term "method" included for mid-nineteenth-century German intellectuals.

Methodologie (methodology) was an established topic of philosophical reflection and also had strong links to pedagogy. A discussion of *Methodologie* was a standard part of works that offered introductory overviews of a given academic field. For example, Carl Schmid described his 1810 work *Allgemeine Encyklopädie und Methodologie der Wissenschaften* as a contribution to "the science of the sciences [*die Wissenschaft der Wissenschaften*]." This "science of the sciences" had two branches. *Encyklopädie* was its "objective side"; that is, it described the objects of scientific inquiry. *Methodologie* was its "subjective side," or "the *Wissenschaft* of the study of the *Wissenschaften*." In other words, methodology described how scientific knowledge was produced. The "subjective" aspects of a science, in this context, meant those things that had to with the knowing subject, the person who possessed or made knowledge.[36] *Encyklopädie* dealt with the "what" (what external object does this science explore?), *Methodologie* with the "how" (how does one, as a knowing subject, create knowledge about this object?). In addition to works like Schmid's, which covered all of the sciences, similar books were common in university disciplines like medicine and theology. Lectures on *"Encyklopädie und Methodologie"* were also offered as university courses, both within the professional faculties and as a general introduction to university study.[37]

Works of *Encyklopädie und Methodologie* defined "method" broadly. *Methodologie* was a science of personas as well as a science of processes. In addition to a description of knowledge-making practices, *Methodologie* included information about the personality traits and skills needed in a given field. For example, in a rare (and so far as I know exceptional) treatment of the natural sciences in this genre, Gustav Suchow divided his discussion of *Methodologie* into two sections: one on "General Methodology" and one on "Requirements for Natural Scientific Study." The latter included things such as a healthy body and good sense organs, a gift for observation, and a good memory.[38]

Methodologie dealt extensively with personal qualities, the concrete competencies and character traits necessary to practice a given science or profession. What kind of person was best suited for a certain kind of *Wissenschaft*? And how, in turn, did the practice of that *Wissenschaft* shape the individual personality? These were all questions that *Methodologie* sought to answer. Discussions of method and discussions of *Bildung*, in other words, covered overlapping intellectual ground, and the issues raised in the school debates blended easily into debates about method. *Bildung* and *Methodologie* both linked the structure of knowledge to the structure of the individual personality.

In earlier discussions of *Methodologie*, one does not find the assertion that the natural sciences shared a single method, common to them but not to other sciences. In the early nineteenth century, proper method might be conceived in very general terms, as, for example, when one spoke of correct philosophical or *wissenschaftliche* method. In this formulation, all fields of knowledge shared a single method. "Methods" could also refer to the specific techniques of a discipline; botanists, for example, spoke of a natural method of classification. An 1826 essay by Carl Gustav Carus on the teaching of anatomy provides a useful example of how the term was typically used. Carus gave a history of "the difference between descriptive, historical, comparative, and philosophical anatomy" and outlined the order in which a beginning student should encounter "each of these different methods" in his studies. For Carus, each of the different methods also represented a historical stage, and the progression from "descriptive" to "philosophical" anatomy was one toward the best method of knowing about the world in general. Significantly, none of the methods Carus mentioned was peculiar to natural science—one could treat any subject descriptively, historically, comparatively, or philosophically.[39]

References to a natural scientific method had also not played an important part in earlier scientific societies' descriptions of their collective character. Common tastes and common sentiments had bound together eighteenth-century "friends of nature," and if they possessed special qualities as a result of their interests, it was Nature who ultimately received the credit, not *Naturwissenschaft*. In the early nineteenth century, the unity of nature, which also dictated the need for unity among nature's students, was the most common philosophical argument advanced in support of a unified community of *"wissenschaftliche"* natural researchers. The school debates of the 1830s and 1840s, in contrast, pushed natural researchers of all stripes to emphasize the supposed epistemic particularity of the natural sciences.

The Natural Scientific Method in Local Debate

Among humanist scholars, one of the clearest early formulations of this new methodological distinction came from a classical philologist, Hermann Köchly. In the 1840s, Köchly was a young schoolteacher at a Dresden Latin school, and he published a number of widely read works on the *Gymnasium* question. Köchly took a particularly strong stance on the differences between the natural and the historical sciences; as a result, the responses inspired by his proposals offer a useful survey of the various intellectual options on the table in *Vormärz* debates over the classification of knowledge. Köchly's proposed reforms, which took as their starting point an assumed methodological difference between the natural and the historical sciences, were both controversial and widely discussed. By the late 1840s, the young Saxon schoolteacher had a national reputation, and in 1848, he was elected the provisional chair of the first General Meeting of German Teachers (Allgemeine Deutsche Lehrerversammlung). After the revolutions of 1848–49, his stature as a school reformer helped win him the chair of classical philology at the University of Zurich.[40]

Unlike many of his fellow philologists, Köchly argued that there were two distinct forms of education that could prepare students for future work in the world of *Wissenschaft*, not just one. Consequently, there ought to be two separate types of secondary schools. The traditional *Gymnasium* could train students for later careers in the historical sciences, which included philology, history, theology, and law, while the *Realschule* would prepare students for university study in the natural sciences and medicine. Both kinds of schools, contrary to contemporary practice, should be accorded equal prestige and status.[41] Köchly's version of the *Realschule*, in other words, elevated this kind of school to the same rank as the *Gymnasium* and assigned it a parallel function.

In the early 1840s, Köchly had been part of the same Dresden literary circle as the liberal Young Hegelian Arnold Ruge, and Köchly drew on Hegelian categories to justify his plan for a new, dual-track school system.[42] While *Wissenschaft* itself was unified, Köchly argued, it had two primary objects, *"Natur"* and *"Geist."* Nature was characterized by "its cyclically recurring states"; the realm of spirit, in contrast, by "progressive development."[43] These were the same distinctions later used by Johann Gustav Droysen, traditionally seen as a seminal figure in the history of historical methodology, in his famous essay *"Natur und Geschichte* [Nature and history]." This parallel usage was one of several congruencies between the careers of these two historians. Droysen also had ties to Young Hegelian circles in the 1830s and 1840s, and during the same period, he criticized classical philology's excessive focus on language at the expense of historical development—another central issue for Köchly, as we shall shortly see.[44]

Like Droysen several decades later, Köchly argued that the two branches of *Wissenschaft* were *methodologically*, and not just topically, distinct. The natural sciences had "a strictly observational and demonstrative method" that was unique to them, and extensive historical training would be wasted on future *Naturforscher*.[45] Philological training provided such students with the wrong kind of "gymnastics of the spirit" (Köchly was an enthusiastic gymnast). *Naturforscher* needed a bodily and sensory education that textual study could not provide. The study of nature involved learning to move from concrete sensory experience to the abstractions of natural law. In contrast, future historians, theologians, and lawyers needed to learn to bring the abstractions presented to them in texts into dynamic relationship with their own concrete inner lives. As Wilhelm Dilthey would later argue, *understanding* was the defining characteristic of the historical sciences.[46]

Köchly's ideas about historical method were at the heart of his criticism of the contemporary *Gymnasium*. Standard teaching methods, he claimed, failed to provide a proper classical education. Instead of imbuing students with the spirit of ancient Greece and Rome, as early neohumanists had intended, the classical high schools simply provided language training. The classical curriculum had lost its way in linguistic minutiae, and it needed to be reformed to provide students with a deep historical understanding of the ancient world. Language study should be a means to an end, not an end in itself. Properly handled, the classical languages provided the conduit through which students came to know the great cultures of Greece and Rome, but if Greek and Latin were taught with dead, philological precision rather than living historical understanding, they were dull tools for training young minds.[47] Köchly's historical approach to teaching the classics was widely discussed in reform circles and his suggestions generally well received.[48]

The suggestion that students bound for the medical faculty should attend *Realschulen*, however, was much more controversial. Köchly's reform plans presented, in a particularly stark form, a distinction that was already common in much that had been previously written about school reform. In describing the internal divisions of *Wissenschaft*, earlier pedagogical writers had often used the opposition between *Geist* and *Natur*, or between an inner, spiritual and an outer, material world, to analyze the different components of the school curriculum.[49] Köchly's suggestion that the two halves of *Wissenschaft* needed to be taught in two separate schools gave institutional flesh and blood to an older philosophical division; the differences between the two kinds of *Wissenschaft* were so great, his reform plan assumed, that they required two separate kinds of *Bildung* from very early in life.

Although Köchly's use of this distinction was not unique, he assigned it a practical significance that went beyond the norm, and reviews of his early books acknowledged his suggestions' relative novelty. Commentators saw Köchly's work as part of a more general innovation, not only in school policy but also in the categorization of knowledge. Karl Mager's review of Köchly's first book, for example, grouped it with two others, both by small-town school directors, who made similar arguments about the internal differences between the *Wissenschaften*. Mager opposed Köchly's plan for two separate kinds of secondary schools (he felt that *Naturforscher* also needed a measure of erudition), but he accepted the epistemological division Köchly had proposed and considered the use of such categories a helpful recent development. "It is nice," Mager wrote, "that the distinction between the *Naturwissenschaften* and the ethical *Wissenschaften* is beginning to be common."[50] Other reviewers mentioned Köchly's new system of classification as noteworthy and argued with him over the appropriate labels for the two branches.[51]

Despite its roots in idealist philosophy, then, this binary division of *Wissenschaft* was considered relatively new in the mid-1840s. The label "*Naturwissenschaft*" was used fairly universally, but there was disagreement about what to call the disciplines on the other side of the divide. Köchly himself preferred the term "historical sciences" because it emphasized the "unique method of the *Geisteswissenschaften*," which he considered the most important source of their distinctiveness. The precise label used, however, was something that Köchly considered inessential—the division was clear, whatever one chose to call the two fields. At one point, he listed six different possible sets of terms that might be used to describe the two main branches of *Wissenschaft* (*Geisteswissenschaften* and *Naturwissenschaften*, the ethical and the physical sciences, the humanistic and the realistic, the historical and the exact,

CHAPTER 8

the spiritual and the sensual, the traditional and the experimental). Whatever labels one used, the fields of history, law, and theology belonged on one side of a divide; the sciences that dealt with nature lay on the other.[52]

Not everyone agreed with Köchly's specific proposals, but given the divisions produced by the school debates, the basic distinctions he used were beginning to seem increasingly like common sense to educated Germans by the mid-1840s. Köchly's relative nonchalance about terminology makes it clear that more than a philosophical distinction was at stake. The categories *Naturwissenschaft* and *Geisteswissenschaft* (or their various equivalents) functioned as placeholders for the interests of different groups of scholars, teachers, and students; they were practical tools to address issues of status and identity in a key social institution.

Köchly's views on natural science developed in close conversation with several local *Naturforscher*, particularly his friend Hermann Richter. His proposals also came under intense scrutiny in his hometown of Dresden, and we have an unusually complete record of the conversations they produced. In 1846, Köchly helped found a society for school reform, the Gymnasialverein. His reform proposals offered a starting point for the Gymnasialverein's work, and one of the group's members made a stenographic record of all the ensuing discussions. Köchly summarized the question the society faced as follows: "Depending on whether they deal with the development of the spirit itself, or the objects of external nature, the sciences break apart into historical or ethical sciences on the one hand and natural sciences on the other." Whether it also made sense to have two different forms of the *Gymnasien* to correspond to these two different branches of *Wissenschaft*, Köchly thought, was the next question the group needed to answer.[53]

The three most prominent and vocal natural scientific members of the society were Hermann Richter, a professor at the local medical academy; Ludwig Reichenbach, the director of Dresden's botanical garden; and Emil Roßmäßler, a forestry professor from nearby Tharandt. None of these men was among the most eminent of German researchers, but all went on to positions of measured national importance in the second half of the nineteenth century. Reichenbach was later elected head of the Leopoldina, Roßmäßler was one of the foremost scientific popularizers of the nineteenth century, and Richter was a leading figure in the national movement for medical reform.[54]

In the Gymnasialverein's debates, the collective sense of grievance that the status quo inspired among *Naturforscher* was abundantly clear.

They complained about the low standing of school mathematics instructors; they griped about ill-prepared medical students and about arrogant humanist peers who considered themselves educated despite knowing nothing about the natural sciences. Dresden's *Naturforscher* had differing views on the specifics of Köchly's reform proposal and varying philosophical commitments. What they shared, however, was a common sense of injury and a common desire for redress.

The positions these men took on the question of school reform had important similarities, but also equally important differences, both in tone and content. Richter was perhaps the most acerbic and combative member of the Gymnasialverein, and also a strident defender of Köchly's proposed split. Richter had been interested in epistemological issues from early on in his career; his dissertation had examined the question of certainty in medical thought, a problem he tackled in conjunction with a study of Francis Bacon (he recommended the creation of a *codex empiricus* in which medical propositions could be collated with relevant observations).[55] At the Dresden medical academy, he was an aggressive spokesman for a reformed, "rational medicine." An admirer of the Younger Viennese school and the methods of physiologist Johannes Müller, he antagonized some of his older colleagues, men he derided as aging medical romantics.[56]

Richter agreed with Köchly that there was a clear methodological distinction between the two branches of *Wissenschaft*, but he had less than flattering things to say about Köchly's own field of classical philology. "The natural sciences have a completely different teaching and research method," he stated, "and it stands in relation to the humanistic method as oil does to water."[57] The natural sciences allowed the student to learn to follow the logic of "things that can speak for themselves"; the traditional humanist curriculum only taught them to follow authority. Studying the classics cultivated, both literally and figuratively, a kind of blindness. The natural sciences, in contrast, taught students to see the world clearly.

Richter—who, like Köchly, was active in Dresden's left-liberal circles—made the political implications of these epistemological differences clear. Philology was a prop for conservative authority; natural science, in contrast, was the perfect foundation for a new political order, one where independent thought and individual freedom mattered more than inherited tradition. An education in the natural sciences would "give every student the ability to look around with a trained eye and an independently thinking spirit at all of the living relations around him," and give him the courage to sweep away all that was dead and sickly. It would also stoke Saxony's productive energies; natural science furthered prosperity as well as freedom.[58]

Fellow society member Ludwig Reichenbach, in contrast, blamed the current epidemic of "pauperism" on a perversion of true humanist values, and on the blurred line between the learned and the productive estates. The disorder of the Hungry Forties, he thought, was the work of an old enemy—an insufficient appreciation for the disinterested worth of learned *Wissenschaft*. Reichenbach held many of the romantic commitments that were anathema to Richter; the botany professor still spoke of Oken and Schelling with approval. Reichenbach also drew very different conclusions than did Richter about the meaning of science for social and political order. A confidant of the Saxon king Friedrich August II, Reichenbach believed the natural sciences would restore faith in traditional monarchy. He had argued in the early 1840s that the study of nature would create "devoted, peace-loving citizens," convinced of the beneficent rights of "the strong and the powerful" to rule, and he repeated this argument in his speeches in the Gymnasialverein.[59]

Reichenbach and Richter also disagreed about the value of the classical tradition. Reichenbach, in addition to promoting the virtues of natural science, also praised the classical curriculum, complaining about the excessive pull of "realism," neohumanism's pedagogical enemy, in current public discussions.[60] Richter, in contrast, considered the traditional curriculum more or less useless.[61] The forestry professor Emil Roßmäßler, a childhood schoolmate of Hermann Richter's, fell somewhere in the middle of the other two men. He shared Richter's political sympathies, but, like Reichenbach, was more conciliatory in his discussion of the humanist tradition. Roßmäßler's main aim was to defend natural science as an activity that could be pursued in a humanist spirit, a subject that provided sound *Bildung* in the best humanist tradition.[62]

In arguing for the pedagogical value of natural science, however, all three *Naturforscher* shared several common reference points. Richter, Roßmäßler, and Reichenbach all claimed that natural science developed unique mental capacities that could be created no other way. Richter was most explicit in using the concept of method to make a case for the particularity of natural science, but all three of the society's *Naturforscher* defended the epistemic distinctiveness of science in some form.

Furthermore, the society's *Naturforscher* all held that the peculiar benefit of a natural scientific education was a sound training of the senses; the end result of this training was the ability to perceive universal laws. For Richter, the natural scientific method began with clear *"sinnliche Anschauung,"* or sensory perception. From there, he thought, the *Naturforscher* continued on through a process of induction to the formulation first of sound scientific descriptions and then of explanatory laws.[63] Similarly, Reichenbach characterized science as a particular kind of "practical logic," grounded in solid sensory training, that led to the discovery of laws. "The spirit learns what law is through nature," he stated, citing earlier author Karl Snell with approval.[64]

All three men also insisted that *Naturwissenschaft* was the sort of unified body of knowledge that offered general, formal *Bildung*. Even the obstreperous Richter, despite his disdain for the classics, used the concept of *Bildung* in ways that would have been recognizable to his neohumanist contemporaries. All three of these men defended an "organically unified *Naturwissenschaft*" that included natural history. They often used the word "*Naturwissenschaft*" in the singular rather than the plural form, in part to forestall competing proposals for reform that included the mathematical parts of physics as *Gymnasium* subjects but left out the natural historical sciences, the fields in which their own scientific interests lay.[65] Despite their political and philosophical differences, both Richter and Reichenbach found in Alexander von Humboldt a worthy embodiment of their scientific principles, praising Humboldt's *Kosmos* as an example of the organically unified natural science they defended.[66]

In discussing natural science's unique qualities, several members of the Gymnasialverein made mention of natural science's "exact methods."[67] What did they mean by this term? For the group's members, "exact" did not just mean quantitative. For Richter, properly scientific expressions came in qualitative forms, too, in "word, line, and number."[68] Indeed, for people interested in arguing for the equal status of the natural historical sciences, keeping mathematics in its proper place was an important concern. Reichenbach hoped to dispel "the misunderstanding that physics [*Naturlehre*] was the entirety of *Naturwissenschaft*," and Richter wrote that *Naturwissenschaft* must be taught as "a great organically coherent whole."[69] "The exact sciences" included botany, zoology, and geology. In addition, the term "method" covered broad ground. The members of the Gymnasialverein slid back and forth among descriptions of pedagogical methods, methods of knowledge production, and common personality traits—a generous use of the term in keeping with the standard topics covered by *Methodologie*.

The stenographic records of the society's meetings also note repeated flares of temper between the *Naturforscher* and the philologists in the group. The competing interests of these two camps sparked tensions even among this group of reformers, all of whom were critical of the neohumanist curriculum in its standard form. In the second meeting of the Gymnasialverein, for example, Richter gave a brief, polemical speech summarizing his views and roundly criticizing the classical curriculum. The main question, according to Richter, was to observe "the difference between the spiritual [*geistigen*] *Bildung* that takes place through sensory perception and that which takes place through word and text." The rest of his comments came close to dismissing the latter kind of *Bildung* completely. Not only were the methods of natural science and humanism completely different, but humanist training, he implied, might be actively harmful.[70]

Richter's claims left many of his fellow society members unnerved. Köchly commented that he felt like the sorcerer's apprentice, someone who had called forth forces he could no longer control; another society member said that Richter had only confirmed his fear that the classical *Gymnasien* were on the defensive against the merciless onslaught of the *Realien*, subjects like natural science and modern languages. When Köchly pressed Richter on whether or not he had really meant to imply that all branches of *Wissenschaft* that dealt with "word and text" were without value, Richter answered that of course he had not meant to question the legitimacy of the historical sciences, or of law, theology, or philosophy; he had only been speaking about educational methods.[71] (Judging from his other writings, this distinction was disingenuous. Richter later argued that what Köchly's field of classics really needed to do was to bring the *"naturwissenschaftliche Methode"* to bear on historical questions.)[72] Reichenbach and Roßmäßler gave more conciliatory speeches at later meetings, and in the end, the society agreed on a compromise curriculum—a unified *Gymnasium* that prepared students for future work in both branches of *Wissenschaft* by including two different courses of study within a single school.[73]

Not everyone in the Gymnasialverein agreed that the methods of the natural and the historical sciences were unique; some members of the society argued that all *Wissenschaften* shared a single method. The day after Richter's controversial speech in the Gymnasialverein, fellow society member Eduard Calinich sat down to write the preface for his recently completed book, *Philosophische Propädeutic für Gymnansien, Realschulen und höhere Bildungsanstalten*, an overview of philosophy for a secondary school audience. The reference to philosophy in his title, he emphasized in the preface, was to philosophy as a set of empirically grounded *Geisteswissenschaften* (as distinct from philosophy as "purely speculative philosophy"). In his opening paragraph, Calinich penned several sentences aimed directly at Richter's speech from the previous day. He complained about those who believed that one could only gain "true and positive knowledge [*ein wahres Wissen und positive Kenntnisse*]" from the natural sciences. The *Geisteswissenschaften* did not merely "sway in the air" without any grounding. Both the *Geisteswissenschaften* and the *Naturwissenschaften* shared a common method, the method of "the *Wissenschaften* in general"—"sensory understanding, experiment, and observation."[74]

Positive Knowledge, Realism, and Natural Science at Midcentury

Calinich's criticisms of Richter, in particular his use of the term "positive knowledge," raise a number of questions. General histories of nineteenth-century Germany have often presented the "rise of natural science" as a defining feature of German culture in the 1840s, 1850s, and 1860s.[75] Natural science's new confidence has sometimes been attributed to an emergent positivism or realism.[76] But how valuable are these two labels—"positivism" or "realism"—in capturing what was novel in discussions of natural science at midcentury? At first glance, neither term adequately describes the range of positions adopted by the *Naturforscher* examined above, either in Dresden or in the national scene more broadly. For one thing, the heritage of German idealism, neohumanism, and *Naturphilosophie* was still evident in a variety of ways, in forms too diverse to be described in terms of simple acceptance or rejection. If some *Naturforscher*, men like Justus Liebig or Hermann Richter, raged against speculative philosophy and the foibles of aging romantics, those same aging romantics (say, Lorenz Oken or Ludwig Reichenbach) were still prominent public figures, and they defended the particularity of natural science with their own set of intellectual resources.

Both "positive knowledge" and "realism" were important categories in the school debates, but one needs to be careful in analyzing what these concepts meant in this particular context. In the terminology of the period, the defenders of both the human and the natural sciences understood themselves to be dealing, at least in part, with "positive" knowledge. Ernst Calinich's angry reaction to Richter's attempt to appropriate this privilege solely for the natural sciences was typical.[77]

More important, positive knowledge could be defined in contradistinction to speculative philosophy, but it was also used to mean the opposite of "formal" kinds of knowledge like grammar or mathematics. For *Naturforscher* and historians who defended the value of positive knowledge, the main target was the excessive formalism of the classical neohumanist position, the idea that "formal *Bildung*" should not be diluted by the addition of too much detailed empirical content. Defending the value of positive knowledge, however, did not necessarily mean rejecting the ideal of formal *Bildung*.[78] More commonly, authors assigned value to formal *Bildung* as well, and the approbatory use of the term "positive knowledge" revealed little about what other epistemological or metaphysical positions a writer might hold.

Talking about a simple shift from idealism to realism at midcentury is similarly problematic, because, like defenses of positive knowledge, defenses of realism were rarely one-sided. The term *"Realismus,"* as it appeared most often in public discussion, referred to a pedagogical tradition, a curriculum centered around the *Realien*, the modern languages, the natural sciences, and modern history, along with the more worldly, practical values that tradition embodied. It did not denote a coherent set of philosophical commitments. Just as support for positive knowledge did not entail a complete rejection of formal *Bildung*, *"Realismus"* did not usually receive unqualified support at the expense of the spiritual or the ideal.[79] Both sets of terms could, and usually did, function as binaries that stood in a productive relationship with each other.

The cumulative result of the school debates was a decided shift in the rhetoric *Naturforscher* employed when speaking about their own forms of knowledge. The debates added a new emotional force to defenses of natural science, and they also made assertions of natural science's epistemic particularity more common. Given the diversity of figures who came forward to defend natural science in the face of neohumanist criticism, however, there is little to suggest that this new stridency (sometimes glossed as a rise in scientific "confidence") was the side effect of some general shift in first principles. In making the case for natural science's epistemic distinctiveness, many *Naturforscher* did not break significantly with ideas that had been common earlier in the nineteenth century. For example, the prominence of *Anschauung*, or sensory apprehension, in the Dresden society's debates reflected a consensus among *Naturforscher* that was not new to the 1840s. In the first third of the nineteenth century, there was widespread agreement among *Naturforscher* of otherwise varying commitments that *"An-*

schaulichkeit," or intuitive clarity, was the best criterion for judging the soundness of a scientific theory.[80] More generally, the idea that refined sensory perception was the hallmark of the *Naturforscher* (and by extension the medical doctor) was a commonplace in introductory textbooks, both in the natural sciences proper and in medicine.[81] Pedagogical writers also regularly praised natural science as a "school for the senses."[82] Among leading researchers, the consensus around the concept of *Anschauung* was beginning to shift by the mid-nineteenth century. Developments like Georg Simon Ohm's mathematical treatment of electricity and the new strategies of measurement taught in Franz Neumann's Königsberg seminar were beginning to push other views to the fore, at least in the physical sciences.[83] But older arguments also still worked perfectly well to bolster claims about the unique qualities of the natural sciences.

What German *Naturforscher* shared at midcentury was not a new set of fundamental commitments but a common position in the debate over *Bildung*. Beyond the unifying effects of this common cause, however, diversity remained the rule. A new emphasis on the uniqueness of "the natural scientific method" did not necessarily signify any strong philosophical agreement on what exactly that method was, and this remains true if one looks ahead to the second half of the nineteenth century. In the numerous memorials for Alexander von Humboldt held after his death, for example, "the natural scientific method" appeared in many of the speeches, but in inconsistent form, with the unique properties accorded to natural science differing according to the disciplinary allegiances of the speaker. The ethnologist Adolph Bastian, speaking in 1869 to a joint meeting of all of Berlin's natural scientific societies, said that the essence of the natural scientific method was rigorous, disciplined comparison, and he considered Humboldt's work the epitome of its use. For the physiologist and physicist Emil Du Bois-Reymond, in contrast, real *Naturwissenschaft* involved the mathematization of phenomena along the lines of Newtonian physics, a project Humboldt's work embodied only very imperfectly.[84]

The school debates continued for the rest of the nineteenth century, forming one of the perennial themes of German cultural politics through the early twentieth century.[85] Indeed, as Norton Wise has noted, a number of key texts in later versions of the *Naturwissenschaft/Geisteswissenschaft* debate were written within this context. Ernst Mach and Wilhelm Ostwald, for example, both wrote on the education debate.[86] Key themes from the school debates can be found throughout Hermann von Helmholtz's famous 1862 speech on the relationship of natural science to other areas of *Wissenschaft*. For Helmholtz, mathematical laws were the hallmark of a mature natural science, and this commitment gave him a particularly elegant counterargument to the standard neohumanist claims about the educational power of grammar. Grammar rules were only approximations, Helmholtz pointed out, rules that were violated all the time in actual speech. Natural law, in contrast, was inviolate.[87] What subject could be better suited to create orderly young minds? Some of Emil Du Bois-Reymond's most strident rhetoric also came in reference to *Gymnasium* reform ("Conic sections! No more Greek composition!").[88] At the inaugural celebration for the University of Tübingen's natural scientific faculty (the first of its kind in Germany), Hugo von Mohl explained the occasion's significance in terms of the debate over *Bildung*. "In the creation of a new natural scientific faculty," he said, "one sees a break with the medieval idea that *Bildung* is only to be found in the humanistic studies."[89]

While the school debates can hardly explain all the permutations of later debates over scientific methods, they did provide a basic cultural scaffolding that made methodological questions a recurring issue for German academics. The German educational tradition, as embodied in the preparatory institution of the *Gymnasium*, linked epistemology to broader questions of moral and intellectual self-development. Because the individual personality supposedly reflected back the characteristics of the knowledge that formed it, claims about method were a necessary accompaniment to claims about the social value of knowledge.

Two Camps or Two Cultures?

In the 1850s, the chemist Justus Liebig and the philhellene Friedrich Thiersch became next-door neighbors in Munich. Thiersch was one of the most strident defenders of the classical curriculum, author of the infamous Bavarian school plan of 1829. Liebig was one of natural science's most famous and acerbic defenders; he had mocked philologists like Thiersch mercilessly in print. In 1855, Liebig's favorite daughter, Johanna, married Thiersch's second son, Karl. The young groom had studied medicine in Munich, Berlin, and Vienna, and he had recently been named to the chair of surgery in Erlangen. He would later move on to a more prestigious appointment in Leipzig. The marriage of the philologist's son and the chemist's daughter was reportedly a very happy one; the couple had four daughters and two sons.

This chapter has focused primarily on the emergence of a heightened sense of division within the world of German *Wissenschaft*, but the preceding anecdote should help us refrain from assuming that this epistemological divide represented the creation of two alien "cultures," one of scholarship and one of science.[90] As Andreas Daum has shown, this rubric is not very useful in helping us position the natural sciences within German public culture. The conflicts between *Naturforscher* and their neohumanist colleagues were real and often bitter, but the intermingling of science and scholarship in the Thiersch and Liebig families was entirely typical; many professors of the natural sciences had familial connections to academics in other fields, and the great academic families of the nineteenth century often had members on both sides of this divide.[91] These intimate alliances shaped the networks of German intellectual life in important ways, as Deborah Coen has shown in her study of the Exner family. For this powerful Austrian academic dynasty, the family circle was a site of rich intellectual exchange among people with an array of disciplinary allegiances.[92]

If relations between *Naturforscher* and their scholarly colleagues sometimes seemed strained, university-educated *Naturforscher* remained part of a cultural world that overlapped heavily with that of their humanistic peers. University-educated *Naturforscher* shared central reference points with their neohumanist rivals. In the context of the school debates, these two camps disagreed on curriculum and methods, but both held up broadly similar notions of *Bildung* as the defining aim of a sound education. Furthermore, to love natural science more was not necessarily to love Greece less, and many university-educated *Naturforscher* continued to place great value on classical erudition. Helmholtz went out of his way to make his respect for the classical tradition clear. Ludwig Reichenbach's eulogy made special mention of his love of the classical tradition.[93] The opposition between the natural sciences and their humanistic counterpart (whatever the latter might be called) was a metadistinction that faded markedly in significance at the level of individual lives and thought.[94]

The older ideal of a unified learned world was certainly under strain by the mid-nineteenth century, and some older unifying forums were fading into irrelevance. Two of the most important general review journals of the early nineteenth century, German-speaking Europe's two *Allgemeine Literaturzeitungen* (one published in Halle and the other in Jena), stopped publication in the late 1840s. A few other general review journals lasted longer (Heidelberg's appeared until 1872), but others had not even made it to midcentury. Leipzig's general review journal, for example, folded in 1834. These *Literaturzeitungen* (still often called *Gelehrtenzeitungen*, or learned papers, in common parlance) had included in their pages not only reviews of books and reports from learned societies but also notices of the official promotions that marked the major turning points in a learned man's life—new educational appointments and bureaucratic posts, both prestigious and more modest. The demise of these review organs marked an important moment in the long, slow transformation of the early modern learned estate into the modern *Bildungsbürgertum*, with its wider educated public and its narrower cadre of specialized, professional scientific and scholarly experts.[95]

The unified ideal of *Wissenschaft* also continued to play a central role in German intellectual life, and the universities and academies of science still stood as the guardians of this cause. The one new state academy founded in this period, Saxony's Academy of Sciences, upheld the tradition of keeping all the fields of *Wissenschaft* together in a single institution.[96] Indeed, the abiding power of this unifying ideal is marked by the lexical difference that now separates the German word *Wissenschaft* and the English "science."[97] Within associational life, history and natural history sometimes moved closer together over the course of the nineteenth century; these pursuits were often joined together in the *Heimat* associations of the later nineteenth century.[98]

To the ties of family and shared cultural values, one can also add the links forged through political allegiances. The *Naturforscher* Hermann Richter and philologist Hermann Köchly, despite their differing intellectual commitments, were clear allies in the left-liberal circles of *Vormärz* Dresden. They belonged to the same liberal political and literary circles, and both were leaders in the local *Turnverein* (gymnastic society), a focal point of liberal sentiment. Building on this earlier activism, both would go on to play roles in the revolutions of 1848–49. When the revolutions failed, Köchly was forced to flee Dresden to avoid arrest, and Richter spent several years in prison.[99] This shared caesura in their careers offers a valuable reminder of the many things that joined these erstwhile friends and opponents. Their quarrels are better understood not as clashes between alien cultures but as family squabbles between learned siblings who fought so vehemently in part because they shared so much. When Germans wrote about the natural and the human sciences, they often invoked the Humboldt brothers, Wilhelm the scholar and Alexander the *Naturforscher*, as symbols for the two halves of *Wissenschaft*. This fraternal metaphor is both telling and apt.

In the late eighteenth century, "*naturwissenschaftlich*" had been a simple—and rarely used—descriptive adjective, designating a topical area of study. By the mid-nineteenth century, it was being wielded as a weapon in any number of intellectual and cultural debates. This little adjective's migration onto the center stage of German modernity was a gradual one. The educational debates of the 1830s and 1840s represented a crucial moment in its history. In these debates, philologists claimed that studying natural objects would spoil the intellectual development of German boys; *Naturforscher* responded that the natural sciences taught intellectual skills that could not be learned by studying books. These two different fields, in other words, had two different methods. Considering the terms of the school debates, it is perhaps not so surprising after all that we can find an early use of the lofty philosophical opposition between the *Geistes-* and the *Naturwissenschaften* in an obscure mid-nineteenth-century pedagogical tract.

Conclusion

In September of 1844, Julius Petzholdt, philologist and Saxon royal librarian, married Hermine Reichard, daughter of a chemical entrepreneur, in the scenic Plauen Gorge just outside of Dresden. In honor of the occasion, the groom's friend Heinrich Dittrich published a brief essay, "Contributions to the Explanation and Criticism of Theokritos." The choice of the bucolic poet, Dittrich explained, was a tribute to Petzholdt's love of country life and his "pleasure in the enjoyment of Nature and her beauties." In the introduction to his scholarly essay, Dittrich celebrated his friendship with the young groom, a friendship built on their common experience of being "raised in Greece and Rome's great spirits." Dittrich went on to reminisce about the cheerful conversations he and Petzholdt had shared as they wandered together through the "romantic Plauen landscape"—conversations, Dittrich was careful to mention, that often took place in Latin.[1]

From the perspective of the late nineteenth century looking backward, the marriage of Julius Petzholdt and Hermine Reichard appears as the incipient merging of the two main segments of the nineteenth-century German middle classes, the joining of *Bildung* and *Besitz*, education and property. The classically educated royal librarian and the daughter of a manufacturer meet each other from opposite sides of a cultural divide. According to Thomas Nipperdey, natural science would have sat with the bridal party at this wedding, an ally of industry and practical life, an enemy of unworldly philology and romantic aestheticism.[2]

CONCLUSION

A different picture emerges if we take a closer look at the people assembled in the gorge on that September day. The lives of both the Reichards and the Petzholdts intersected at multiple points with the developments recounted in this book. The groom was not as distant from the natural sciences as some images of nineteenth-century German culture might lead us to expect; his love of nature could also plausibly have been described as an affection for *Naturwissenschaft*. As was often the case among Germany's university-educated elite, Julius had relatives who studied nature rather than texts. Julius's brother Alexander was a chemist and a professor at the local technical academy, and despite the differences in their primary disciplinary allegiances, the Petzholdt siblings were close allies in Dresden civic life in the early 1840s. Julius was the secretary of Dresden's Natural Scientific Society, and both brothers belonged to the Manufacturing Association, too.[3]

The bride's parents, Gottfried and Wilhelmine Reichard, made a brief appearance in chapter 2, as purveyors of public scientific and technical spectacles. Wilhelmine had achieved widespread fame as the German lands' first female hot-air balloonist.[4] Her husband, Gottfried, who gave public chemical lectures in a number of German cities in the 1810s, did not have a university degree. He likely got his start in chemistry from his father, who had run a porcelain manufactory.

The Reichards were the kind of experts, or *"Naturkundigen,"* left at best on the margins, and often completely outside, the early nineteenth-century's narrowed definition of natural research. When Julius Petzholdt wrote his mother-in-law's obituary, he felt compelled to lament the fact that the Reichards' ballooning had not done more to benefit *Wissenschaft*, since the miscellaneous observations made on their ascents had never been systematically ordered.[5] In the late eighteenth century, however, people like Hermine's father and mother, with their prodigious chemical knowledge, might also have been considered "learned" alongside their new son-in-law. Of course, there was only so much in a name, and the division between the educated and the propertied middle classes had deep material and social roots in the pre-1800 period.[6] Nonetheless, in the late eighteenth century, the networks of natural research had created a significant gray area at the border between the "learned" and the "practical." Men like Gottfried Reichard's father, people who oversaw things like porcelain manufactories, were sometimes also considered members of the learned world. The cultural logic that placed the Reichards outside Dresden's core community of *Naturforscher*, which made them listeners instead of participants in a wider conversation about the properties of nature, was relatively new.

CONCLUSION

A year after her marriage, Hermine Reichard appeared as a subscriber to the Natural Scientific Society's lecture series as Frau Dr. Petzholdt, along with two of her male relatives and two of her female in-laws. Her mother's public feats had been described as *"wissenschaftlich"* in their day; Hermine's assigned place in the audience of the Natural Scientific Society is not a bad symbol for what the progressive social enclosure of elite science meant for women's relationship to it.[7]

Yet there is little to suggest that the Reichard family suffered much from being placed outside the new, narrower world of *Wissenschaft*. The university-educated elite of the early nineteenth century may have looked down their noses at performers like the Reichards, but their performances had still been financial successes. The money the couple raised from lectures and ballooning allowed them to set up a small sulfuric acid manufactory in Plauen. By the 1840s, this was a flourishing concern with twenty-four employees. In the early 1840s, Gottfried Reichard was even planning to publish a scholarly work on the Plauen Gorge. Though his book never appeared (he died in 1844), he nevertheless benefited from the *Vormärz*'s expanding print culture. Print houses were some of the best customers for the chemical products he made. Reichard won a royal medal for his contribution to Saxon industry, and he was also a prominent figure in the Dresden manufacturing society.[8]

There is even evidence that the Petzholdt brothers looked to Reichard as a potential ally in their efforts to raise the status of science among Dresden's businessmen. Both of the ambitious young Petzholdt brothers were trying to make themselves useful to Dresden's more practical citizens in the 1840s. Alexander, as we saw in chapter 7, completed a study of electromagnetic plating techniques in 1842 at the request of the Manufacturing Association. The society's members dismissed the chemist's efforts as useless and impractical. Julius published a guide to the coal-rich Plauen Gorge in 1842, which was similarly derided by practical critics as a waste of paper.[9] A year later, Julius, still the secretary of the Manufacturing Association, published a pamphlet about the group's library. He dedicated this modest work to his future father-in-law, praising Reichard as a man who could help combat "the mistaken opinion that theory counts for nothing, and only practice matters in industrial life." He also called Reichard, as an erstwhile public lecturer, by the honorary title "Professor."[10]

Petzholdt's dedication and the form of address he chose may give us hints about his romantic ambitions in the year before his wedding. This act of deference also speaks volumes about the complex position that *Naturwissenschaft*, as both a category and a social cause, occupied within German public life. Here, we find a philologist appealing on behalf of *Theorie* and *Wissenschaft* to a former scientific showman–turned–entrepreneur.

In the different strands of these two family histories, we can trace out the crosscutting developments and traditions that shaped the boundaries of the German category "*Naturwissenschaft.*" In the Petzholdt brothers, we have the fraternal ties that bound together university-educated German men, the social core of the learned world. In the history of the Reichard family, we see almost all of the developments that stretched learned networks of natural knowledge in new directions. From the bride's grandfather, we hear the echoes of state interest in new luxury products like porcelain. Through the bride's parents, we can watch the spread of chemical skills, and see the myriad ways in which natural knowledge might be put to work in the broader cultural and economic marketplace.

In the Petzholdt brothers' complex relationship to their new in-laws, we also get a hint of the many remaining ambiguities that the seeming solidity of the concept "natural science" made less visible. Different connections and alliances come into view depending on the scale that one chooses as a frame. At the level of associational life, where we can watch large groups coming together, "the natural sciences" could seem a well-defined and distinct collective cause. At the level of individual biography, the division between the natural sciences and other kinds of endeavors could be much less well marked out.

In the introduction, we heard Mattias Flurl and Werner Siemens praising the "escape" of the natural sciences from the closed circles of the learned. In a nutshell, this was the development that caused the early modern period's classificatory schemes to strain and break. If taken too literally, this statement is misleading, given that the study of nature was already a socially diverse enterprise well before 1770. But there is also no doubt that the basic habits of learned men—writing for publication, attending protocolled society meetings, pursuing serious reading, experimenting, and collecting—became much more widely distributed beginning in the late eighteenth century. At the same time that less institutionalized kinds of learned behavior were spreading, the German states' formal educational systems also went through marked change. The spread of normal schools, agricultural schools, and technical academies meant that a wider range of people now shared what had previously been a hallmark of the learned—the experience of training for your profession through formal schooling rather than through practical apprenticeship.

CONCLUSION

By the 1840s, the basic coinage of the republic of letters was traded in a much wider market. This led to a certain degree of devaluation; behaviors that had once been clear markers of learned distinction now counted for much less, and this fueled efforts to create new kinds of hierarchies. At the same time, however, the spread of these behaviors also provided one of the most important concrete referents for midcentury talk about the power of *Naturwissenschaft*. It allowed "natural science" to be presented as something that was not just the property of a small elite but the common property of all the educated. Where exactly this "common property" was manufactured, however, was a different question. The older, more integrative ideals of intellectual community inherited from the preceding century were not entirely gone in 1850, but they had weakened considerably. In the coming decades, the distinction between the popular and the professional would become progressively clearer.

Struggles for power, prestige, and resources were not the only things that shaped the history of "natural science" as a category, however. Science, as its acolytes often pointed out, was a passion, and its pursuit was a pleasure. There is a reason this book has been about the growing salience of a concept rather than the hegemonic rise of a new form of ideology. The pull of intellectual curiosity could be socially destabilizing, creating new bonds that cut across established social hierarchies. The pursuit of natural knowledge forged novel ties between men, and sometimes women, of differing backgrounds, and these novel relationships were made stronger by the fact that the sciences also forged new emotional connections between humans and the natural world. People came to care deeply about plants, insects, minerals, chemicals, and other physical phenomena that otherwise would have attracted little notice. Stories of friendships forged through botanizing, of a man's devastation after a fire destroyed the natural history collection that had been "his life"—the emotions described in these tales were transformative in their own right.

Julius Petzholdt and Hermine Reichard were married in one of Dresden's most famously scenic landscapes, the Plauen Gorge. In the history of these two families, the gorge appeared in many guises. It was a site of industrial development, a place for natural historical research, an object of aesthetic contemplation, and an evocation of classical landscapes lost. With its coal deposits and geologically interesting cliffs, the gorge was a rich field for the naturalist or entrepreneur, but also a spot where two young learned men might walk together, conversing in Latin, or where the Saxon royal librarian might court the daughter of a chemical entrepreneur. These multiple uses would begin to seem more difficult to reconcile by the end of the nineteenth century, as workers' tenements encroached on the celebrated beauties of the gorge, but in the 1840s, these different forms of engagement existed side by side with relative ease.[11] To understand what natural science meant to nineteenth-century Germans, it helps to think about it in the full complexity of such local settings.

Notes

INTRODUCTION

1. See Wengenroth, "Science, Technology and Industry."
2. Siemens, "Das naturwissenschaftliche Zeitalter," 144.
3. Flurl, *Rede von dem Einfluße der Wissenschaften*.
4. On nineteenth-century standards of experimental precision, see Wise, *Values of Precision*. Quote from Siemens, "Das naturwissenschaftliche Zeitalter," 144.
5. See, e.g., Cahan, *From Natural Philosophy to the Sciences*; and Cunningham and Williams, "De-centering the 'Big Picture.'"
6. Stichweh, *Zur Entstehung des modernen Systems wissenschaftlicher Disziplinen*, 14–17.
7. Pickstone, "Working Knowledges before and after circa 1800"; Yeo, "Classifying the Sciences."
8. On eighteenth-century definitions of "philosophy," see Schneiders, "Philosophiebegriff des philosophischen Zeitalters." On natural history as a model for other kinds of knowledge making, see Lindenfeld, *Practical Imagination*, 28–33.
9. Koselleck, "Einleitung."
10. See, e.g., the entries for "science" in *Johnson's English Dictionary*; Reid, *Dictionary*.
11. Lloyd and Noehden, *Dictionary*, s.v. "Naturwissenschaft."
12. Translator's footnote in Helmholtz, "Relation of Natural Science to Science in General," 1.
13. Hoelder and Peschier, *Hand-Wörterbuch*, s.v. "Naturwissenschaft" and "Wissenschaft"; Peschier, *Wörterbuch*, s.v. "Naturwissenschaft" and "Wissenschaft."
14. I have used only the masculine pronoun in the preceding sentences because, although the label "*Naturforscher*" was

sometimes applied to women, in its stereotypical form this persona was gendered masculine in this time period. On the historical variety of scientific personae, see Daston and Sibum, "Scientific Personae."

15. Hoelder and Peschier, *Hand-Wörterbuch*, s.v. "Naturforscher"; and Peschier, *Wörterbuch*, s.v. "Naturforscher."
16. Brewster to Brougham, March 14, 1829, in Thackray and Morrell, *Gentlemen of Science: Early Correspondence*, 23.
17. Johnston, "Cultivators of Natural Science."
18. Ross, "Scientist"; Merton, "De-gendering the 'man of science.'" Ross's article is often cited to attest to the advent of "the scientist" as a distinct persona starting in the 1830s, but Ross actually argues that the term was still quite controversial as late as 1900. Merton confirms this picture of a very slow adoption in English and shows that neither Whewell nor any of his contemporaries actually used the word in any meaningful way.
19. Morrell and Thackray, *Gentlemen of Science: Early Years*, 223–96; and Knight, *Age of Science*, 8.
20. See, e.g., Ringer, *Decline of the German Mandarins*.
21. Anderton, *Limits of Science*; Jurkowitz, "Liberal Unification of Science"; and Galison, "Context of Disunity," 3–4.
22. Schnädelbach, *Philosophy in Germany*, 74–75.
23. Porter, "Science, Provincial Culture, and Public Opinion." These societies still described their collective aims in these same broad terms in the late nineteenth century; see the descriptions given in *Yearbook of Scientific and Learned Societies of Great Britain*. If one looks at the work of Jack Morrell and Arnold Thackray, it is easy to miss the fact that the British associational landscape did not, in fact, have very many groups at all that were organized around the idea of "science" in the modern colloquial sense of the word. In deciding which groups from this period should be labeled "scientific societies," Morrell and Thackray took as their standard the different topics included in the BAAS; since the BAAS had sections on agriculture and the mechanical sciences, Britain's field of "scientific associations" looks very large in their overview, since it includes societies devoted to the improvement of agriculture and manufacturing. This is a perfectly appropriate way to define a "scientific association" in the British context, given the standard nineteenth-century British usages of the word "science," but it makes comparisons with Germany difficult. At first glance, it looks impressive to say that Britain had 102 scientific societies by 1850. If one did a similar survey of German associations, however, and defined "science" in this very broad way, the German numbers would be much larger than the British ones. By 1850, Prussia alone had 361 agricultural societies. For the British numbers, see Morrell and Thackray, *Gentlemen of Science: Early Years*, app. 3. For the Prussian statistic, see Nipperdey, *Deutsche Geschichte, 1800–1866*, 150.

24. This generalization holds true, in any case, for the societies that survived from this period into the late nineteenth century. See *Yearbook of Scientific and Learned Societies of Great Britain*.
25. Fox, "Savant Confronts His Peers."
26. *Yearbook of Scientific and Learned Societies of Great Britain*, 1–47; and Müller, *Die wissenschaftlichen Vereine und Gesellschaften Deutschlands*, xv–xvii.
27. See, e.g., Galison and Stump, *Disunity of Science*; and Shapin, *Never Pure*.
28. For example, Daston and Galison, *Objectivity*.
29. For example, McClellan, *Science Reorganized*.
30. On the very gradual emergence of a "scientific community" in the nineteenth century, see Cahan, "Institutions and Communities," 327–28.
31. See, e.g., Liebig, *Studium der Naturwissenschaften*, 46. In comparison to the British case, the German natural theological tradition has received little historical attention. See Brooke, "Functions of Natural Theology." Recent work has begun to fill in this gap. See Daum, "Science, Politics, and Religion"; and Kleeberg, *Theophysis*.
32. On Schweigger, see Phillips, "Science, Myth and Eastern Souls."
33. Even after 1850, when debates about materialism and Darwinism made the question of science's relationship to religion a much more provocative one, the relationship between religion and natural science remained complex and multifaceted. See, e.g., Richards, *Tragic Sense of Life*, chap. 9; Daum, "Science, Politics, and Religion"; Gregory, *Nature Lost?*; and Gregory, *Scientific Materialism*.
34. On German civil society and evolving notions of citizenship in this period, see McNeely, *Emancipation of Writing*; and Hull, *Sexuality, State, and Civil Society*. On understandings of the nation in this period, see Vick, *Defining Germany*; and Levinger, *Enlightened Nationalism*.
35. Cahan, "Institutions and Communities," 313. See also Alter, *Reluctant Patron*, 1–12.
36. Nyhart, *Modern Nature*; Penny, *Objects of Culture*; Zimmerman, *Anthropology and Antihumanism*; and Daum, *Wissenschaftspopularisierung*.
37. See, e.g., Tuchman, "Institutions and Disciplines"; Schubring, *"Einsamkeit und Freiheit" neu besichtigt*; Olesko, *Science in Germany*; Olesko, *Physics as a Calling*; Nyhart, *Biology Takes Form*; Mendelsohn, "Emergence of Science as a Profession"; and Turner, "German Science, German Universities."
38. Hilpert, *Jacob Sturm*, 1.
39. For a summary of this view, see McClellan, "Scientific Institutions."
40. Zimmerman, *Anthropology and Antihumanism*, 4.
41. Daum, *Wissenschaftspopularisierung*, 5–14.
42. Hilpert, *Jacob Sturm*, 3. Hilpert claimed that Sturm was not just loved by those who knew him but also valued by "the learned men of the rest of the world."

43. Within the history of particular disciplines, historians sometimes mention in passing that this community persists into the nineteenth century. See, e.g., David Knight, preface to Knight and Kragh, *Making of the Chemist*, xiv.
44. Goldgar, *Impolite Learning*; and La Vopa, "Conceiving a Public."
45. Goodman, *Republic of Letters*; and Brockliss, *Calvet's Web*.
46. Daston, "Republic of Letters."
47. [Lorenz Oken], *Isis* (1817): 1–2.
48. Virchow, *Bedeutung der Naturwissenschaften*, 7. Cf. Eskildsen, "Republic of Letters."
49. Turner, "Social Patterns of German Science"; Turner, "University Reformers and Professorial Scholarship"; and Turner, "Growth of Professorial Research." Rudolf Stichweh set a similar break around 1800 and also linked it (though within a different theoretical scaffolding) to the transition from the hierarchical society of the Old Regime to the vertically, functionally organized society of the modern period. See Stichweh, *Entstehung des modernen Systems wissenschaftlicher Disziplinen*.
50. Turner, "*Bildungsbürgertum* and the Learned Professions."
51. See Jaumann, "*Respublica Litteraria* / Republic of Letters."
52. Siemens, "Das naturwissenschaftliche Zeitalter," 19.
53. In her excellent dissertation on scientific associational life in Frankfurt am Main, Ayako Sakurai has shown that the language of learnedness retained a prominent place in local associational life through the first part of the nineteenth century, and "the learned" remained a clearly defined group. See Sakurai, "Science, Identity, and Urban Reinvention," esp. chap. 1. However, since work on associational life has begun to move away from framing the changes that took place in this period as part of a general structural transition "from a society of estates to a society of classes," we no longer need to set the language of estates and the language of civil society in opposition to each other, as Sakurai does; we can analyze how these two forms of conceptualizing society interact with each other without assuming that talk of estates necessarily represents a conservative attempt to preserve the conditions found in a previous "stage" of development. In the eighteenth century, discussion of civil society often involved discussions of the different tasks of the various estates. For a summary of the waning interpretive emphasis given to the "estates to classes" rubric, see Zaunstöck, "Wege in der Sozietätsgeschichte."
54. Broman, "Habermasian Public Sphere"; Broman, "Periodical Literature"; and Broman, "Epistemology of Criticism."
55. Broman, *German Academic Medicine*, 202; and Broman, "Preliminary Considerations on Science and Civil Society."
56. Bosse, "Die gelehrte Republik"; and Habermas, *Structural Transformation of the Public Sphere*. For an article that coincides with Bosse's main point in

many ways but looks particularly at natural knowledge, see Wood, "Science, the Universities, and the Public Sphere."
57. Blanning, *Culture of Power*, 111.
58. La Vopa, "Conceiving a Public."
59. See, e.g., Altmayer, *Aufklärung als Popularphilosophie*.
60. Mah, "Phantasies of the Public Sphere."
61. Clark, *Academic Charisma*.
62. Clark's point about the significance of embodied and oral charisma is a valuable one, but his tendency to cast orality as an archaic persistence also poses some problems. Late eighteenth- and nineteenth-century forms of learned sociability were certainly oral, but they also had many novel features; it is not clear how we understand them better by casting them as anachronisms. Indeed, the thrust of the large literature on sociability would seem to cut in precisely the opposite direction. "Modern" social identities, constituted through new (oral) forms of sociability, were very much embodied and enacted.
63. Wittmann, *Geschichte des deutschen Buchhandels*, chap. 6; and Hellmuth and Piereth, "Germany, 1760–1815."
64. On Merian, see Davis, *Women on the Margins*, 140–202.
65. Nipperdey, *Deutsche Geschichte, 1800–1866*, 451–83; and Gispen, *New Profession, Old Order*, pt. 1.
66. On Fraunhofer, see Jackson, "Can Artisans Be Scientific Authors?"; and Jackson, *Spectrum of Belief*.
67. For the argument that "improving landlords" were as prevalent in Germany as in Britain, see Hagen, *Ordinary Prussians*. On agricultural improvement, see Haushofer, *Deutsche Landwirtschaft*; and Kiesewetter, *Industrialisierung und Landwirtschaft*.
68. Ringer, *Education and Society in Modern Europe*, 45–54; and Gispen, *New Profession, Old Order*, pt. 1.
69. The dynamics of this kind of crowding were somewhat different from those typically discussed within the literature on Europe's posited "excess of educated men." See, e.g., O'Boyle, "Excess of Educated Men"; and Titze, "Überproduktion von Akademikern."
70. For overviews of this large literature, see Hoffmann, *Civil Society*; Dann, *Vereinswesen und bürgerliche Gesellschaft*; Van Horn Melton, *Rise of the Public*; Zaunstöck and Meumann, *Sozietäten, Netzwerke, Kommunikation*; and Hardtwig, *Genossenschaft, Sekte, Verein*. In the German context, a seminal early article was Thomas Nipperdey's "Verein als soziale Struktur."
71. In Richard van Dülmen's opinion, for example, learned societies had limited oppositional potential; he considered them important during an early stage in the development of a middle-class public but saw their significance as waning thereafter. Wolfgang Hardtwig advanced a similar view. See van Dülmen, *Society of Enlightenment*; and Hardtwig, "Entwicklungstendenzen des Vereinswesens in Deutschland, 1789–1848."

NOTES TO PAGES 21-25

72. Hoffmann, *Politics of Sociability*; Düding, *Organisierter gesellschaftlicher Nationalismus*; and Dann, *Vereinsbildung und Nationsbildung*.
73. Given that recent work has treated class affiliations as negotiated cultural identities rather than as the straightforward function of social or economic relations, paying attention to the labels people claimed for themselves has taken on added importance. For overviews, see Sperber, "*Bürger, Bürgerlichkeit, Bürgerliche Gesellschaft*"; Breuilly, "Elusive Class"; and Eley and Nield, *Future of Class*.
74. On the history of this social formation, see Conze and Kocka, *Bildungsbürgertum im 19. Jahrhundert*.
75. Ulrich Engelhardt, *"Bildungsbürgertum": Begriffs- und Dogmengeschichte*, 33–96; Vierhaus, "Bildung"; and Koselleck, "Semantic Structure of *Bildung*."
76. See Hull, *Sexuality, State, and Civil Society*; and McNeely, *Emancipation of Writing*.
77. See Stewart, "Feedback Loop"; Lynn, *Popular Science and Public Opinion*; Kim, "'Public' Science"; and Morus, Schaffer, and Secord, "Scientific London."
78. See, e.g., Kanz, *Nationalismus und internationale Zusammenarbeit*; and Brock, *Justus von Liebig*.
79. On the strength of the German university tradition, and the particular social cohesion of the university educated in Germany, see Clark, *Academic Charisma*; McClelland, *State, Society, and University*; and Kocka and Mitchell, *Bourgeois Society*, especially Kocka, "European Pattern."
80. On the challenges faced by eighteenth-century cameralists, see Wakefield, *Disordered Police State*.
81. Fox, "Savant Confronts His Peers"; and Inkster and Morrell, *Metropolis and Province*.
82. This has long been recognized as an important factor in shaping German scientific institutions. See, e.g., Penny, "Wissenschaft in einer polyzentrischen Nation"; and Ben-David, *Scientist's Role in Society*.
83. Work that has looked at the 1850s has argued for a strong connection between these two programs. See Anderton, "Limits of Science"; Jurkowitz, "Liberal Unification of Science"; and Galison, "Context of Disunity," 3–4.
84. See, e.g., Green, *Fatherlands*; and Planert, "Collaboration to Resistance."
85. Broman, *German Academic Medicine*, 199; and Sheehan, *German History*, 173–74.
86. Kronick, *Scientific and Technical Periodicals*, 190–93.
87. Hilpert, *Jacob Sturm*, 8.
88. For a later example of this kind of activity, see Olesko, "Civic Culture and Calling."
89. See Retallack, *Saxony in German History*.
90. Blackbourn and Eley, *Peculiarities of German History*.

CHAPTER ONE

1. This was how it appeared for the first time in a book title. Scheuchzer, *Physica, oder Natur-Wissenschaft*. For other early eighteenth-century examples, see König, "Naturwissenschaften," 641–42.
2. Steinbach, *Deutsches Wörterbuch*.
3. See König, "Naturwissenschaften," 641.
4. See, e.g., Schäffer, *Förderung der Naturwissenschaft*; and Reuß, "Grund zu einer wohleingerichteten Oekonomie," 5–6.
5. Cited in König, "Naturwissenschaften," 644.
6. Stichweh, *Entstehung des modernen Systems wissenschaftlicher Disziplinen*.
7. Several recent works on the late eighteenth century have been interested in tracing out the integrative mechanisms and norms of modern "scientific communities" emerging around 1800. Ute Schneider's study of the *Allgemeine Deutsche Bibliothek*, for example, examines the role of this journal as an "integrating mechanism" of the republic of letters; the end point to the process, for her, is the development of specialized scientific communities at the turn of the century. To my mind, the distinction Schneider draws between "professional" readers and others seems anachronistic, given that many members of the late eighteenth-century republic of letters understood their learned activities as a form of leisure. See Ute Schneider, *Nicolais Allgemeine Deutsche Bibliothek*, 319–66. Heinrich Bosse and Oliver Hochadel posit a similar moment of clear separation between *Wissenschaft* and the public around 1800. Hochadel, *Öffentliche Wissenschaft*; and Bosse, "Das gelehrte Republik." The period around 1800 is clearly one of elite consolidation, but it is anachronistic to talk about "the scientific community" in the early nineteenth century. See Cahan, "Institutions and Communities."
8. "Naturwissenschaft," *Allgemeine Deutsche Real-Encyclopädie*, 740–46.
9. "Physik," 493–573.
10. It should be noted that this was not an exclusive identity; from the perspective of the eighteenth century, there was no reason a *Naturkenner*, a connoisseur of nature, should not know about other fields as well.
11. For an overview of this label's different uses, see Brockliss, *Calvet's Web*. For an example of the broader use of the term, see Kim, "'Public' Science."
12. On the persistence of these practices, see Brockliss, *Calvet's Web*. Brockliss contrasts his own account of the republic of letters with the work of scholars like Thomas Broman, who have used Habermas's concept of the public sphere. The concept of "the learned public" (a phrase that was ubiquitous in the late eighteenth century) suggests that it is worthwhile to think more carefully about how these two things, the learned world on the one hand and "the public" on the other, were intertwined. Particularly in the German context, it seems to me that many of Brockliss's and Broman's

insights can be meaningfully combined rather than set in opposition to each other. Broman, "Habermasian Public Sphere."
13. This argument bears a family resemblance to one made by Anthony La Vopa and Thomas Broman, who have both argued that modern forms of professional authority were built out of the public sphere, through (actually closed) communities of experts whose knowledge was (in theory) openly available. For my own purposes here, however, thinking about processes of professionalization is not very helpful, since science is still only partially "professionalized" through the entire period covered in this book. La Vopa, "Conceiving a Public"; and Broman, *German Academic Medicine*.
14. Skinner, "Language and Social Change."
15. See, e.g., Droysen, "Charakteristik der europäischen Krisis."
16. There are several shorter studies that have looked at the early intellectual history of this category. See König, "Naturwissenschaften"; Ziche, "Naturgeschichte zur Naturwissenschaft"; and Briedbach, "Transformation statt Reihung." One important interpretive tradition has identified two general trends that defined early modern natural knowledge's transformation into modern *Naturwissenschaft*: the study of nature became mathematicized and temporalized. It is hard to see how either of these trends, however, would have given the German concept its particularly unified cast; these processes affected different parts of the natural sciences (temporalization was more important in the earth and life sciences, mathematicization in the physical sciences) and do not necessarily knit these different fields together. Furthermore, neither trend was particular to German-speaking Europe. A defining work in this tradition is Lepenies, *Ende der Naturgeschichte*.
17. Karsten, *Kenntniß der Natur*, xii, ix.
18. Ibid., v; Karsten expanded on these points two years later in another work. Quote from Karsten, *Kurzer Entwurf der Naturwissenschaft*, unnumbered preface.
19. Karsten, *Kenntniß der Natur*, xvi.
20. Kant, *Metaphysical Foundations of Natural Science*. Of course, Kant also wrote extensively and influentially about the natural historical disciplines; for a summary, see Jardine, *Scenes of Inquiry*, 11–55.
21. Klopstock, *Die deutsche Gelehrtenrepublik*, 15; and Lambert, *Deutscher gelehrter Briefwechsel*, iv.
22. Ziche, "Naturgeschichte zur Naturwissenschaft."
23. On Kant's nineteenth-century reception, see, e.g., Friedman and Nordmann, *Kantian Legacy*.
24. Hufbauer, *German Chemical Community*; and Hochadel, *Öffentliche Wissenschaft*.
25. Daston, "Ethos of Enlightenment"; and Daston, "Attention and the Values of Nature in the Enlightenment." On the role of sensibility in eighteenth-century French science, see Riskin, *Age of Sensibility*.

NOTES TO PAGES 33-36

26. Foreword to *Beschäftigungen der Berlinischen Gesellschaft Naturforschender Freunde* 1 (1775): xi.
27. Daston and Galison, *Objectivity*.
28. Martini, "Entstehungsgeschichte der Gesellschaft Naturforschender Freunde in Berlin," S II, Gazelle, Zool. Mus., Historische Bild- und Schriftgutsammlung, MfN d. HUB, xvi, xvii.
29. Hufbauer, *German Chemical Community*, 5.
30. On the mid-eighteenth-century German enthusiasm for natural theology, see Clark, "Death of Metaphysics."
31. Ehrhart, *Beiträge zur Naturkunde*, 84.
32. Leysser, *Beyträge zur Naturkunde*, 5.
33. Aulich, "Anfänge der Naturforschenden Gesellschaft"; and Friedrich Heinrich Wilhelm Martini, "Anrede des bestaendigen Secretairs der Gesellschaft an die beym zweeten [sic] jaehrigen Stiftungs-Tag versammelten Mitglieder den 9ten Juli 1775," S II, Gazelle, Zool. Mus., Historische Bild-u. Schriftgutsammlungen, MfN d. HUB, xxxiv.
34. Martini, "Anrede des bestaendigen Secretairs der Gesellschaft an die beym zweeten [sic] jaehrigen Stiftungs-Tag versammelten Mitglieder den 9ten Juli 1775," MfN d. HUB, xxxvii–xxxviii.
35. Lowood, *Patriotism*.
36. Wakefield, "Apostles of Good Police," 5.
37. For a European-wide list, see Im Hof, *Das gesellige Jahrhundert*, 259–63; Roberts, "Dutch Enlightenment"; and Krueger, *Czech, German, and Noble*, 89–126. Roy Porter has argued that direct utilitarian concerns were not in fact very important in the Literary and Philosophical Societies of late eighteenth-century Britain. Porter, "Science, Provincial Culture, and Public Opinion"; cf. Thackray, "Natural Knowledge in Cultural Context." Enlightened good taste and a concern for utility are harder to tease apart in the German case, where these two ideals were often presented as intricately linked.
38. Lindenfeld, *Practical Imagination*, 14–45; Wakefield, *Disordered Police State*; Wakefield, "Police Chemistry"; and Lisbet Koerner, *Linnaeus*. If cameralism was distinctive in some ways, many of its basic concerns were more widely shared. On Britain and France, see Emma C. Spary, "Political, Natural, and Bodily Economies"; and Drayton, *Nature's Government*, chap. 3.
39. Beckmann, *Anleitung zur Technologie*.
40. Homburg, "Two Factions, One Profession"; and Lowood, *Patriotism*, 202–5.
41. For the emergence of this view of natural "riches," see Cooper, *Inventing the Indigenous*, chap. 4.
42. Lowood, *Patriotism*, 205–90.
43. For example, Flurl, *Rede von dem Einfluße der Wissenschaften*, 12, 18–20.
44. *Festschrift zum 150-jährigen Bestehen der Ökonomischen Sozietät zu Leipzig*, 7.
45. Lowood, *Patriotism*, 83–131, 291–366.

46. Gesetze der hiesigen Privatgesellschaft Naturforschender Freunde nach den Verbesserungen von 3. Mai, 1774, S II, Gazelle, Zool. Mus., Historische Bild- und Schriftgutsammlungen, MfN d. HUB, xxxvii.
47. *Plan und Gesetze nebst dem Verzeichniss der jetztlebenden Mitglieder der Gesellschaft naturforschender Freunde*, 12.
48. Schütz, *Beschreibung des Zinnstockwerks zu Altenberg*.
49. Vierhaus, "'Patriotismus'"; and Hubrig, *Die patriotischen Gesellschaften*, 65–67.
50. Martini, "Entstehungsgeschichte der Gesellschaft," xvi (see chap. 1, n. 28).
51. Foreword to *Beschäftigungen der Berlinischen Gesellschaft Naturforschender Freunde* 1 (1775): vii.
52. "Plan und Gesetze der Naturforschenden Gesellschaft in Halle," xxxvi–xxxvii. See Kleinert, "Die Naturforschende Gesellschaft zu Halle"; and Aulich, "Anfänge der Naturforschenden Gesellschaft."
53. For example, Johann Traugott Müller, *Oekonomische und Physikalische Bücherkunde*.
54. Cobres, *Büchersammlung zur Naturgeschichte*.
55. Bumann, "Wissenschaftsbegriff im deutschen Sprach- und Denkraum," 67.
56. For example: Gleditsch's *Theoretisch-praktische Geschichte*; Löwe's *Handbuch*; Baldinger's *Litterar-Geschichte*; and Suckow's *Anfangsgründe*.
57. Flurl, *Rede von dem Einfluße der Wissenschaften*, 12.
58. For example, ibid., 13.
59. Karsten, *Kenntniß der Natur*, xiv.
60. Lindenfeld, *Practical Imagination*, 14.
61. I see little in eighteenth-century German intellectual life to support Stephen Gaukroger's contention that natural philosophy had already won the exclusive right to dictate "cognitive standards" to other fields of knowledge by the late seventeenth century. See Gaukroger, *Emergence of a Scientific Culture*.
62. Kant, "Über den Gemeinspruch," 41–42.
63. Hansen, "Swiss Community Enlightenment," 148–55. On positive images of the German peasantry in this period, see Gagliardo, *Pariah to Patriot*.
64. Christian Heinrich Schmid, *Abriß der Gelehrsamkeit*, 69.
65. Bumann, "Wissenschaftsbegriff im deutschen Sprach- und Denkraum," 69.
66. Many of these defenses were inspired by Rousseau's controversial essay on the influence of the arts and sciences. The first German translation of Rousseau's *Discours sur les sciences et les arts* appeared in 1752: Rousseau, *Wiederherstellung der Wissenschaften und Künste*. Rousseau's work could still inspire polemic over fifty years later. For example, see Flurl, *Rede von dem Einfluße der Wissenschaften*, 7.
67. German historians typically see "the learned" as a social category relevant to the early modern period but do not consistently follow this

vocabulary into the later eighteenth century. See, e.g., Ulrich Engelhardt, *"Bildungsbürgertum."*
68. Schlosser, *Ueber Pedanterie*, 5–6.
69. Ibid., 19.
70. Turner, *"Bildungsbürgertum* and the Learned Professions."
71. See, e.g., Stichweh, "Universität und Öffentlichkeit."
72. Košenina, afterword to *Ueber Pedanterie*, 26.
73. Schlosser, *Ueber Pedanterie*, 18–20.
74. Similar questions about the relationship between the learned man and the wider public were also at the heart of the debates over *Popularphilosophie*. See, e.g., Altmayer, *Aufklärung als Popularphilosophie*; and Kleinhans, *Der "Philosoph" in der neueren Geschichte*.
75. William Clark's recent work largely recapitulates this older chronology, following the shift from early eighteenth-century erudition to late eighteenth-century research in the changing contours of philology dissertations. Clark, *Academic Charisma*, chap. 6.
76. Haugen, "Academic Charisma"; and Gierl, "Bestandaufnahme im gelehrten Bereich."
77. Jaumann, *"Respublica litteraria* / Republic of Letters." In this chapter, I use the term "the republic of letters," "the learned world," and "the learned public" as if they were synonyms, contravening Herbert Jaumann's warning that it is important to distinguish "the republic of letters" from other possible modes of conceptualizing scholarly communication. It would undoubtedly also be interesting to analyze the specific moments in which the communication of the learned world is styled as "republican"; for the purposes of the current study, however, such fine distinctions between different modalities of conceptualizing learned communication are of less moment. The crucial distinction here is the line late eighteenth-century Germans drew between the "learned" public and other readers and writers, and that is a complicated enough question in its own right.
78. Klopstock, *Deutsche Gelehrtenrepublik*.
79. "Ankündigung," *Kaiserlich privilegirter Reichs-Anzeiger* 1 (1794): 327.
80. *Rheinisches Magazin zur Erweiterung der Naturkunde* 1 (1793), unnumbered preface.
81. Prizelius, *Vollständige Pferdewissenschaft*, 442.
82. Kennedy, "Abhandlung von einigen in Baiern gefundenen Beinen," 8.
83. For example, Marcus Bloch asked for help from the "natural historical public" in gaining more information about a fish called the piper; Borckhausen offered his work to "the botanical public." See Bloch, *Oekonomische Naturgeschichte der Fische*, 264; and Borckhausen, "Ueber die linneischen Gattungen Crataegus," 85.
84. "Ueber den Don Karlos," 184. In criticizing Adrian Johns's arguments about the instability of print as a medium, Roger Chartier also makes the point that the evaluative mechanisms relevant to scientific and schol-

arly texts were different from those relevant to literary or artistic works. Chartier, "Printing Revolution," 404–5.
85. For a particularly strong formulation of this view, see Blanning, *Culture of Power*, esp. 111.
86. On the importance of learned societies in this period, see, e.g., Vierhaus, "Organisation wissenschaftlicher Arbeit"; Voss, "Akademien als Organisationsträger der Wissenschaften"; and Döring and Nowak, *Gelehrte Gesellschaften*.
87. Gesetze der hiesigen Privatgesellschaft, xxvii (see chap. 1, n. 46).
88. Schlosser, *Ueber Pedanterie*, 19.
89. See te Heesen, "Vom naturgeschichtlichen Investor."
90. Gesetze der hiesigen Privatgesellschaft, xxviii (see chap. 1, n. 46).
91. Ursula Klein, "Apothecary-Chemists in Eighteenth-Century Germany"; and Heilbron, *Electricity in the 17th and 18th Centuries*.
92. Müller-Wille, *Botanik und weltweiter Handel*.
93. Brockliss, *Calvet's Web*.
94. Ziche and Bornschlegell, "Wissenschaftskultur in Briefen"; and Böhme-Kassler, *Gemeinschaftsunternehmen Naturforschung*.
95. Barner, "Gelehrte Freundschaft." For an overview of the literature on literary friendship, see Meyer-Krentler, "Freundschaft im 18. Jahrhundert."
96. Friedrich Heinrich Wilhelm Martini, "Anrede des bestaendigen Secretairs der Gesellschaft, an die beym ersten jaehrigen Stiftungs-Tag versammelten Mitgleider den 9ten Juli 1774," S II, Gazelle, Zool. Mus., Historische Bild- und Schriftgutsammlungen, MfN d. HUB, lxii.
97. *Plan und Gesetze nebst dem Verzeichniss der jetzlebenden Mitglieder der Gesellschaft naturforschender Freunde*, 6.
98. See Yeo, *Encyclopaedic Visions*, esp. 57.
99. Karsten, *Kenntniß der Natur*, ii–xiii.
100. Roth, *Natur-Geschichte in Schulen*, 14.
101. Flurl, *Rede von dem Einfluße der Wissenschaften*, 4.
102. Bernoulli, "Nachricht an die Gelehrten," 292.
103. Götze, "Nachricht an das Publikum," 422.
104. Klopstock, *Deutsche Gelehrtenrepublik*, 1.
105. See, e.g., Curio, "An das Publikum," 189.
106. On the challenges the commercial book market presented for the authorial persona, see La Vopa, *Fichte*, 407–24.
107. Commodification plays a prominent role in William Clark's analysis of the construction of academic charisma. See Clark, *Academic Charisma*.
108. Reddy, "Structure of a Cultural Crisis."
109. Götze, "Nachricht an das Publikum," 420.
110. Varrentrapp and Wenner, book ad, 170–71.
111. Woodmansee, *Author, Art, and the Market*.
112. Bernoulli, "Lamberts Schriften," 290.

113. te Heesen, "Vom naturgeschichtlichen Investor"; and Müller-Wille, "Nature as a Marketplace."
114. Jablonsky, *Natursystem*.
115. Schrank, *Allgemeine Anleitung, die Naturgeschichte zu studiren*.
116. Jablonsky, *Natursystem*; and Smellie and Lichtenstein, *Philosophie der Naturgeschichte*. Richard Yeo has argued that subscriber lists functioned much as dedications to individual patrons had in earlier periods; they were a way for the author to advertise the support he enjoyed. Yeo, *Encyclopaedic Visions*, 46–49.
117. Broman, "Habermasian Public Sphere"; Broman, "Periodical Literature"; and Broman, "On the Epistemology of Criticism."
118. Clark, "Medieval Universitas Scholarium to the German Research University," 431.
119. Selwyn, *Everyday Life in the German Book Trade*.
120. Ute Schneider, *Nicolais Allgemeine Deutsche Bibliothek*.
121. In the late eighteenth century, anonymity was often a thin veil anyway; see Terrall, "Uses of Anonymity."
122. La Vopa, *Grace, Talent, and Merit*.
123. Varrentrapp and Wenner, book ad, 173.
124. Bernoulli, "Lamberts Schriften," 290.
125. *Allgemeine Literatur-Zeitung* 3–4 (1787): 276.
126. Johann Gottlob Schneider, "Vorwort," in Hunter, *Naturgeschichte der Wallfischarten*, 2.
127. Nau and Zinner, *Tabellarischer Entwurf der Naturgeschichte*, 22; and Bergmann, *Anfangsgründe der Naturgeschichte*, 268.
128. Volkmann, *Italienische Bibliothek*, unnumbered preface.
129. *Auserlesene Bibliothek der neuesten deutschen Literatur* 8 (1775): 239; and *Leipziger Magazin zur Naturkunde, Mathematik und Oekonomie* 1 (1781): 393.
130. Selwyn, *Everyday Life in the German Book Trade*, 15–28.
131. Curio, "An das Publikum," 189.
132. See Goldgar, *Impolite Learning*; and Erlin, *Berlin's Forgotten Future*, 88–90.
133. Terrall, "Frederick the Great's Berlin."
134. Sutton, *Science for a Polite Society*; and Goodman, *Republic of Letters*.
135. Schlosser, *Ueber Pedanterie*, 5.
136. Schiebinger, *The Mind Has No Sex?* This chronology seems to fit best with the French case. See Spary, "Forging Nature in the Republican Museum"; and Goodman, *Republic of Letters*. If one looks at the official state academies, it makes some sense for Germany as well. See Mommertz, "Invisible Economy of Science."
137. See Weckel, "'Der mächtige Geist der Assoziation'"; and Tolkemitt, "Beziehungsnetz der Gebildeten." For a critique of the idea that the late eighteenth century saw the appearance of clearly defined separate spheres for women and men, see Vickery, "Golden Age to Separate Spheres?"
138. Kleinert, "Naturforschende Gesellschaft zu Halle," 254.

139. On the long homosocial tradition of the university, see Noble, *World without Women*; and Van Horn Melton, *Rise of the Public in Enlightenment Europe*, 215.
140. Lowood, *Patriotism*, 77.
141. See Weckel, "Fiberfrost des Freiherrn."
142. Arenswald, *Galanterie-Mineralogie*, 3–6.
143. La Vopa, "Herder's Publikum"; and Umbach, "Culture and *Bürgerlichkeit*."
144. Blumenbach, *Handbuch der Naturgeschichte*, unnumbered preface.
145. Ehrhart, *Beiträge zur Naturkunde*, unnumbered preface.
146. Ochsenheimer, *Schmetterlinge Sachsens*, 1:8.
147. On the persistence of Latin-language periodicals in German-speaking Europe through the 1780s, see Waquet, *Latin*, 84.
148. Jentzsch, *Der deutsch-lateinische Büchermarkt*, 333.
149. Blumenbach, *Handbuch der Naturgeschichte*, unnumbered preface.
150. Ibid. On connections between history, natural philosophy, and natural history in Göttingen, see Heilbron, "Physics and Its History at Göttingen."
151. Fryer, "Jacob Christian Schäffer."
152. Schäffer, *Förderung der Naturwissenschaft*, 2.
153. Ibid.
154. This was a recurring theme in the essays in a recent edited volume on this theme. Nyhart and Broman, *Science and Civil Society*.
155. Broman, "Habermasian Public Sphere."
156. For an overview of important eighteenth-century scientific controversies, see, e.g., Hankins, *Science and the Enlightenment*.
157. Porter, "Science, Provincial Culture and Public Opinion."
158. See, e.g., Möller, *Vernunft und Kritik*; Porter, *Creation of the Modern World*; and Porter and Teich, *Enlightenment in National Context*.
159. Wakefield, *Disordered Police State*.
160. On the distinctive features of the German patriotic and economic societies, see Lowood, *Patriotism*, 10–11. This is not to claim that the useful "civic expert" was a uniquely German phenomenon; it clearly was not. See Shapin, "The Image of the Man of Science."
161. Roche, "Natural History in the Academies."
162. For an account of some of the more ambitious late eighteenth-century projects for transforming nature, see Blackbourn, *Conquest of Nature*.

CHAPTER TWO

1. Haymann, *Schriftsteller und Künstler*, 111–12, 116–17.
2. Ibid.
3. Cooper, *Inventing the Indigenous*; and Daston and Park, *Wonders and the Order of Nature*, 267–73.
4. See, e.g., Nummedal, *Alchemy and Authority*; Pamela H. Smith and Findlen, *Merchants and Marvels*; Pamela H. Smith, *Body of the Artisan*; and Pamela H. Smith, *Business of Alchemy*.

5. Mencke, *Gelehrten-Lexicon*, unnumbered preface.
6. On Dresden's expanding leisure culture, see Rosseaux, *Freiräume*. Many of the figures in the following pages had links to medicine; on consumer culture and health in this period, see Mary Lindemann, *Health and Healing*, 158–64. More generally, see Brewer and Porter, *Consumption and the World of Goods*.
7. Watanabe-O'Kelly, *Court Culture in Dresden*.
8. Turner, "*Bildungsbürgertum* and the Learned Professions."
9. Kaspar Risbeck, quoted in Haenel and Kalkschmidt, *Das alte Dresden*, 74–76.
10. Engelhardt, "*Bildungsbürgertum*," 66–70; Vierhaus, "Bildung"; and Koselleck, "Semantic Structure of *Bildung*."
11. Walker, *German Home Towns*; James Sheehan referred to the common-born members of this group as "non-noble elites." See Sheehan, *German History*, 132–43.
12. Haymann, *Schriftsteller und Künstler*, unnumbered preface.
13. Ibid., 97–152, 217–28.
14. On nobles and the universities, see McClelland, *State, Society and University*, 50–51.
15. Hansen, "Swiss Community Enlightenment."
16. Lowood, *Patriotism*, 80.
17. Haymann, *Schriftsteller und Künstler*, 138–44.
18. Hagen, *Ordinary Prussians*.
19. Haymann, *Schriftsteller und Künstler*, 123–24.
20. Schnieber, *Entwicklung des Zierpflanzenbaues*, 9–17.
21. Ibid., 18–19.
22. Ursula Klein, "Apothecary-Chemists."
23. Haymann, *Schriftsteller und Künstler*, 113.
24. Ibid., 116–17, 118–20.
25. Ibid., 119–21.
26. Haenel and Kalkschmidt, *Das alte Dresden*, 48.
27. Hansen, "Swiss Community Enlightenment," 148–49.
28. Böhme, "Die Gesellschaft Naturforschender Freunde zu Berlin."
29. Martini, "Entstehungsgeschichte der Gesellschaft," vi, xiii (see chap. 1, n. 28); and Böhme-Kassler, *Gemeinschaftsunternehmen Naturforschung*, 38–42, 91–104.
30. Martini, "Vorbericht," x.
31. Martini, "Entstehungsgeschichte der Gesellschaft," v–vi (see chap. 1, n. 28).
32. "Fortgesetzte Nachrichten von der hiesigen Gesellschaft naturforschender Freunde," *Neue Mannigfaltigkeiten* 3 (1776): 73.
33. Martini, "Entstehungsgeschichte der Gesellschaft," xxi (see chap. 2, n. 29).
34. Kleinert, "Die Naturforschende Gesellschaft zu Halle," 259.
35. te Heesen, "Vom naturgeschichtlichen Investor."

36. Kurt Becker, "Geschichte der Gesellschaft Naturforschender Freunde zu Berlin."
37. Lowood, *Patriotism*, 79.
38. Ibid., 78–79.
39. Viereck, *"Zwar sind es weibliche Hände,"* 70.
40. Schäffer, *Förderung der Naturwissenschaft*, 1–2.
41. Leysser, *Beyträge zur Naturkunde*, 8.
42. Johann Samuel Schröter, *Abhandlungen über verschiedene Gegenstände der Naturgeschichte*, vol. 2 (Halle, 1776–77), 48, quoted in te Heesen, "Vom naturgeschichtlichen Investor," 63.
43. Hufbauer, "Social Support for Chemistry," 228.
44. See Böning, "Der 'gemeine Mann' als Adressat."
45. Jenisch, *Cultur-Charakter und Cultur-Geschichte*, 131.
46. Lawrence Eliot Klein and La Vopa, *Enthusiasm and Enlightenment*.
47. La Vopa, *Grace, Talent, and Merit*.
48. Schäffer, *Förderung der Naturwissenschaft*, 19.
49. Röhling, *Deutschlands Flora*, unnumbered preface.
50. On this cameralist ideal, see Walker, *German Home Towns*.
51. Wakefield, *Disordered Police State*.
52. On the challenges that utilitarian ideals could pose for the status of learned physicians, see Broman, *Transformation of German Academic Medicine*, 67–72.
53. "Vorrede," *Neue Schriften der Gesellschaft Naturforschender Freunde Westphalens* 2 (1805): v–vi.
54. "Ist es zweckmäßig gehandelt, wenn man bey Fragen über irgend einen Zweig der Naturkunde nur lateinische Abhdandlungen als Antworten verlangt?," *Neue Schriften der Gesellschaft Naturforschender Freunde Westphalens* 2 (1805): 19.
55. Ibid.
56. Cobres, *Büchersammlung zur Naturgeschichte*, ii.
57. Ehrhart, "Verzeichniss der um Hannover wild wachsenden Pflanzen," in *Beiträge zur Naturkunde*, 1:84–85.
58. Quoted in Haymann, *Schriftsteller und Künstler*, 140.
59. "Biographie des Herrn D. Bernhard Feldmann," 510.
60. Martini, "Entstehungsgeschichte der Gesellschaft," xv (see chap. 1, n. 28).
61. Reuß, "Grund zu einer wohleingerichteten Oekonomie," 9.
62. On the late eighteenth- and early nineteenth-century education of Saxony's leading midcentury natural researchers, see Jungnickel, "Royal Saxon Society of Sciences," 35–41.
63. Leysser, *Beförderung der Naturkunde*, vol. 1, unnumbered preface.
64. In the 1760s, out of the twenty-two botanical textbooks published in German-speaking cities, nineteen were in Latin (86 percent); this percentage dropped sharply over the next decade to 50 percent. Latin's share of the botanical book market continued to decline, but not without bumps.

According to von Miltitz, *Handbuch der botanischen Literatur*, 36 percent of botanical handbooks were in Latin in the 1780s, 47 percent in 1790s, and finally, 26 percent between 1800 and 1810.

65. See, e.g., Röhling, *Deutschlands Flora*.
66. On the German reception of Linnaeus, see Stafleu, *Linnaeus and the Linnaeans*, 241–65. On Linnaeus's role in facilitating botanical communication, see Lisbet Koerner, "Carl Linnaeus in His Time and Place"; and Müller-Wille, "Walnuts in Hudson Bay, Coral Reefs in Gotland." For qualifications of the argument that new classificatory systems made communication simple and transparent, see Ritvo, *Platypus and the Mermaid*. On the development of natural systems, see Stevens, *Development of Biological Systematics*.
67. Ehrhart, "Nachricht an das Publikum, betreffend die Herausgabe meines Phytophylaciums" [1779], *Beiträge zur Naturkunde*, 70.
68. Dann, "Lesegesellschaften des 18. Jahrhunderts"; and Martino and Stützel-Prüsener, "Publikumsschichten, Lesegesellschaften und Leihbibliotheken."
69. Wittmann, *Geschichte des deutschen Buchhandels*, 121–218; and Engelsing, *Bürger als Leser*.
70. Jentzsch, *Deutsch-lateinische Büchermarkt*, 136, 333.
71. Kirchner, *Das deutsche Zeitschriftenwesen*, 1:151, 1:156; and Kronick, *Scientific and Technical Periodicals*.
72. Lorenz, "Funktionsbestimmungen von Naturlehre"; and Lorenz, "Physik im Hamburgischen Magazin."
73. See, e.g., Sloan, "Buffon-Linnaeus Controversy."
74. Guardians or parents sometimes taught children Latin in the home; this was how the Latinate women of the eighteenth century typically learned the language. See Mommertz, "Invisible Economy," 164.
75. Reuß, *Dictionarium botanicum*, 1st ed., 1781; 2nd ed., 1786. Frege, *Versuch eines allgemeinen botanischen Handwörterbuchs*; Hassler, *Lateinisch-teutsches und teutsch-lateinisches Pflanzen-Lexicon*; Borkhausen, *Botanisches Wörterbuch*; and Römer, *Versuch eines möglichst vollständigen Wörterbuchs der botanischen Terminologie*.
76. Koch, *Botanisches Handbuch*.
77. Golinski, *Science as Public Culture*; and Lynn, *Popular Science and Public Opinion*.
78. Haymann, *Schriftsteller und Künstler*, 113, 115, 116, 132.
79. Rosseaux, *Freiräume*, 198–209; Hochadel, *Öffentliche Wissenschaft*; and Rüger, "Populäre Naturwissenschaft in Nürnberg." On similar events in Britain, see Schaffer, "Natural Philosophy and Public Spectacle"; and Stewart, "Meaning for Machines."
80. Monjau, "'Auffahrt und Rettung.'"
81. See Hochadel, *Öffentliche Wissenschaft*, 302–3.
82. Rosseaux, *Freiräume*, 208–9.

83. Andrea van Dülmen, *Das irdische Paradies*, 14–23; Maurer, "Die pädagogische Reise"; and Claudia Becker, "Natürliche Erziehung."
84. Mitchell, *German Landscape Painting*, 29–30.
85. Kehn, "Schönheiten der Natur"; and Kaschuba, "Die Fußreise."
86. Andrea van Dülmen, *Das irdische Paradies*.
87. On walking habits, see Lempa, *Beyond the Gymnasium*. On urban expansion, see Engeli, "Stadterweiterungen in Deutschland."
88. Hermand, "Nationalisierung des Harzes." Most of the secondary literature on travel in this period has focused on long-distance travel. In addition to the preceding volume, see also Jäger, *Europäisches Reisen*; and Griep and Jäger, *Reisen im 18. Jahrhundert*.
89. Rosseaux, *Freiräume*, chap. 5.
90. Quoted in Jäckel, "Entdeckung der Stadtlandschaft," 23.
91. For example, Hasse, *Dresden*; Lindau, *Gegend um Dresden*; and F. D. Reichel, *Dresdens Umgebung*.
92. Kleist an Wilhemine v. Zenge, September, 3, 1800, *Heinrich von Kleist*, 103.
93. Jäckel, "Entdeckung der Stadtlandscaft"; and *Merkwürdigkeiten Dresdens*, 50–57, 193–216.
94. *Stadtlexikon Dresden*, s.v. "Plauenscher Grund"; and Julius Petzholdt, *Der Plauensche Grund*, iii–xl. On the Sächsische Schweiz, see Paul Emil Richter, *Verzeichniss selbständiger Werke*.
95. Jäckel, "Entdeckung der Stadtlandschaft," quote 18.
96. Joseph Leo Koerner, *Caspar David Friedrich*, 5–14.
97. Otto Richter, *Dresdens Umgebung*.
98. *Merkwürdigkeiten Dresdens*, 10.
99. Ibid., 52.
100. Ibid., 50–57.
101. Lustig, "Cultivating Knowledge"; Gothein, *Geschichte der Gartenkunst*, 2:415–19; and van Dülmen, *Das irdische Paradies*, 23–30.
102. Schnieber, "Entwicklung des Zierpflanzenbaues," 170–87; and Naumann, *Dresdens Gartenbau*, 41–49.
103. van Dülmen, *Das irdische Paradies*, 23–30, 110–31.
104. Mitchell, *German Landscape Painting*, 27–29.
105. Ibid., 11–29.
106. Daßdorf, *Beschreibung der vorzüglichsten Merkwürdigkeiten*, 713.
107. Wilhelm Gottlieb Becker, *Der Plauische Grund*.
108. Lang, *Beschreibung des Plauenschen Grundes*.
109. Götzinger, *Schandau und seine Umgebungen*.
110. Ibid., unnumbered introduction.
111. *Merkwürdigkeiten Dresdens*, 210, 196, 214.
112. *Merkwürdigkeiten Dresdens*, 211. On the mechanics of natural history collecting, see Larson, "Equipment for the Field."
113. Haymann, *Schriftsteller und Künstler*, 133.
114. *Merkwurdigkeiten Dresdens*, 54.

115. Wilhelm Gottlieb Becker, *Der Plauische Grund*, xi–xii.
116. August Harzer, "Kleine Fußwanderungen," 35 (1912): 73–74.
117. Ibid., 76. On the promotion of clover, see Achilles, *Deutsche Agrargeschichte*, 57–58.
118. August Harzer, "Kleine Fußwanderungen," 35 (1912): 74. On the basalt controversy, see Rudwick, "Minerals, Strata, and Fossils," 270.
119. August Harzer, "Kleine Fußwanderungen," 35 (1912): 75, 77, 76, 89.
120. Ibid., 125.
121. August Harzer, "Kleine Fußwanderungen," 35 (1912): 101, and 37 (1914): 207.
122. August Harzer, "Kleine Fußwanderungen," 37 (1914): 208.
123. Ficinus, *Flora*, unnumbered preface.
124. Ibid.; and Ruppius, *Flora Jenensis*.
125. Schröter, "Einige Bemerkungen für die Sammler der Papilionen," 490.
126. Ochsenheimer, *Schmetterlinge Sachsens*, 3–4.
127. Ibid.

CHAPTER THREE

1. Kleinert, "Die Naturforschende Gesellschaft zu Halle," 256.
2. These two examples are taken from *Schriften der Berlinischen Gesellschaft Naturforschender Freunde* 6 (1785): 1–71, 158–84.
3. On unity as an important strain within Romantic science, see Cunningham and Jardine, "Age of Reflection."
4. For a summary, see Olesko, introduction to *Science in Germany*.
5. For an overview of this diversity, see, e.g., Engelhardt, "Romantic *Naturforschung* around 1800"; Brain, Cohen, and Knudsen, *Hans Christian Ørsted*; and Caneva, "Physics and *Naturphilosophie*." On the role of "the general" in nineteenth-century science, see Hagner and Laubichler, *Hochsitz des Wissens*.
6. See Schaffer, "Genius in Romantic Natural Philosophy."
7. *Bulletin der Naturwissenschaftlichen Sektion der Schlesischen Gesellschaft für vaterländische Cultur*, published from 1821 to 1832.
8. Engelhardt, *Zwei Jahrhunderte Wissenschaft und Forschung*; and Pfetsch, *Entwicklung der Wissenschaftspolitik*, 252–313.
9. Parthier, *Leopoldina: Bestand und Wandel*, 26–31.
10. Broman, *German Academic Medicine*.
11. Steffens, *Grundzüge der philosophischen Naturwissenschaft*, vii.
12. Steffens, "Ueber die Bedeutung eines freien Vereins," 161.
13. Kleine-Natrop, *Das heilkundige Dresden*, 139–41.
14. On this shift, see Engelhardt, *Historisches Bewußtsein*, 152–53. Indeed, as a unifying program, *Naturphilosophie* was in considerable crisis long before the 1820s. See Broman, "University Reform in Medical Thought," 52–53.
15. Lowood, *Patriotism*, 291–366.

16. Lindenfeld, *Practical Imagination*, chap. 2 and 3; Lundgreen, *Techniker in Preussen*, 7–40; and Brose, *Politics of Technological Change*, 34–38.
17. Nipperdey, *Deutsche Geschichte*, 150.
18. Oken, *Ueber den Werth der Naturgeschichte*, 3.
19. Ibid., 4, 7.
20. Humpert, *Bibliographie der Kameralwissenschaften*.
21. Engelmann and Enslin, *Bibliotheca oeconomica*.
22. Laurop, *Handbuch der Forst- und Jagdliteratur*.
23. Dierbach, *Repertorium botanicum*.
24. Lindenfeld, *Practical Imagination*, 46–141; Brose, *Politics of Technological Change*, 34–38; and Schmiel, "Landwirtschaftliches Bildungswesen," 306–10.
25. On state support for the agricultural sciences in this period, see Haushofer, *Die deutsche Landwirtschaft*, 15–38; and Klemm, *Agrarwissenschaften in Deutschland*, 52–143.
26. *Denkschriften der Königlichen Akademie der Wissenschaften zu München* 1 (1809): iv.
27. *Neue Statuten der Königlich Preußischen Akademie gemeinnütziger Wissenschaften zu Erfurt*, 1.
28. Ibid., 2.
29. "Verzeichniss der Mitglieder der Königl. Academie gemeinnütziger Wissenschaften zu Erfurt," *Bericht über die Arbeiten der Academie gemeinnütziger Wissenschaften zu Erfurt* (Erfurt: August Stenger, [1847]), 7–8.
30. On Saxon officials' long-term power struggle with the burghers of Leipzig, see Beachy, *Soul of Commerce*.
31. *Festschrift zum 150-jährigen Bestehen der Ökonomischen Gesellschaft*, 27–34, 106; and am Ende, *Die Oekonomische Gesellschaft im Königreich Sachsen*, 14–15.
32. For a list of the officers, see *Schriften und Verhandlungen der ökonomischen Gesellschaft im Königreich Sachsen* 1 (1818): 10–11. On the reform and conservative factions in the Saxon administration, see Gross, *Die bürgerliche Agrarreform*, 56–103.
33. Groß, *Geschichte Sachsens*, 187–93.
34. *Festschrift zum 150-jährigen Bestehen der Ökonomischen Gesellschaft*, 106.
35. "Statuten der ökonomischen Gesellschaft," *Schriften und Verhandlungen der ökonomischen Gesellschaft im Königreich Sachsen* 1 (1818): 3–4.
36. "Bekanntmachung der Gesellschaft," *Schriften und Verhandlungen der ökonomischen Gesellschaft im Königreich Sachsen* 1 (1818): 55–56.
37. Lowood, *Patriotism*, 80.
38. "Verzeichniß sämmtlicher ordentlichen- und Ehrenmitglieder der ökonomischen Gesellschaft im Königreich Sachsen," *Schriften und Verhandlungen der ökonomischen Gesellschaft im Königreich Sachsen* 1 (1818): 10–48.
39. Zaunstöck, *Sozietätslandschaft und Mitgliederstrukturen*, 165.
40. *Physikalische-Oekonomische Bibliothek* 1 (1770), unnumbered preface.

41. Oken, *Naturgeschichte*.
42. Many studies of social categorization in this period convey the impression that "the learned" had largely disappeared as an important social referent by the early nineteenth century. See, e.g., Ulrich Engelhardt, *"Bildungsbürgertum,"* 64–114.
43. Oken, *Naturgeschichte*, 16.
44. Whaley, "Transformation of *Aufklärung*."
45. For example, Lenz, *Geschichte der Universität zu Berlin*. In the 1970s, R. Steven Turner buttressed this intellectual argument with a sociological one, analyzing the spread of *Wissenschaftsideologie* as a tool of professionalization. Neohumanist philology, he argued, became the model of a modern professional "research discipline" in the early nineteenth century, and the natural sciences then later followed its lead. Turner, "Prussian Universities and the Research Imperative."
46. McClelland, *State, Society, and University*.
47. For example, see Olesko, *Physics as a Calling*.
48. Wehler, *Deutsche Gesellschaftsgeschichte*, 2:506.
49. Baumgarten, *Professoren und Universitäten*; and Paletschek, "Verbreitete sich ein Humboldt'sches Modell?"
50. Thomas Broman has argued that the ideals of *Wissenschaft* and *Bildung* gave university-educated physicians new opportunities for building reputations in the public sphere; these ideals also spoke to learned physicians' misgivings about the implication of Enlightenment utilitarianism for their status. It seems worthwhile to expand this argument to other kinds of *Naturforscher* as well. Broman, *German Academic Medicine*, 67–101.
51. For summaries of this position, see Schindling, *Bildung und Wissenschaft*, 69; and Turner, "German Science, German Universities."
52. Schleiermacher, "Universitäten in deutschem Sinn," 235; and Steffens, "Ueber das Verhältniß unserer Gesellschaft," 139.
53. Ziolkowski, *German Romanticism*, 218–308; Shaffer, "Romantic Philosophy"; and Anrich, *Idee der deutschen Universität*.
54. Jacobi, *Ueber gelehrte Gesellschaften, ihren Geist und Zweck*, 14.
55. H. Otto Sibum has argued that these issues were especially acute for experimentalists in the early nineteenth century, given that their close relationship to instruments and technical invention made them particularly vulnerable to artisanal associations. Sibum, "Experimentalists in the Republic of Letters." Many of the figures examined in this chapter, however, came from the natural historical disciplines, and they expressed similar anxieties; the unstable status of learned natural researchers seems to have been a problem felt across the entire range of early nineteenth-century scientific disciplines.
56. Wucherer, *Naturlehre*, 29.
57. Jacobi, *Ueber gelehrte Gesellschaften*, 11–12; similarly, see Carus, *Bearbeitung der Naturwissenschaften*, 9.

NOTES TO PAGES 100–107

58. Wucherer, *Naturlehre*, 7.
59. Lindner, "Die Linnéische Sozietät zu Leipzig." The final volume of Jena's *Nachricht von dem Fortgange der Naturforschenden Gesellschaft zu Jena* was published in 1802.
60. Hansen, "Swiss Community Enlightenment."
61. Böhme, "Die Gesellschaft Naturforschender Freunde."
62. H. St. a 2481, Acta, die zu Dresden zusammengetretene Gesellschaft für Natur- und Heilkunde betr., Geheimes Kabinett, Sozietäten, SHSA.
63. "Verzeichniss der Mitglieder der naturforschenden Gesellschaft."
64. *Personalbestand und Büchersammlung der Gesellschaft für Natur- und Heilkunde*. This source organized the membership according to the year that members joined, and it also lists members who had died by 1841, so it gives an historical overview of the membership in addition to its state in 1841.
65. Ficinus, *Flora*, vol. 1, preface.
66. Schnieber, "Entwicklung des Zierpflanzenbaues," 170–87; and Naumann, *Dresdens Gartenbau*.
67. "Johann Heinrich Seidel und seine Pflanzen."
68. *Personalbestand und Büchersammlung der Gesellschaft für Natur- und Heilkunde*, 11, 12; and "Verzeichniss der Mitglieder der Naturforschenden Gesellschaft," 11.
69. Jackson, *Spectrum of Belief*, 88–97, quote 91.
70. Sakurai, "Science, Identity, and Urban Reinvention," chap. 1.
71. On the Leipzig commercial elite, see Beachy, *Soul of Commerce*.
72. "Verzeichniss der Mitglieder der naturforschenden Gesellschaft." On the other dense social connections among the commercial and learned elites of Leipzig, see Beachy, "Club Culture and Social Authority."
73. Jacobi, *Ueber gelehrte Gesellschaften*, 6–7.
74. Oken, *Naturgeschichte*, 7.
75. Oken, *Isis* 1 (1817): 6.
76. H. St. a 2481, Acta, die zu Dresden zusammengetretene Gesellschaft für Natur- und Heilkunde betr., Geheimes Kabinett, Sozietäten, SHSA, Dresden.
77. Ibid.
78. Kirchner, *Das deutsche Zeitschriftenwesen*, 40.
79. Meinel, "Die wissenschaftliche Fachzeitschrift."
80. For an example, see Brain and Wise, "Muscles and Engines."
81. For example, the aforementioned Laurop, *Forst- und Jagdliteratur*.
82. Dierbach, *Repertorium botanicum*; and Krüger, *Bibliographia botanica*.
83. Starke, "Von der Residenzstadt zum Industriezentrum," 3.
84. On the cultural significance of water cures, see Blackbourn, "Fashionable Spa Towns"; and Stewart, "Culture of the Water Cure."
85. Flechsig, *Die Struve'schen Mineralwässer*, 11–14.

86. Annalen der Gesellschaft für Natur- und Heilkunde, 1818–29, Gesellschaft für Natur- und Heilkunde, SHSA, Dresden.
87. Ibid.
88. The protocol from the directorial meeting preceding Struve's presentation includes a note to the effect that Struve ought to receive a special invitation to present his work to the society. Annalen der Gesellschaft für Natur- und Heilkunde, 1818–29, Gesellschaft für Natur- und Heilkunde, SHSA, Dresden.
89. Ibid.
90. Struve, *Nachbildung der natürlichen Heilquellen*; and Kreyssig, *De l'usage des eaux minérales*.
91. von Ammon, *Brunnendiätetik*.
92. Marchand, *Down from Olympus*; Grafton, "*Polyhistor* into *Philolog*"; La Vopa, "Specialists against Specialization"; and O'Boyle, "Klassische Bildung und soziale Struktur."
93. For an example, take the following passage from William Clark: "Up to the 1830s, and perhaps up to 1848 or even later, natural scientists in the German lands had to play on the social-cultural stage erected by philologists and philosophers, by Romantics and Idealists." Clark, *Academic Charisma*, 446. *Naturforscher* themselves actually worked quite hard to build the stage to which Clark refers.
94. *Neue Statuten der Königlich Preußischen Akademie gemeinnütziger Wissenschaft zu Erfurt*, 6.
95. Berg and Parthier, "Die 'kaiserliche' Leopoldina," 47.
96. Grulich, *Akademie der Naturforscher*, 120–85.
97. See Turner, "Social Patterns of German Science"; Houghton, *Scientific Periodicals*; and Pörksen, "Übergang vom Gelehrtenlatein." For the declining place of Latin in the German book market over the eighteenth century, see Jentzsch, *Der deutsch-lateinische Büchermarkt*.
98. Waquet, *Latin*, 80–99.
99. Ibid., 95–96.
100. Ijsewijn, *Companion to Neo-Latin Studies*, 1:190–91; and Ijsewijn and Sacré, *Companion to Neo-Latin Studies*, 2:345.
101. Pörksen, "Übergang vom Gelehrtenlatein," 246; and Felix Klein, "Gauss' wissenschaftliches Tagebuch."
102. Snelders, "Oersted's Discovery of Electromagnetism," 228.
103. For example, Schultes and Römer, *Systema vegetabilium*; and Candolle, *Regni vegetabilis systema naturale*.
104. *Journal für Chemie und Physik* 33 (1821): 1.
105. Ibid.
106. Reichenbach, *Catechismus der Botanik*, 20.
107. Mössler, *Handbuch der Gewächskunde*, x.
108. Jahn and Krausse, *Geschichte der Biologie*, 905.

109. A. F. Schweigger's collaboration with Bekker was mentioned by J. S. C. Schweigger in *Journal für Chemie und Physik* 33 (1821): 2.
110. Schweigger, *Einleitung in die Mythologie*.
111. The Schweigger brothers, for example, received early training in classical languages from their father. See Martius, "Johann Salomo Christoph Schweigger," 348.

CHAPTER FOUR

1. "Naturwissenschaft," *Allgemeine Deutsche Real-Encyclopädie*, 740–46.
2. See, e.g., Richards, "Nature Is the Poetry of Mind"; Beiser, *German Idealism*; Gregory, "Kant's Influence on Natural Scientists"; Lenoir, *Strategy of Life*; and Kamphausen and Schnelle, *Romantik als naturwissenschaftliche Bewegung*, chap. 1.
3. Breidbach, "Transformation statt Reihung."
4. Dettelbach, "Humboldtian Science."
5. Strickland, "Galvanic Disciplines"; and Knight, *Age of Science*, 56.
6. On the Romantic era's aesthetic and ontological commitment to the unity of nature, see Mitchell, *German Landscape Painting*, chap. 2 and 3. On the *Naturphilosophen* and other strands of Romantic science, see Jardine, "*Naturphilosophie* and the Kingdoms of Nature"; Beiser, "Kant and *Naturphilosophie*"; and Dietrich von Engelhardt, "Romanticism in Germany." The ambitious philosophical projects of this generation were once maligned as dangerously speculative diversions, but more recent historical research has recognized their importance to later nineteenth-century science. See, for example, Richards, *Romantic Conception of Life*; and Cunningham and Jardine, "Age of Reflexion."
7. Wucherer, *Naturlehre*.
8. Carus, *Bearbeitung der Naturwissenschaften*.
9. Link, *Ideen zu einer philosophischen Naturkunde*; and Steffens, *Grundzüge der philosophischen Naturwissenschaft*.
10. On *Bildung*, see Vierhaus, "Bildung"; and Koselleck, "Semantic Structure of *Bildung*."
11. Carus, "Von den Naturreichen."
12. See, e.g., Stichweh, *Entstehung des modernen Systems wissenschaftlicher Disziplinen*; and Daum, "*Wissenschaft* and Knowledge."
13. Martini, "Entstehungsgeschichte der Gesellschaft," vi (see chap. 1, n. 28).
14. *Beschäftigungen der Berlinischen Gesellschaft Naturforschender Freunde* 1 (1775): vii.
15. Olesko, *Physics as a Calling*; and Schubring, "Bonn Natural Science Seminar."
16. For a typical formulation of the argument that specialized societies can be used as an index of the functional differentiation of the sciences, see Hardtwig, "Verweinswesen in Deutschland," 15.

17. Böhme-Kaßler, *Gemeinschaftsunternehmen Naturforschung*.
18. Lorenz Oken, "Vorlesungen: Dresdens Winterhalbjahre, 1816–1817," *Isis* 1 (1817): 195.
19. "Vorwort," *Schriften der Gesellschaft zur Beförderung der gesammten Naturwissenschaften zu Marburg* 1 (1823): v.
20. "Gesellschafts-Nachrichten," *Abhandlungen der naturforschende Gesellschaft zu Görlitz* 1 (1827): 165–72, quotes pp. 165, 168.
21. Anderton, "Limits of Science"; Jurkowitz, "Liberal Unification of Science"; and Galison, "Context of Disunity," 3–4.
22. For example, see Jackson, *Harmonious Triads*, chap. 3.
23. Penny, *Objects of Culture*. For an older argument about the importance of competing centers to the shape of German university development, see Ben-David, *Scientist's Role in Society*.
24. Johnston, "Cultivators of Natural Science."
25. Green, *Fatherlands*.
26. Sakurai, "Science, Identity, and Urban Reinvention," chap. 1.
27. For an overview of the process of political and territorial consolidation, see Sheehan, *German History*, 235–50. On the university closings, see McClelland, *State, Society, and University*, 101–2. On the longer-term convergence of the university system with the consolidation of central places, see Blotevogel, "Kulturelle Stadtfunktionen und Urbanisierung," 146, 149.
28. On the eighteenth-century founding of the Göttingen academy, see Toellner, "Entstehung und Programm."
29. Johnston, "Cultivators of Natural Science," 202–3.
30. Vick, *Defining Germany*.
31. Johnston, "Cultivators of Natural Science," 241.
32. Daston, "Republic of Letters"; and Helmut Walser Smith, *Continuities of German History*, chap. 2.
33. Cosmopolitanism (or at least a broader sense of common European identity) remained an important value for other nationalist Romantic thinkers, too. See Scheuner, "Staatsbild und politische Form."
34. Steffens, "Ueber die Bedeutung eines freien Vereins," 161; and Jacobi, *Ueber gelehrte Gesellschaften*.
35. Immermann, *Epigonen*, 503.
36. Fritsch, Mencke, and Jöcher, *Gelehrten-Lexicon*, quote from title page; and Baader, *Das gelehrte Baiern*. In his preface, Baader lists fourteen similar guides to different regions, cities, and states in German-speaking Europe and ends the long list with "and so on," to indicate that it is not complete.
37. Wehler, *Deutsche Gesellschaftsgeschichte*, 2:524.
38. Kronick, *Scientific and Technical Periodicals*, 190–93.
39. Ibid., 88–95.
40. Schweigger, "Leopoldinisch-Carolinischen Akademie," 16.
41. Baader, *Das gelehrte Baiern*, v.
42. Gradmann, *Das gelehrte Schwaben*, unnumbered preface.

43. te Heesen, "Vom naturgechichtlichen Investor."
44. Gesetze der hiesigen Privatgesellschaft Naturforschender Freunde nach den Verbesserungen vom 3ten Mai, 1774, S II, Gazelle, Zool. Mus., Historische Bild- und Schriftgutsammlungen, MfN d. HUB; and H. St. A. 2481, Acta, die zu Dresden zusammengetretene mineralogische Gesellschaft betr. Aq. 1817, Sozietäten, Geheimes Kabinett, SHSA.
45. On correspondence networks, see Goldgar, *Impolite Learning*; and Goodman, *Republic of Letters*.
46. For a brief exploration of the continued importance of letter writing in early nineteenth-century German science, see Kanz, "Briefwechsel Goethes."
47. "Einladung," *Zeitgenossen* 1 (1816), printed on inner front cover (emphasis in the original).
48. Bücking, "Beireis," 87.
49. Ibid., 88.
50. See Hohendahl, *Geschichte der deutschen Literaturkritik*; and Rowland and Fink, *Eighteenth-Century German Book Review*. On the importance of the institution of *Kritik* in constituting the public authority of the sciences, see Broman, "Epistemology of Criticism."
51. Kirchner, *Das deutsche Zeitschriftenwesen*, 201–4.
52. Lorenz Oken, *Isis* 1 (1817): 1–2.
53. Broman, "Epistemology of Criticism."
54. My reading of Oken's views on authorship differs somewhat from the one offered by Helmut Müller-Sievers, who has argued that the *Isis* ascribed to a depersonalized view of authorship, considering it as nothing more than timed publication. I agree with Müller-Sievers's claim that Oken's primary concern in crafting his editorial policy was to create a forum of free circulation that resisted anything that resembled censorship, but I think that a closer look at Oken's own critical practice and his projects of social organizing suggests that he was also concerned with preserving a more robust link between the author's personality and his work, albeit according to a different logic than Goethe. See Müller-Sievers, "Skullduggery."
55. Oken, "Link mit Sprengel."
56. Lorenz Oken, "Ankündigung: Selbst-Recension [gilt nicht], Bitte an Botaniker . . . ," *Isis* 1 (1817): 460.
57. Goldgar, *Impolite Learning*; and Turner, "Historicism, *Kritik*, and the Prussian Professoriate," 453–57. On eighteenth-century French solutions to analogous problems, see Goodman, *Republic of Letters*.
58. Goldfuß, "Die in Vorschlag gebrachte Versammlung," 41–42.
59. Oken, *Isis* 1 (1817): 4.
60. Fichte, "Deduzierter Plan," vii–ix. On the role of embodied oral authority in the modern research university, see Clark, *Academic Charisma*.
61. Goldfuß, "Die in Vorschlag gebrachte Versammlung," 42.
62. Lorenz Oken, "Eine Versammlung deutscher Naturforscher," *Isis* (1821): 40.
63. On Oken's politics, see Breidbach, Fliedner, and Ries, *Lorenz Oken*.

64. Oken, *Isis* (1821), inside the covers of nos. 7 and 8 of this year.
65. On the reciprocal relationship between local societies and the GDNA, see Daum, *Wissenschaftspopularisierung*, 119–25.
66. Steffens, "Ueber die Bedeutung eines freien Vereins," 155.
67. Ibid., 162, 164.
68. Jacobi, *Ueber gelehrte Gesellschaften*, 17.
69. Schleiermacher, "Universitäten im deutschen Sinn."
70. See, e.g., Gordin, "Importation of Being Earnest"; Daston and Park, *Wonders*, 241–46; Biagioli, "Etiquette, Interdependence, and Sociability"; Shapin, *Social History of Truth*; Shapin and Schaffer, *Leviathan and the Air-Pump*; and Dear, "*Totius in verba*." For the late eighteenth and nineteenth centuries, the best-developed work along these lines is on England and Scotland. For an overview, see Spary, "Identity Problems"; and Cahan, "Institutions and Communities."
71. H. St. A. 2481. Acta, die zu Dresden zusammengetretene mineralogische Gesellschaft betr. Aq. 1817, Sozietäten, Geheimes Kabinett, SHSA.
72. H. St. a 2481, Acta, die zu Dresden zusammengetretene Gesellschaft für Natur- und Heilkunde betr. Aq. 1820, Sozietäten, Geheimes Kabinett, SHSA.
73. H. St. A. 2481, Acta, die zu Dresden zusammengetretene mineralogische Gesellschaft betr. Aq. 1817, Sozietäten, Geheimes Kabinett, SHSA.
74. H. St. a 2481, Acta, die zu Dresden zusammengetretene Gesellschaft für Natur- und Heilkunde betr. Aq. 1820, Sozietäten, Geheimes Kabinett, SHSA.
75. "Königliche Bestätigung der Gesellschaft und ihrer Statuten," *Schriften der Naturforschenden Gesellschaft zu Leipzig* 1 (1818), unnumbered page.
76. Daston and Park, *Wonders*, 241–46; Shapin, *Social History*; and Shapin and Schaeffer, *Leviathan and the Air-Pump*.
77. Schweigger, "Leopoldinisch-Carolinischen Akademie," 15.
78. Trumpler, "Verification and Variation."
79. Statuten des Vereines für Natur- und Heilkunde im Voigtlande, 1831, Beilage zu den Annalen, vol. 2, 1828–37, Gesellschaft für Natur- und Heilkunde, SHSA.
80. H. St. a 2481, Acta, die zu Dresden zusammengetretene Gesellschaft für Natur- und Heilkunde betr. Aq. 1820, Sozietäten, Geheimes Kabinett, SHSA.
81. Schweigger, "Leopoldinisch-Carolinischen Akademie," 16.
82. Statuten, Beilagen zu den Annalen, vol. 1, Gesellschaft für Natur- und Heilkunde, SHSA.
83. H. St. a 2481, Acta, die zu Dresden zusammengetretene Gesellschaft für Natur- und Heilkunde betr. Aq. 1820, Sozietäten, Geheimes Kabinett, SHSA.
84. 15. Juni, 1819, Annalen der Gesellschaft für Natur- und Heilkunde, 1818–29, Gesellschaft für Natur- und Heilkunde, SHSA.

85. 15. October, 1828, Annalen der Gesellschaft für Natur- und Heilkunde, 1818–29, Gesellschaft für Natur- und Heilkunde, SHSA.
86. Jahn, "Gesellschaft Naturforschender Freunde."
87. Kronick, *History of Scientific Periodicals*, 118–70.
88. "Vorwort," *Schriften der Gesellschaft zur Beförderung der gesammten Naturwissenschaften zu Marburg* 1 (1823): vii.
89. For a bibliography of society publications, see Johannes Müller, *Die wissenschaftlichen Vereine*.
90. Dierbach, *Repertorium botanicum*, title page; and *Nachricht von dem Leben und Wirken Dr. Burkhard Wilhelm Seiler*, 13–14.
91. On the shift in publishing patterns in this period, see Turner, "Social Patterns of German Science."
92. "Vorrede," *Magazin für die neusten Entdeckungen in der gesammten Naturkunde* 1 (1807): iii–iv.
93. Nyhart, "Writing Zoologically."
94. H. St. a 2481, Acta, die zu Dresden zusammengetretene Gesellschaft für Natur- und Heilkunde betr. Aq. 1820, Sozietäten, Geheimes Kabinett, SHSA.
95. "Auszüge aus den Protokollen der Sitzungen der naturforschenden Gesellschaft zu Leipzig," *Schriften der Naturforschenden Gesellschaft zu Leipzig* 1 (1822): 226, 229.
96. "Königliche Bestätigung," *Schriften der Naturforschenden Gesellschaft zu Leipzig*, 4.
97. "Auszüge," *Schriften der Naturforschenden Gesellschaft zu Leipzig*, 211, 214, 227, 228, 214.
98. Ibid., 213, 224.
99. Ibid., 211, 218, 220, 226.
100. On the role of seventeenth-century societies in the constitution of credit and authorship, see Biagioli, "Etiquette, Interdependence, and Sociability." On scientific authorship in the early modern period, see Biagioli and Galison, *Scientific Authorship*, pt. 1.
101. See Goldgar, *Impolite Learning*; and te Heesen, "Vom naturgeschichtlichen Investor."
102. *Merkwürdigkeiten Dresdens*, 79.
103. See, e.g., Beachy, "Club Culture and Social Authority"; Stegmann, *Harmonie zu Dresden*, 39. Two members of the Gesellschaft für Natur- und Heilkunde, an Oberkonsistorialrat Scheider and Heinrich Ficinus, are mentioned in the society's history, pp. 33 and 37, respectively. Kranke, *Geschichte der Freimaurerei*, includes membership lists for two Dresden Freemason lodges, the Loge zu den drei Schwerten and the Loge zum goldenen Apfel.
104. See, e.g. Inkster and Morrell, *Metropolis and Province*; Fox, "Savant Confronts His Peers"; and Lynn, *Popular Science and Public Opinion*.

105. For example, Nipperdey, "Verein als soziale Struktur," 177–78.
106. Jardine, *Scenes of Inquiry*.

CHAPTER FIVE

1. Carus, "Von den Naturreichen."
2. See, e.g., Winter, "Between Louis and Ludwig"; and Barclay, *Frederick William IV*.
3. Cf. Sakurai, "Science, Identity, and Urban Reinvention," chap. 1.
4. See, e.g., Clark, *Academic Charisma*. Clark's descriptions of German (and especially Prussian) bureaucracies as machinelike and relentlessly rationalizing also does not mesh very well with more recent research, which has emphasized the continued importance of personal alliances and patronage relationships. See, e.g., Brewer and Hellmuth, eds. *Rethinking Leviathan*; and Brose, *Politics of Technological Change*. For examples of the importance of personal patronage in early nineteenth-century Oriental scholarship, see Marchand, *German Orientalism*, 95–97. On Alexander von Humboldt's importance as a patron in Prussia, see Barclay, "Mäzenatentum am Preußischen Hof."
5. On Goethe's status in literary critical discourse in this period, see Berghahn, "Von der klassizistischen zur klassischen Literatur."
6. "Versammlung der Flora am 28. August 1829," *Mittheilungen aus dem Gebiete der Flora und Pomona* 1 (1829): 57.
7. In the secondary literature on early nineteenth-century science, the label "Romantic" has typically been used as a broad period term; that is also how I am using it here. This usage unfortunately tends to obscure an aesthetic label that early nineteenth-century figures often would have found relevant—"classicism"—and I have also used this designation when it seemed appropriate, such as when a tie to Goethe and Schiller was particularly obvious. On Romanticism and *Naturphilosophie*, see Engelhardt, "Naturgefühl, Naturwissenschaft, und Naturphilosophie"; Poggi and Bossi, *Romanticism in Science*; and Richards, *Romantic Conception of Life*.
8. Bödeker, "Die 'gebildeten Stände'"; Engelsing, *Bürger als Leser*, 259–76; and Bollenbeck, *Bildung und Kultur*, 148–59. On salon culture, see Seibert, *Der literarische Salon*; and Hertz, *Jewish High Society*.
9. On analogous concerns in Britain, see Secord, "Botany on a Plate."
10. Kanz, "Briefwechsel Goethes," 206.
11. Rudolphi, *Gemälde weiblicher Erziehung*, 12, 16–17, 48–49, 224–25, 276, quote 63.
12. See, e.g., Schiebinger, *Mind Has No Sex?* On eighteenth-century philosophical discussions of sex and gender, see Jordanova, "Sex and Gender."
13. Reill, *Vitalizing Nature*, 220–36.

14. Harleß, *Verdienste*, unnumbered dedication. On the position of noblewomen in the early nineteenth century, see Paletschek, "Adelige und bürgerliche Frauen."
15. Harleß, *Verdienste*, 1.
16. On the different qualities commonly ascribed to the sexes in the nineteenth century, see Hausen, "Polarisierung der 'Geschlechtscharaktere'"; and Frevert, *Mann und Weib*.
17. Harleß, *Verdienste*, 1–2.
18. Daston, "Quantifizierung der weiblichen Intelligenz"; and Schiebinger, "Gender and Natural History."
19. Harleß, *Verdienste*, 2.
20. These inclusions were not unprecedented; as part of *Oekonomie* broadly conceived, household management could be placed within the broad tent of natural research as the eighteenth century had defined it. In addition, in the early modern period, cookery had links to medicine, given that many remedies were prepared in the home. The cameralist literature was also important, however, in promoting new gender dichotomies. See Gray, *Productive Men, Reproductive Women*. On women's ties to medical cookery and pharmacology in the early modern period, see Schiebinger, *Mind Has No Sex?*, chap. 4.
21. Harleß, *Verdienste*, 256.
22. Herminghouse, "Women and the Literary Enterprise," 79; Tebben, *Beruf—Schriftstellerin*; and Weckel, *Zwischen Häuslichkeit und Öffentlichkeit*.
23. Harleß, *Verdienste*, 295.
24. Ibid.
25. The Cotta–Nees von Esenbeck correspondence has been published in Dorothea Kuhn, "Nees von Esenbeck."
26. Kirchner, *Das deutsche Zeitschriftenwesen*, 1:260–61.
27. Johann Cotta to Goethe, in Dorothea Kuhn, "Nees von Esenbeck," 70.
28. Nees von Esenbeck to Goethe, July 26, 1816, in Nickol, *Johann Wolfgang von Goethe und Christian Gottfried Daniel Nees von Esenbeck*, 8–9.
29. Nees von Esenbeck to Cotta, August 16,1816, in Dorothea Kuhn, "Nees von Esenbeck," 73.
30. Kirchner, *Das deutsche Zeitschriftenwesen*, 1:261.
31. Nees von Esenbeck, "Von der Metamorphose."
32. Engelhardt, *Historisches Bewußtsein*, 103–58.
33. Huber to Cotta, April 29, 1817, in Dorothea Kuhn, "Nees von Esenbeck," 76.
34. Huber to Cotta, December 30, 1817, in Dorothea Kuhn, "Nees von Esenbeck," 81.
35. Huber to Cotta, February 20, 1818, in Dorothea Kuhn, "Nees von Esenbeck," 83.
36. Huber to Cotta, August, 1818, in Dorothea Kuhn, "Nees von Esenbeck," 89.

37. Huber to Cotta, February 20, 1818, in Dorothea Kuhn, "Nees von Esenbeck," 83.
38. Ibid.
39. Huber to Nees von Esenbeck, December 18, 1817, in Dorothea Kuhn, "Nees von Esenbeck," 78.
40. See, for example, Dettelbach, "Humboldtian Science"; and Dettelbach, "Face of Nature."
41. Broman, *German Academic Medicine*, 90–96.
42. Nees von Esenbeck, "Von der Metamorphose," 996.
43. Nees von Esenbeck to Redaktion (Th. Huber), December 27, 1817, in Dorothea Kuhn, "Nees von Esenbeck," 79.
44. Nees von Esenbeck to Redaktion (Th. Huber), June 11, 1817, in Dorothea Kuhn, "Nees von Esenbeck," 76.
45. Nees von Esenbeck to Redaktion (Th. Huber), December 27, 1817, in Dorothea Kuhn, "Nees von Esenbeck," 79.
46. Huber to Nees von Esenbeck, February 20, 1818, in Dorothea Kuhn, "Nees von Esenbeck," 84.
47. Nees von Esenbeck to Redaktion (Th. Huber), March 1, 1818, in Dorothea Kuhn, "Nees von Esenbeck," 84.
48. Nees von Esenbeck to Redaktion (Th. Huber), May 6, 1818, in Dorothea Kuhn, "Nees von Esenbeck," 87.
49. Goldfuß to Cotta, August 16, 1818, in Dorothea Kuhn, "Nees von Esenbeck," 88.
50. Huber to Cotta, August, 1818, in Dorothea Kuhn, "Nees von Esenbeck," 89.
51. Other examples include Dierbach, *Anleitung zum Studium*; and Candolles and Sprengel, *Grundzüge der wissenschaftlichen Pflanzenkunde*.
52. "Nekrolog [H. G. Ludwig Reichenbach]," 1–3.
53. Reichenbach and Mössler, *Mösslers Handbuch*, 1:xiv; and Reichenbach, *Catechismus der Botanik*.
54. Reichenbach and Mössler, *Mösslers Handbuch*, 1:xv.
55. Advertisement included in *Mösslers Handbuch*, vol. 3.
56. Reichenbach, *Iconographia botanica exotica*; and Reichenbach, *Iconographia botanica*.
57. Lisbet Koerner, *Linnaeus*; and Stevens, *Development of Biological Systematics*.
58. Eisnerova, "Botanische Disziplinen," 304–5.
59. Cf. Lisbet Koerner, "Goethe's Botany."
60. Reichenbach, *Botanik für Damen*, v.
61. Ibid., vi.
62. Ibid., 529–67, quote 552–53.
63. Ibid., 534.
64. Ibid., 535.
65. Ibid., 566–67.
66. Ibid., 302–6; and Zunck, *Die natürlichen Pflanzensysteme*.

67. On Humboldtian plant geography, see Dettelbach, "Humboldtian Science."
68. Reichenbach and Mössler, *Mösslers Handbuch*, 1:xiii.
69. Ibid., 1:xv.
70. H. G. L. Reichenbach, "Vorwort," *Magazin der aesthetischen Botanik* 1 (1822): unnumbered page.
71. Weckel, "Fiberfrost des Freiherrn."
72. Schön, "Weibliches Lesen."
73. H. G. L. Reichenbach, "Vorwort," *Magazin der aesthetischen Botanik* (1824): vii, ix.
74. Reichenbach, *Botanik für Damen*, 540.
75. Ibid., 4.
76. The names of the initial directors are listed in Dänhardt, *Festschrift*, 12.
77. *Mittheilungen aus dem Gebiete der Flora und Pomona* 1 (1829): 1.
78. "Versammlung der Flora am 28. August 1829," *Mittheilungen aus dem Gebiete der Flora und Pomona* 1 (1829): 57–60, 61–63; and Carus, *Briefe über Landschaftsmalerei*.
79. "Versammlung der Gesellschaft Flora, am 21. October 1829," *Mittheilung aus dem Gebiete der Flora und Pomona* 1 (1829): 74–76, 79; quote from Falkenstein to Reichenbach, October 22, 1829, MISC. DRESDEN V 1: Verhandlungsschriften, 1826–30, SLB.
80. MISC. DRESDEN V 1: Verhandlungsschriften, 1826–30, SLB.
81. Dänhardt, *Festschrift*, 13–20.
82. Schnieber, "Entwicklung des Zierpflanzenbaues"; and "Johann Heinrich Seidel und seine Pflanzen."
83. *Verzeichniß der Mitglieder der Flora*.
84. *Mittheilungen aus dem Gebiete der Flora und Pomona* 1 (1829): 1.
85. On the *Abendzeitung*, see Doering-Manteuffel, *Dresden und sein Geistesleben*, 9–14; Haenel and Kalkschmidt, *Das alte Dresden*, 242–53; and Kirchner, *Das deutsche Zeitschriftenwesen*, 1:263.
86. *Mittheilungen aus dem Gebiete der Flora und Pomona* 1 (1829): 1.
87. *Einheimisches, Beiblatt zur Abendzeitung* 12 (1828): 52.
88. Doering-Manteuffel, *Dresden und sein Geistesleben*, 9.
89. Reprinted in Haenel and Kalkschmidt, *Das alte Dresden*, 246.
90. *Hundert Jahre Geschichte der Arnoldischen Buchhandlung zu Dresden*.
91. Doering-Manteuffel, *Dresden und sein Geistesleben*, 25.
92. Weckel, "Lost Paradise"; Weckel, "'Der mächtige Geist der Assoziation'"; and Tolkemitt, "Beziehungsnetz der Gebildeten."
93. Haenel and Kalkschmidt, *Das alte Dresden*, 256–60; quote: Karl Falkenstein, *C. A. Tiedges Leben* (Leipzig, 1841), in Haenel and Kalkschmidt, *Das alte Dresden*, 260.
94. Jäckel, "Von den 'sächsischen Frauenzimmer.'"
95. Doering-Manteuffel, *Dresden und sein Geistesleben*, 15.
96. Weckel, "Fiberfrost des Freiherrn."

97. Ernst Scherzlieb [Wilhelm von Lüdemann], *Dresden, wie es ist* (Zwickau, 1830), excerpt reprinted in Haenel and Kalkschmidt, *Das alte Dresden*, 248–52.
98. Weckel, "Lost Paradise."
99. See Kaschuba, "German *Bürgerlichkeit*."
100. "Vorwort," *Magazin für aesthetische Botanik* 1 (1822): ix. On Englishwomen's participation in botany in this period, see Shteir, *Cultivating Women, Cultivating Science*.
101. Mscr. Dresd. v. 1: Verhandlungschriften, 1826–30, SLB.
102. *Merkwürdigkeiten Dresdens*, 55.
103. Dänhardt, *Festschrift*.
104. Flora, 1841–43, Verhandlungsschriften, SLB.
105. Friedrich Schlegel, *Kritische Fragmente*, 22.
106. On the salon as a site of literary production, see Seibert, *Literarische Salon*, 204–36.
107. Daum, *Wissenschaftspopularisierung*, 273–80.
108. Rupke, *Alexander von Humboldt*, 43–47.
109. Daum, *Wissenschaftspopularisierung*; and Richards, *Tragic Sense of Life*.
110. Schleiden, *Grundzüge der wissenschaftlichen Botanik*, 52.
111. Ibid., xv–xvi.
112. Woodmansee, *Author, Art and the Market*.

CHAPTER SIX

1. Zaunick, "Gründung und Gründer."
2. Cited in Zaunick, "Gründung und Gründer," 30.
3. Andreas Daum has pointed out the ways in which organizations like the GDNA continued to cut across the boundaries between these two worlds. See Daum, *Wissenschaftspopularisierung*, 119–37.
4. For other examples of such men, see Lynn Nyhart's work on the taxonomist Philipp Leopold Martin and several other midcentury figures: Nyhart, *Modern Nature*, chap. 2; see also Daston, "Glass Flowers."
5. Fraas, *Klima und Pflanzenwelt*, iv.
6. General natural historical or natural scientific societies were founded during this period in: Blankenburg (1831), Mannheim (1833), Dresden (1833), Mainz (1834), Bamberg (1834), Detmold (1835), Kassel (1836), Hamburg (1837 and 1844), Posen (1837), Dessau (1839), Karlsruhe (1840), Dürkheim (1840), Erfurt (1842), Darmstadt (1844), Stuttgart (1844), Bautzen (1845), Meissen (1845), Augsburg (1846), Elberfeld (1846), Güstrow (1847), Clausthal (1848), and Groß-Schönau (1849). The divide between the pre- and post-1830 period is not absolutely rigid; a few provincial centers already had societies earlier in the nineteenth century (the town of Görlitz in the Oberlausitz, for example). These groups continued to proliferate through the second half of the nineteenth century, with 101 founded in German-

speaking Europe between 1850 and 1900. Daum, *Wissenschaftspopularisierung*, 91–95.
7. Nyhart, *Modern Nature*.
8. Nyhart, "Natural History"; cf. Lepenies, *Ende der Naturgeschichte*.
9. Stevens, *Development of Biological Systematics*, 199–218, quote 218.
10. On the continued importance of the natural history tradition in post-1848 popular understandings of science, see Daum, *Wissenschaftspopularisierung*.
11. *Yearbook of Scientific and Learned Societies of Great Britain*; and Fox, "Savant Confronts His Peers."
12. "Mitglieder des Königlichen Sächsischen Alterthums-Vereins."
13. Green, *Fatherlands*.
14. Sheehan, *German History*, 139–41.
15. On the complexities of German political identities in this period, see Green, *Fatherlands*; and Vick, *Defining Germany*.
16. On later ties between representations of nature and German identities, see Lekan, *Imagining the Nation*; Schama, *Landscape and Memory*; Lekan and Zeller, *Germany's Nature*; and Blackbourn and Retallack, *Localism, Landscape and the Ambiguities of Place*, especially the essays by David Blackbourn, Thomas M. Lekan, Caitlan Murdock, and Pieter M. Judson.
17. Applegate, *Nation of Provincials*; and Confino, *Nation as a Local Metaphor*.
18. Rita Krueger shows a similar relationship between scientific activity and the formation of new kinds of Czech identity during the eighteenth century; the link between intellectual sociability, cosmopolitan standards, and political identity was of course not new in the nineteenth century. Krueger, *Czech, German, and Noble*.
19. Zaunick, "Gründung und Gründer."
20. *Personalbestand und Büchersammlung der Gesellschaft für Natur- und Heilkunde*, 4.
21. *Statuten der Isis*, 5.
22. On state building in these decades, see Green, *Fatherlands*; and Walker, *German Home Towns*, pt. 3. On educational expansion, see Jeismann and Lundgreen, *Handbuch der deutschen Bildungsgeschichte*, vol. 3.
23. Starke, "Von der Residenzstadt zum Industriezentrum," 3.
24. Mitglieder der Gesellschaft [1841]. Hist. Saxon 6 315, L8, SLB.
25. "Vorwort," *Correspondenzblatt des naturhistorischen Vereins für die preussischen Rheinlande* 1 (1843): 3.
26. "Vorwort," *Abhandlungen der Naturwissenschaftlichen Gesellschaft "Saxonia" zu Gross- und Neuschönau* 1 (1851–52): iii–vi.
27. "Verzeichniss der Mitglieder," *Verhandlungen des naturforschenden Vereines in Brünn* 1 (1862): xiii–xix.
28. See Harwood, *Technology's Dilemma*; and Gispen, *New Profession, Old Order*, chap. 1.

29. Hermann Richter, "Denkschrift der Gesellschaft Isis," and Ludwig Reichenbach, "Denkschrift der Gesellschaft für Natur- und Heilkunde," in Richter and Reichenbach, *Der naturwissenschaftliche Unterricht auf Gymnasien*, 2, 18.
30. Bescherer, *Methodik des naturwissenschaftlichen Unterrichts*.
31. Carl Traugott Sachse, "Welches sind die Aufgaben der naturhistorischen Gesellschaften und Vereine Deutschlands?" and "Jahresbericht der Isis in Dresden," *Allgemeine Deutsche Naturhistorische Zeitung* 2 (1847): 169, 212.
32. Speyer, *Thätigkeit des Vereins für Naturkunde*.
33. Nägeli, "Aufgabe der Naturgeschichte," 4–5.
34. On the program of "scientific zoology," see Nyhart, *Biology Takes Form*, 90–102. Quote from *Zeitschrift für wissenschaftliche Zoologie* 1 (1848), cited in Nyhart, *Biology Takes Form*, 94.
35. *Jahreshefte des Vereins für vaterländische Naturkunde in Württemberg* 1 (1845): 3–4.
36. Sachse, "Bericht," *Allgemeine deutsche naturhistorische Zeitung* 1 (1846): 185–86.
37. *Jahresbericht der Pollichia* 1 (1843): 9.
38. Reichel, *Pflanzen in der Umgegend von Dresden*; Reichel and Dietrich, *Darstellungen der Heilquellen*; and Reichel, *Dresdens Umgebung*.
39. te Heesen, "Vom naturgeschichtlichen Investor"; and Hamm, "Goethes Sammlungen."
40. Reichenbach, *Das naturhistorische Museum*, vii.
41. Hoffmannsegg to GNHK, Beilagen der Annalen, vol. 1: 1818–27, Gesellschaft für Natur- und Heilkunde, SHSA, Dresden.
42. Zaunick, "Gründung und Gründer," 41.
43. Sachse, "Bericht der Isis," *Allgemeine Deutsche Naturhistorische Zeitung* 2 (1847): 208–9.
44. Ibid., 209.
45. "Organische Bestimmungen des Vereins für vaterländische Naturkunde in Württemberg," *Jahreshefte des Vereins für vaterländische Naturkunde* 1 (1845): 13–14.
46. See the reports of attendance in the *Jahresbericht der Pollichia* 1 (1843) through 6 (1848).
47. Volkens, "Die Geschichte des Botanischen Vereins," 1.
48. Reichenbach, *Flora Saxonica*.
49. Sachse, "Bericht der Isis," *Allgemeine Deutsche Naturhistorische Zeitung* 2 (1847): 210–11.
50. Carl August Friedrich Harzer, *Naturgetreue Abbildungen*, 3.
51. "Vorrede," *Neueste Schriften der Naturforschenden Gesellschaft in Danzig* 1 (1843): iv.
52. Drechsler, "Kurzgefasste Geschichte," 71.

53. Phillips, "Friends of Nature."
54. On the diverse patronage networks of midcentury natural history, see Nyhart, *Modern Nature*, 35–49.
55. Carl August Friedrich Harzer, *Naturgetreue Abbildungen*, 3.
56. Rep 17, Nr 500, Journal des naturwissenschaftlichen Vereines in Halle, AUHW.
57. Fiedler, *Physikalische Gesellschaft*.
58. Rep 17, Nr 501 and Nr 503, AUHW.
59. Vorstand des Naturwissenschaftlichen Vereins an die Naturforschende Gesellschaft. 18 09/5/3: Archiv 1841–50 [Briefe], AUHW.
60. Kieferstein an die Naturforschende Gesellschaft, January 8, 1851, 09/5/3: Archiv 1841–50 [Briefe], AUHW.
61. Keferstein, *Erinnerungen*.
62. Drechler, "Kurzgefasste Geschichte."
63. Nyhart, "Civic and Economic Zoology."
64. Sakurai, "Science, Identity, and Urban Reinvention," chap. 1.
65. Hennig, *Romantik bis zur Gründerzeit*, 11–16.
66. *Jahresbericht über die königlichen Studien-Anstalten zu Bamberg*; and *Alt-Bamberg* 7 (1904–5): 39–51.
67. For descriptions of the Theresienfest, see Hennig, *Von Romantik*, 21–34. On invented traditions, see Hobsbawm and Ranger, *Invention of Tradition*.
68. Kunz, *Verortete Geschichte*, 119–32.
69. *Statuten der naturforschenden Gesellschaft zu Bamberg*.
70. Crown Prince Maximilian to the Society for Natural Research, October 24, 1842, Allgemeine Korrespondenz, 1834–60, Naturforschende Gesellschaft, SAB; and *Tagblatt der Stadt Bamberg* (1835), no. 346.
71. *Tagblatt der Stadt Bamberg* (1836), no. 89.
72. *Tagblatt der Stadt Bamberg* (1837), no. 292.
73. Ibid., no. 341.
74. *Tagblatt der Stadt Bamberg* (1840), no. 349.
75. Das Central-Comité des Theresien Volksfestes zu Bamberg an die nf. G., May 31, 1834, Allgemeine Korrespondenz, 1834–60, Naturforschende Gesellschaft, SAB; and *Tagblatt der Stadt Bamberg* (1839), no. 83.
76. Sitzungsberichte, 1833–56, Naturforschende Gesellschaft, SAB. To determine whether or not society members had been ordained, I used Wachter, *General-Personal-Schematismus*.
77. *Alt-Bamberg*, 4 (1901): 21; and *Tagblatt der Stadt Bamberg* (1836), no. 89.
78. *Alt-Bamberg*, 12 (1912–13): 177; and 2 (1918–19): 117.
79. Weis, *Säkularisation der bayerischen Klöster*.
80. Oskar Kuhn, "Naturalienkabinett," 32.
81. Seit, "Michael von Deinlein."
82. Lösch, *Adam Gengler*.
83. Wehler, *Deutsche Gesellschaftsgeschichte*, 2:476.

84. Sitzungsberichte, 1833–56, Naturforschende Gesellschaft, SAB.
85. *Tagblatt der Stadt Bamberg* (1838), no. 212; (1837), no. 289; and (1841), no. 228.
86. *Tagblatt der Stadt Bamberg* (1836), no. 290; (1837), no. 254; (1840), no. 259.
87. For example, Jäck, *Pantheon der Literaten*; and Jäck, *Leben und Werke der Künstler Bambergs*.
88. Penny, "Fashioning Local Identities."
89. Applegate, *Nation of Provincials*.
90. "Zur Geschichte des Vereins," *Jahresbericht der Pollichia* 1 (1843): 2.
91. Ibid., 9–10.
92. Ibid., 10–11.
93. *Jahresbericht der Pollichia* 6 (1848): 11–16.
94. *Jahresbericht der Pollichia* 9 (1851): 3.
95. "Prospectus," *Allgemeine deutsche naturhistorische Zeitung* 1 (1846): 2.
96. Ibid.

CHAPTER SEVEN

1. Kreutzberg, "Einfluß der Naturwissenschaften," 286. On middle-class enthusiasm for these new technologies, see Blackbourn, *Long Nineteenth Century*, 271–73.
2. Droysen, "Charakteristik der europäischen Krisis."
3. *Allgemeines Organ für Handel und Gewerbe und damit verwandte Gegenstände* 4 (1837).
4. Lips, *Nürnberg-Fürther Eisenbahn*, 11.
5. See Wengenroth, "Science, Technology and Industry." Cf. Jacob and Stewart, *Practical Matter*.
6. Langewiesche, *Liberalismus in Deutschland*, 12–60; Sheehan, *German Liberalism*, pt. 1; and Sheehan, "German States."
7. Kreutzberg, "Einfluß der Naturwissenschaften," 286.
8. Ibid., 294.
9. Freese, *Das deutsche Gymnasium*, 12.
10. "Handwörterbuch der reinen und angewandten Chemie," *Technisches Litteratur-Blatt* 1 (1838): 2.
11. For overviews, see Retallack, "Society and Politics in Saxony"; and Zeise, "Bürgerliche Umwälzung," 332–80. On Saxon liberalism, see Muhs, "Zwischen Staatsreform und politischem Protest."
12. Kiesewetter, *Industrialisierung und Landwirtschaft*.
13. Cf. Lenoir, *Instituting Science*. Lenoir has argued that the natural sciences (specifically, the organic physics of Helmholtz and others) helped forge a new ideology of "material interest" that was crafted as an alternative to the supposedly bankrupt ideal of *Bildung*; this ideology of material interest was designed to appeal to the "rising industrial bourgeoisie." Two elements of this argument are out of step with the picture that other German

historians have painted of the middle third of the nineteenth century. Other historians have not seen *Bildung* as bankrupt at midcentury—quite the opposite. Historians looking at the propertied middle classes around midcentury also generally no longer conceptualize this group as a "rising" industrial bourgeoisie. See, e.g., Bollenbeck, *Bildung und Kultur*; and Sheehan, *German History*, 509–13.

14. On the horizons of expectation characteristic of early industrial promotion, see Brose, *Politics of Technological Change*.
15. Wehler, *Deutsche Gesellschaftsgeschichte*, 2:499–504; Gispen, *New Profession, Old Order*; and Brose, *Politics of Technological Change*, chap. 3.
16. Harwood, *Technology's Dilemma*, 79.
17. Beger, *Idee des Realgymnasiums*.
18. Wehler, *Deutsche Gesellschaftsgeschichte*, 2:502–3.
19. *Statuten des Gewerbevereins in Dresden*, 1.
20. Ibid.
21. Ibid., 2.
22. Schleiden and Schmid, *Encyclopädie der gesammten theoretischen Naturwissenschaften*, 1:v–viii.
23. Geinitz, "Verzeichnis der Mitglieder."
24. See, e.g., Julius Petzholdt, *Nachricht von der Bibliothek*.
25. Alexander Petzholdt, *Galvanische Vergoldung*, 3.
26. "Professor Schubert und die Dresdner Dampfschiffahrts-Kompagnie," *Gewerbe-Blatt für Sachsen* 3 (1838): 392.
27. See Harwood, *Technology's Dilemma*.
28. Preusker, *Sophien-Ducaten*, 63.
29. Liebig, *Studium der Naturwissenschaften*, 9.
30. "Mitgliederverzeichniß der Gesellschaft 'Isis' in Dresden," "Mitglieder des 'naturhistorischen Vereins' für das Erzgebirge in Schneeberg," and "Mitglieder des Vereines für Naturkunde in Meissen," *Allgemeine deutsche naturhistorische Zeitung* 1 (1846): 1–11.
31. For a study of these patterns in a later period, see Augustine, "Arriving in the Upper Class."
32. See the annual *Geschäftsbericht der Naturwissenschaftlichen Gesellschaft in Dresden* (1843–49).
33. *Gesetze der naturwissenschaftlichen Gesellschaft*.
34. Alexander Petzholdt, *Populäre Vorlesungen über Naturwissenschaft*. Lists appear in the yearly *Geschäftsbericht der Naturwissenschaftlichen Gesellschaft in Dresden*, published from 1845 to 1849.
35. *Gesetze der naturwissenschaftlichen Gesellschaft*.
36. "Mitgliederverzeichniß," *Geschäftsbericht der naturwissenschaftlichen Gesellschaft in Dresden* (Dresden: n.p., [1845]), 14–17.
37. Daum, *Wissenschaftspopularisierung*, 436–58.
38. Geyer, "Ueber das Verhältniss der Philosophie zu den Naturwissenschaften."

39. Schubert, "Ueber Beharrungsvermögen oder Trägheit der Masse."
40. Arleen Tuchman, *Science, Medicine, and the State*; and Lenoir, *Instituting Science*.
41. On the institutionalization of organic chemistry, see Borscheid, *Naturwissenschaft, Staat und Industrie*.
42. "Vorwort," *Bilder-Conversations-Lexikon*, vol. 1.
43. See, e.g., Pinkard, *German Philosophy*, 356–59; and Sheehan, *German History*, 802–3.
44. "Naturwissenschaft," *Bilder-Conversations-Lexikon*, 3:247–48.
45. See Vick, *Defining Germany*, 15–17.
46. Cited in Kocka, *Industrial Culture and Bourgeois Society*, 6.
47. Woodmansee, *Author, Art, and the Market*.
48. "Gedankenaustellungen," *Gewerbe-Blatt für Sachsen* 4 (1839): 373.
49. See Brose, *Politics of Technological Change*.
50. "Schutzworte für die Gewerbe-Verein," *Gewerbe-Blatt für Sachsen* 3 (1838): 73–74.
51. Preusker, "Gewerbegeist," 28.
52. *Technisches Litteratur-Blatt: Beilage zum Gewerbe-Blatt für Sachsen* (1838): 12.
53. *Gewerbe-Blatt für Sachsen* 4 (1839): 283.
54. "Grotesken," *Gewerbe-Blatt für Sachsen* 4 (1839): 70.
55. Beger, *Idee des Realgymnasiums*.
56. "Ein Blick in unser industrieles Leben," *Mittheilungen für das Erzgebirge und Voigtland* 1 (1836): 291.
57. Engels, *Dialektik der Natur*, 195.
58. *Mittheilungen für das Erzgebirge und Voigtland* 4 (1839): 284.
59. Preusker, *Bürgerhalle*, inside cover.
60. Cf. Lenoir, *Instituting Science*.
61. Beger, *Idee des Realgymnasiums*, 14.
62. Preusker, *Gutenberg und Franklin*, 16.
63. Brophy, "Public Sphere," 191; and Wittmann, *Geschichte des deutschen Buchhandels*, 218–56.
64. "Ueber die wahre Zentralization der Kräften und Interessen," *Gewerbe-Blatt für Sachsen* 4 (1839): 268.
65. See, for example, "Die Eröffnung der Gewerbeschule zu Plauen, nebst einem Worte über die dasige Sonntagsschule," *Mittheilungen für das Erzgebirge und Voigtland* 1 (1836): 165.
66. Clauss, *Chronik des Gewerbe-Vereins*, 7–11.
67. For an examination of the ways in which the writing practices associated with the state helped constitute civil society in an earlier period, see McNeely, *Emancipation of Writing*.
68. "Über die Einfluss der Wissenschaften auf die Gewerbe," *Mittheilungen für das Erzgebirge und Voigtland* 1 (1836): 372.
69. "Bücherscheu der Gewerbsleute," *Gewerbe-Blatt für Sachsen* 4 (1839): 251.

70. "Schlußworte für die Gewerb-Vereine," *Gewerbe-Blatt für Sachsen* 3 (1838): 73–74.
71. Preusker, *Sonntags- und Gewerbeschulen*; Preusker, *Bürgerhalle*; and Preusker, *Sophien-Ducaten*. On Preusker, see Heerde, *Karl Benjamin Preusker*.
72. Tenfelde, "Lesegesellschaften und Arbeiterbildungsvereine"; and Dräger, *Volksbildung in Deutschland*.
73. Preusker, "Gewerbegeist," 27–29, 36–37. This speech was given at the 1838 anniversary celebration of the Großenhain *Gewerbeverein*.
74. Ibid., 29.
75. Preusker, *Sophien-Ducaten*, 57–58.
76. On Preusker's mixed success in promoting civility, science, and morality among young journeymen through intellectual socializing, see Hunziker, *Associations, Demonstrations, Recreations*.
77. Preusker, *Sophien-Ducaten*, 65.
78. In this vein, Emil Roßmäßler offered science popularization as an answer to the "Social Question" in the 1850s and 1860s. Daum, *Wissenschaftspopularisierung*, 154–61.
79. On the place of *Handwerk* in German liberalism more generally, see Haupt and Lenger, "Liberalismus und Handwerk"; and Thamer, "Emanzipation und Tradition."
80. F. G. Wieck, "Die Richtungen der Gewerbevereine und die Technik," *Gewerbe-Blatt für Sachsen* 4 (1839): 195.
81. *Gewerbe-Blatt für Sachsen* 3 (1848): 94.
82. "Ueber die wahre Zentralization der Kräften und Interessen," *Gewerbe-Blatt für Sachsen* 4 (1839): 267–68.
83. "Ein Blick in unser industrielles Leben," *Mittheilungen für das Erzgebirge und Voigtland* 1 (1836): 291. *Aktiengesellschaften* also occasionally provided a model for the internal organization of scientific societies. Württemberg's natural history society did not have yearly dues; its members bought stock shares in the group. "Organische Bestimmungen des Vereins für vaterländische Naturkunde in Württemberg," *Jahreshefte des Vereins für vaterländische Naturkunde in Württemberg* 1 (1845): 10.
84. See Gispen, *New Profession, Old Order*; and Harwood, *Technology's Dilemma*.
85. See, e.g., Nyhart, *Modern Nature*, 155–57.
86. Schleiden and Schmid, *Encyclopädie der gesammten theoretischen Naturwissenschaften*, vi.
87. "Ueber den Einfluß der Wissenschaften auf die Gewerbe," *Mittheilungen für das Erzgebirge und Voigtland* 1 (1836): 372.
88. "Gewerbliches," *Mittheilungen für das Erzgebirge und Voigtland* 1 (1836): 136.
89. [Advertisements], *Illustrirte Zeitung* 3 (1844): 175, 79, 238.
90. Hardtwig, "Vereinswesen in Deutschland"; and Tenfelde, "Arbeiterbildungsvereine."
91. *Annual Report of the Board of Regents*, 90–94.

CHAPTER EIGHT

1. Schmaus, "Understanding and Explanation"; and Roger Smith, "Natural Sciences and the Humanities." On later debates in the British context, see Collini, introduction to *Two Cultures*.
2. Brief overviews can be found in Connolly and Keutner, *Hermeneutics versus Science?*, 1–2; and Engelhardt, *Historisches Bewußtsein*, 161–69. On the early stages of the *Methodenstreit*, see MacLean, "History in a Two-Cultures World."
3. Calinich, *Philosophische Propädeutik*.
4. Flint, *Philosophy as Scientia Scientiarum*, 139ff. On Calinich, see Diemer, "Differenzierung der Wissenschaften, 187.
5. Eckert, *Die schulpolitische Instrumentalisierung*.
6. Yeo, *Science in the Public Sphere*, chap. 1 and 4. More generally, see Schuster and Yeo, *Politics and Rhetoric*.
7. Snyder, *Reforming Philosophy*.
8. Whitman, "Roman Lawyers in Germany." On Bastian, see Zimmerman, *Anthropology and Antihumanism*. "Naturwissenschaftliche Medizin" (natural scientific medicine) is another composite that would have sounded strange to eighteenth-century ears. For an early use of this language to talk about medicine, see Virchow, "Naturwissenschaftliche Methode."
9. Marchand, *Down from Olympus*, 4–35.
10. Jäger, "Philologisch-historische Fächer."
11. Nipperdey, *Deutsche Geschichte*, 459–60.
12. The founding theorists of this new school type also used the language of *Bildung* to defend their curriculum. See Eckert, *Schulpolitische Instrumentalisierung*; and Daum, *Wissenschaftspopularisierung*, 43–84.
13. Schöler, *Geschichte des naturwissenschaftlichen Unterrichts*, chap. 4.
14. Jeismann, "Knabenschulwesen"; Schubring, "Mathematisch-naturwissenschaftliche Fächer"; Olesko, *Physics as a Calling*, 21–60; and Bonnekuh, *Naturwissenschaft als Unterrichtsfach*. On Saxony, see Jungnickel, "Teaching and Research."
15. Grossmann, "Separatvotum," 180–81.
16. Friedrich Lindemann, *Mängel des Gelehrtenschulwesens*, 26.
17. Marchand, *Down from Olympus*, 31.
18. Ibid., 28.
19. See, e.g., Thiersch, *Jesuitismus und Obscurantismus*; and Raschig, "Sendschreiben."
20. Thiersch, *Jesuitismus und Obscurantismus*; Grossmann, "Separatvotum"; and Raschig, "Sendschreiben."
21. Friedrich Lindemann, *Mängel des Gelehrtenschulwesens*, 22. See also Snell, *Begründung des Gymnasial-Unterrichts*.
22. Jeismann, "Knabenschulwesen," 154–55.

23. Ibid., 167–68.
24. Raschig, "Sendschreiben," 2.
25. Friedrich Lindemann, *Mängel des Gelehrtenschulwesens*, 13, 34.
26. Köchly, *Princip des Gymnasialunterrichtes*, vi.
27. Oken, "Für die Aufnahme der Naturwissenschaften."
28. Freese, *Das deutsche Gymnasium*.
29. Mager, "Schriften zur Pädagogik."
30. Köchly, *Princip des Gymnasialunterrichtes*, 2.
31. Oken, "Aufnahme der Naturwissenschaften."
32. Oken, *Werth der Naturgeschichte*, 13–14.
33. Heuser and Zott, *Die streitbaren Gelehrten*.
34. Liebig, *Studium der Naturwissenschaften*, 11–21, 43–47.
35. Turner, "Historicism, *Kritik*, and the Prussian Professoriate," 460–62.
36. Schmid, *Encyklopädie und Methodologie*, iv.
37. See, e.g., Wagner, *Encyklopädie und Methodologie*; and Heusinger, *Encyklopädie und Methodologie*. For examples of regularly offered courses in this vein, see *Verzeichniß der auf der vereinigten Hallischen und Wittenbergischen Friedrichs-Universität . . . zu haltenden Vorlesungen* (1829–40).
38. Gustav Suchow, *Encyklopädie und Methodologie der theoretischen Naturwissenschaft*.
39. Carus, "Unterschied."
40. Böckel, *Köchly*, 53, 91–102.
41. Köchly, *Princip des Gymnasialunterrichtes*, 4.
42. On Köchly's ties to Ruge, see Böckel, *Köchly*, 29–31.
43. Köchly, *Princip des Gymnasialunterrichtes*, 4; Köchly, *Zur Gymnasialreform*, 47; on Hegel's original use of these categories, see Pinkard, *German Philosophy*, 266–304.
44. On Droysen, see Wise, "Relation of Physical Science"; and MacLean, "Development of Historical Hermeneutics," 352–53.
45. Hermann Köchly, "Zwei und siebzig Sätze zur Gymnasialreform," *Vermischte Blätter zur Gymnasialreform* 1 (1847): 25.
46. Köchly, *Zur Gymnasialreform*, 59–62.
47. This criticism was not an uncommon one; see Grafton, "*Polyhistor* into *Philolog*"; and La Vopa, "Specialists against Specialization."
48. Mager, "Schriften zur Pädagogik"; Fuchs, "Schriften zur Pädagogik"; and "Geschichte und Verhandlungen des Dresdener Gymnasialvereins von Michaelis bis Weihnachten 1846," *Vermischte Blätter zur Gymnasialreform* 1 (1847): 50.
49. See, e.g., Snell, *Begründung des Gymnasial-Unterrichts*, 19; and Friedrich Lindemann, *Mängel des Gelehrtenschulwesens*, 21–22, 41–43.
50. Mager, "Schriften zur Pädagogik," 58–62, 66.
51. "Ueber Gymnasialreform"; "Schul- und Unterrichtswesen"; and A. A., *Über das Gymnasialwesen*, 2.

52. Hermann Köchly, "Polemik und Kritik," *Vermischte Blätter zur Gymnasialreform* 1 (1847): 113, 129–30.
53. "Geschichte und Verhandlungen des Dresdener Gymnasialvereins," 59–60.
54. "Nekrolog [H. G. Ludwig Reichenbach]," *Sitzungsberichte der Naturwissenschaftlichen Gesellschaft Isis* (1879): 1–8; Daum *Wissenschaftspopularisierung*, 203–9; Daum, "Science, Politics and Religion"; and G. M. Richter, "Hermann Eberhard Richter."
55. Grosse, *Hermann Eberhard Richter*, 14; and Böckel, *Köchly*, 51.
56. G. M. Richter, "Hermann Eberhard Richter," 7, 25–30.
57. Hermann Richter, "Allgemeiner Bericht über Naturwissenschaft," *Vermischte Blätter zur Gymnasialreform* 2–3 (1848): 105; and Hermann Richter, "Vorrede," in Richter and Reichenbach, *Der naturwissenschaftliche Unterricht*, viii.
58. Richter, "Denkschrift der Gesellschaft Isis," 55.
59. Reichenbach, *Leben der Thierwelt*, 88; and Reichenbach, "Denkschrift," 13.
60. "Geschichte und Verhandlungen des Dresdener Gymnasialvereins," 61–67; and Reichenbach, "Denkschrift," 16.
61. "Geschichte und Verhandlungen des Dresdener Gymnasialvereins," 53–57.
62. Ibid., 70–76.
63. Richter, "Allgemeiner Bericht," 101–2.
64. Reichenbach, "Denkschrift," 9.
65. Ibid.; Richter, "Allgemeiner Bericht," 102; Reichenbach, "Gymnasium und die Naturkunde," 71; and "Geschichte und Verhandlungen des Dresdener Gymnasialvereins," 70–76.
66. Richter, "Vorrede," in Richter and Reichenbach, *Der naturwissenschaftliche Unterricht*, vi; Reichenbach, "Gymnasium und die Naturkunde."
67. See, e.g., Hermann Köchly, "Erwiderung des Herausgebers an Herrn Rector M. Raschig," *Vermischte Blätter zur Gymnasialreform* 1 (1847): 24; and Hermann Richter, "Die Vorbildung der Aerzte zuf Gymnasien," *Vermischte Blätter zur Gymnasialreform* 1 (1847): 174.
68. Richter, "Allgemeiner Bericht," 101–2.
69. Reichenbach, "Gymnasium," 71; and Richter, "Allgemeiner Bericht," 102.
70. "Geschichte und Verhandlungen des Dresdener Gymnasialvereins," 55.
71. Ibid., 58–60.
72. Richter, "Denkschrift," 41–42.
73. Hermann Richter, Hermann Köchly, and H. Herz, "Gesammtbericht," *Vermischte Blätter zur Gymnasialreform* 2–3 (1848): 233–68.
74. Calinich, *Philosophische Propädeutik*, iv, 3.
75. See, e.g., Sheehan, *German History*, 802–20.
76. Cf. Lenoir, *Instituting Science*, 131–78.
77. MacLean, "Two-Cultures World," 474–75.

78. On the continued commitment to the ideal of *Bildung* among *Naturforscher* later in the century, see Engelhardt, "Bildungsbegriff der Naturwissenschaft"; Daum, *Wissenschaftspopularisierung*, 52–57; and Cahan, "Helmholtz and the Civilizing Power of Science."
79. Engelhardt, "Bildungsbegriff der Naturwissenschaft," 106–16; and Daum, *Wissenschaftspopularisierung*, 52–57.
80. Caneva, "Galvanism to Electrodynamics"; and Heidelberger, "Patterns of Change," 8–9.
81. See, e.g., Ficinus, *Physik*, 1–2; Choulant, *Anleitung zum Studium*, 116–19; and Wagner, *Encyclopädie und Methodologie*, viii.
82. See, e.g., Freese, *Das deutsche Gymnasium*, 57–58; and Beger, *Idee des Realgymnasiums*, 44, 117.
83. Caneva, "Galvanism to Electrodynamics"; Heidelberger, "Patterns of Change"; and Olesko, *Physics as a Calling*.
84. Bastian, *Alexander von Humboldt*; and Du Bois-Reymond, "Humboldt Denkmäler."
85. Daum, *Wissenschaftspopularisierung*, 51–57.
86. Wise, "Relation of Physical Science," 28–29.
87. Helmholtz, "On the Relation of Natural Science"; and Jurkowitz, "Liberal Unification of Science."
88. Cited in Lenoir, *Instituting Science*, 173–74.
89. von Mohl, "Eröffnung der naturwissenschaftlichen Facultät," 208.
90. See, e.g., MacLean, "Two-Cultures World"; and Hörz, *Brückenschlag zwischen zwei Kulturen*, 10–11, 15. Andreas Daum has already pointed out the ways in which the idea of "two cultures" fails to capture the natural sciences' place in public culture. See Daum, *Wissenschaftspopularisierung*.
91. Baumgarten, *Professoren und Universitäten*, 130, 134–36.
92. Coen, *Age of Uncertainty*.
93. "Nekrolog [H. G. Ludwig Reichenbach]."
94. Recent work on Helmholtz, for example, has emphasized his ties to artists, philosophers, and humanist intellectuals. See Hörz, *Zwei Kulturen*; Kruger, *Universalgenie Helmholtz*; and Lenoir, *Instituting Science*, 135–36. On the interconnections between musicians and physicists, see Jackson, *Harmonious Triads*.
95. For the publication runs of these journals, see Kirchner, *Das deutsche Zeitschriftenwesen*, 201–4.
96. Jungnickel, "Royal Saxon Society of Sciences."
97. Daston, "Unity of Knowledge."
98. Applegate, *Nation of Provincials*.
99. Böckel, *Köchly*, 45, 71–135; and G. M. Richter, "Hermann Eberhard Richter," 12–18.

CONCLUSION

1. Dittrich, *Zur Feier der Vermählung*, unnumbered preface.
2. Nipperdey, *Deutsche Geschichte*, 497–98.
3. Julius Petzholdt, *Nachricht von der Bibliothek*; *Geschäftsbericht der naturwissenschaftlichen Gesellschaft*, 14; and *Erster Hauptrechenschaftsbericht des Gewerbe-Vereins*.
4. Monjau, "'Auffahrt und Rettung.'"
5. Julius Petzholdt, "Johanne Wilhelmine Siegmundine Reichard."
6. For an overview of this large literature, see Sperber, *"Bürger, Bürgerlichkeit, Bürgerliche Gesellschaft."*
7. *Geschäftsbericht der naturwissenschaftlichen Gesellschaft in Dresden*, 16.
8. Julius Petzholdt, *Der Plauensche Grund*, i, 49–50.
9. Dittrich mentioned these critics in his preface. See Dittrich, *Zur Feier der Vermählung*, 5.
10. Julius Petzholdt, *Nachricht von der Bibliothek*, iii.
11. On Plauen in the late nineteenth century, see *Stadtlexikon Dresden*, s.v. "Plauenscher Grund." It is important not to overstate the degree to which these different forms of engaging nature diverged later in the century, however. For an argument of the continued multivalencies of modern nature, see Nyhart, *Modern Nature*, 368.

Selected Bibliography

This bibliography includes all major works cited. A handful of short or unusually formatted sources have been excluded, with full citations given in the notes.

Archives Consulted

Archiv der Martin-Luther-Universität Halle-Wittenberg [AUHW], Rep 17 Nr 500–505, Naturwissenschaftlichen Vereines in Halle.
Museum für Naturkunde der Humboldt-Universität zu Berlin [MfN d. HUB], Historische Bild-u. Schriftgutsammlungen, Bestand Gesellschaft Naturforschender Freunde.
Sächsische Landesbibliothek [SLB], Dresden: Handschriften-sammlung, Flora Gesellschaft.
Sächsiches Hauptstaatsarchiv [SHSA], Dresden: Geheimes Kabinett, Sozietäten; Gesellschaft Natur- und Heilkunde.
Stadtarchiv Bamberg [SAB], Naturforschende Gesellschaft.

Published Sources

A., A. *Zur Verständigung über das Gymnasialwesen*. Dresden: Arnold, 1847.
Achilles, Walter. *Deutsche Agrargeschichte im Zeitalter der Reformen und der Industrialisierung*. Deutsche Agrargeschichte. Stuttgart: E. Ulmer, 1993.
Adelung, Johann Christoph. *Grammatisch-kiritisches Wörterbuch der hochdeutschen Mundort*. Leipzig: Johann Gottlob Immanuel Breitkopf, 1793–1806.
Alter, Peter. *The Reluctant Patron: Science and the State in Britain, 1850–1920*. Translated by Angela Davies. Oxford: Berg, 1987.

Altmayer, Claus. *Aufklärung als Popularphilosophie: Bürgerliches Individuum und Öffentlichkeit bei Christian Garve*. St. Ingbert: W. J. Röhrig, 1992.

Am Ende, C. G. E. *Die Oekonomische Gesellschaft im Königreich Sachsen in ihrer geschichtlichen Entwicklung seit 120 Jahren*. Dresden: G. Schönfeld, 1884.

Ammon, Friedrich August von. *Brunnendiätetik*. Dresden: Paul Gottlob Hilscher, 1825.

Anderton, Keith. "The Limits of Science: A Social, Political and Moral Agenda for Epistemology in Nineteenth-Century Germany." PhD diss., Harvard University, 1993.

Annual Report of the Board of Regents of the Smithsonian Institution. Washington, DC: Government Printing Office, 1865.

Anrich, Ernst, ed. *Die Idee der deutschen Universität*. Darmstadt: Hermann Gentner, 1956.

Applegate, Celia. *A Nation of Provincials: The German Idea of Heimat*. Berkeley: University of California Press, 1990.

Arenswald, C. F. von. *Galanterie-Mineralogie und Vorschläge zur Naturwissenschaft für die Damen*. Halle: Johann Jacob Gebauer, 1780.

Augustine, Dolores L. "Arriving in the Upper Class: The Wealthy Business Elite of Wilhelmine Germany." In *The German Bourgeoisie: Essays on the Social History of the German Middle Class from the Late Eighteenth to the Early Twentieth Century*, edited by David Blackbourn and Richard J. Evans, 56–73. London: Routledge, 1991.

Aulich, Reinhard. "Die Anfänge der Naturforschenden Gesellschaft zu Halle." In *Europa in der frühen Neuzeit: Festschrift für Günter Mühlpfordt*, edited by Erich Donnert, 155–65. Weimar: Böhlau, 1997.

Baader, Klement Alois. *Das gelehrte Baiern*. Nuremberg: In der Johann Esaias Seidelschen Kunst- und Buchhandlung, 1804.

Baldinger, Ernst Gottfried. *Ueber Litterar-Geschichte der theoretischen und praktischen Botanik*. Marburg: in der neuen akademischen Buchhandlung, 1794.

Barclay, David E. *Frederick William IV and the Prussian Monarchy, 1840–1861*. Oxford: Oxford University Press, 1995.

———. "Mäzenatentum am Preußischen Hof: Christian Carl Josias Bunsen, Alexander von Humboldt, und die Umgebung Friedrich Wilhelm IV." *Jahrbuch Stiftung Preußische Schlösser und Gärten Berlin-Brandenburg* 1 (1995–96): 11–17.

Barner, Wilfried. "Gelehrte Freundschaft im 18. Jahrhundert: Zu ihren traditionalen Voraussetzungen." In *Frauenfreundschaft, Männerfreundschaft: Literarische Diskurse im 18. Jahrhundert*, edited by Wolfram Mauser and Barbara Becker-Cantarino, 23–46. Tübingen: M. Niemeyer, 1991.

Bastian, Adolph. *Alexander von Humboldt*. Berlin: Wiegandt & Hemp, 1869.

Baumgarten, Marita. *Professoren und Universitäten im 19. Jahrhundert: Zur Sozialgeschichte deutscher Geistes- und Naturwissenschaftler*. Göttingen: Vandenhoeck & Ruprecht, 1997.

Beachy, Robert. "Club Culture and Social Authority: Freemasonry in Leipzig, 1741–1830." In *Paradoxes of Civil Society: New Perspectives on Modern German and British History*, edited by Frank Trentmann, 157–75. New York: Berghahn, 2000.

———. *The Soul of Commerce: Credit, Property, and Politics in Leipzig, 1750–1840*. Leiden: Brill, 2005.

Becker, Claudia. "Natürliche Erziehung—Erziehung zur Natur? Kontroverses um Rousseau." In *Idealismus und Aufklärung: Kontinuität und Kritik der Aufklärung in Philosophie und Poesie um 1800*, edited by Christoph Jamme and Gerhard Kurz, 152–73. Stuttgart: Klett-Cotta, 1988.

Becker, Kurt. "Abriß einer Geschichte der Gesellschaft Naturforschender Freunde zu Berlin." *Sitzungsberichte der Gesellschaft Naturforschender Freunde zu Berlin*, N.F. 13 (1973): 61–80.

Becker, Wilhelm Gottlieb, ed. *Der Plauische Grund bei Dresden: Mit Hinsicht auf Naturgeschichte und schöne Gartenkunst*. Nuremberg: In der Frauenholzischen Kunsthandlung, 1799.

Beckmann, Johann. *Anleitung zur Technologie*. Göttingen: Im Verlag der Wittwe Vandenhoeck, 1777.

Beger, August. *Die Idee des Realgymnasiums*. Leipzig: J. C. Hinrich, 1845.

Beiser, Frederick C. *German Idealism: The Struggle against Subjectivism, 1781–1801*. Cambridge, MA: Harvard University Press, 2002.

———. "Kant and *Naturphilosophie*." In *The Kantian Legacy in Nineteenth-Century Science*, edited by Michael Friedman and Alfred Nordmann, 7–26. Cambridge, MA: MIT Press, 2006.

Ben-David, Joseph. *The Scientist's Role in Society: A Comparative Study*. Englewood Cliffs, NJ: Prentice-Hall, 1971.

Berg, Wieland, and Benno Parthier. "Die 'kaiserliche' Leopoldina im Heiligen Römischen Reich Deutscher Nation." In *Gelehrte Gesellschaften im mitteldeutschen Raum, 1650–1820*, edited by Detlef Döring and Kurt Nowak, 39–52. Leipzig: Verlag der Sächsischen Akademie der Wissenschaften, 1999.

Berghahn, Klaus J. "Von der klassizistischen zur klassischen Literatur." In *Geschichte der deutschen Literaturkritik, 1730–1980*, edited by Peter Uwe Hohendahl, 10–75. Stuttgart: J. B. Metzler, 1985.

Bergmann, Joseph. *Anfangsgründe der Naturgeschichte*. Frankfurt am Main: Andreä, 1787.

Bernoulli, Johann. "Nachricht an die Gelehrten von Joh. Heinrich Lamberts Schriften." *Leipziger Magazin zur Naturkunde, Mathematik und Oekonomie* 1 (1781): 290–92.

Bescherer, Julius. *Methodik der naturwissenschaftlichen Unterrichts für Schulen*. Dresden and Leipzig: Arnold, 1838.

Biagioli, Mario. "Etiquette, Interdependence, and Sociability in Seventeenth-Century Science." *Critical Inquiry* 22 (1996): 193–238.

Biagioli, Mario, and Peter Galison, eds. *Scientific Authorship: Credit and Intellectual Property in Science*. London: Routledge, 2003.

BIBLIOGRAPHY

Bilder-Conversations-Lexikon. 3 vols. Leipzig: F. A. Brockhaus, 1837–41.
"Biographie des Herrn D. Bernhard Feldmann zu Ruppin." *Beschäftigungen der Berlinischen Gesellschaft Naturforschender Freunde* 3 (1777): 505–18.
Blackbourn, David. *The Conquest of Nature: Water, Landscape, and the Making of Modern Germany.* New York: W. W. Norton, 2006.
———. "Fashionable Spa Towns in Nineteenth-Century Europe." In *Water, Leisure and Culture: European Historical Perspectives*, edited by Susan C. Anderson and Bruce H. Tabb, 2–23. Oxford: Berg, 2002.
———. *The Long Nineteenth Century: A History of Germany, 1780–1918.* New York: Oxford University Press, 1998.
Blackbourn, David, and Geoff Eley. *The Peculiarities of German History: Bourgeois Society and Politics in Nineteenth-Century Germany.* Oxford: Oxford University Press, 1984.
Blackbourn, David, and James N. Retallack, eds. *Localism, Landscape, and the Ambiguities of Place: German-Speaking Central Europe, 1860–1930.* German and European Studies. Toronto: University of Toronto Press, 2007.
Blanning, T. C. W. *The Culture of Power and the Power of Culture: Old Regime Europe, 1660–1789.* Oxford: Oxford University Press, 2003.
Bloch, Marcus Eliezer. *Oekonomische Naturgeschichte der Fische Deutschlands.* Vol. 2. Berlin: Buchhandl. d. Realschule, 1784.
Blotevogel, Hans Heinrich. "Kulturelle Stadtfunktionen und Urbanisierung: Interdependente Beziehungen im Rahmen der Entwicklung des deutschen Städtesystems im Industriezeitalter." In *Urbanisierung im 19. und 20. Jahrhundert: Historische und geographische Aspekte*, edited by Hans Jürgen Teuteberg, 143–85. Cologne: Böhlau, 1983.
Blumenbach, Johann Friedrich. *Handbuch der Naturgeschichte.* 1st ed. Göttingen: J. C. Dietrich, 1779.
Böckel, Ernst. *Hermann Köchly: Ein Bild seines Lebens und seiner Persönlichkeit.* Heidelberg: Carl Winter, 1904.
Bödeker, Hans Erich. "Die 'gebildeten Stände' im späten 18. und frühen 19. Jahrhundert: Zugehörigkeit und Abgrenzungen, Mentalitäten und Handlungspotentiale." In *Bildungsbürgertum im 19. Jahrhundert: Politischer Einfluß und gesellschaftliche Formation*, 4:21–53. Stuttgart: Klett-Cotta, 1989.
Böhme, Katrin. "Die Gesellschaft Naturforschender Freunde zu Berlin: Bestand und Wandel einer gelehrten Gesellschaft." *Berichte zur Wissenschaftsgeschichte* 24 (2001): 271–83.
Böhme-Kaßler, Katrin. *Gemeinschaftsunternehmen Naturforschung: Modifikation und Tradition in der Gesellschaft Naturforschender Freunde zu Berlin, 1773–1906.* Stuttgart: Franz Steiner, 2005.
Bollenbeck, Georg. *Bildung und Kultur: Glanz und Elend eines deutschen Deutungsmusters.* Frankfurt: Insel, 1994.
Böning, Holger. "Der 'gemeine Mann' als Adressat aufklärerischen Gedankengutes: Ein Forschungsbericht zur Volksaufklärung." *Das 18. Jahrhundert* 12, no. 1 (1989): 52–82.

Bonnekuh, Werner. *Naturwissenschaft als Unterrichtsfach: Stellenwert und Didaktik des naturwissenschaftlichen Unterrichts zwischen 1800 und 1900.* Frankfurt am Main: Peter Lang, 1992.

Borckhausen, Moritz Balthasar. "Ueber die linneischen Gattungen Crataegus." *Archiv für die Botanik* 1 (1798): 85–91.

Borkhausen, Moritz Balthasar. *Botanisches Wörterbuch.* Giessen: G. F. Heyer, 1797.

Borscheid, Peter. *Naturwissenschaft, Staat und Industrie in Baden, 1848–1914.* Stuttgart: Klett, 1976.

Bosse, Heinrich. "Die gelehrte Republik." In *Öffentlichkeit im 18. Jarhundert,* edited by Hans-Wolf Jäger, 51–76. Göttingen: Wallstein, 1997.

Brain, Robert M., Robert S. Cohen, and Ole Knudsen. *Hans Christian Ørsted and the Romantic Legacy in Science: Ideas, Disciplines, Practices.* Dordrecht, Holland: Springer, 2007.

Brain, Robert M., and M. Norton Wise. "Muscles and Engines: Indicator Diagrams and Helmholtz's Graphical Methods." In *Universalgenie Helmholtz: Rückblick nach 100 Jahren,* edited by Lorenz Krüger, 124–45. Berlin: Akademie, 1994.

Breidbach, Olaf. "Transformation statt Reihung—Naturdetail und Naturganzes in Goethes Metamorphosenlehre." In *Naturwissenschaften um 1800: Wissenschaftskult in Jena-Weimar,* edited by Olaf Breidbach and Paul Ziche, 46–64. Weimar: Hermann Böhlaus Nachfolger, 2001.

Breidbach, Olaf, Hans-Joachim Fliedner, and Klaus Ries, eds. *Lorenz Oken (1779–1851): Ein politischer Naturphilosoph.* Weimar: Verlag Hermann Böhlaus Nachfolger, 2001.

Breuilly, John. "The Elusive Class: Some Critical Remarks on the Historiography of the Bourgeoisie." *Archiv für Sozialgeschichte* 38 (1998): 385–95.

Brewer, John, and Eckhart Hellmuth. *Rethinking Leviathan: The Eighteenth-Century State in Britain and Germany.* London: German Historical Institute, 1999.

Brewer, John, and Roy Porter, eds. *Consumption and the World of Goods.* London: Routledge, 1993.

Brock, William H. *Justus von Liebig: The Chemical Gatekeeper.* Cambridge: Cambridge University Press, 1997.

Brockliss, L. B. W. *Calvet's Web: Enlightenment and the Republic of Letters in Eighteenth-Century France.* Oxford: Oxford University Press, 2002.

Broman, Thomas H. "The Habermasian Public Sphere and 'Science in the Enlightenment.'" *History of Science* 36 (1998): 123–49.

———."Introduction: Some Preliminary Considerations on Science and Civil Society." In *Science and Civil Society.* Vol. 17. Osiris, 2002.

———. "On the Epistemology of Criticism: Science, Criticism, and the German Public Sphere, 1760–1800." In *Literaturwissenschaft und Wissenschaftsgeschichte,* edited by Jörg Schönert. Stuttgart: J. B. Metzler, 2000.

———. "Periodical Literature." In *Books and the Sciences in History,* 225–37. Cambridge: Cambridge University Press, 2000.

———. *The Transformation of German Academic Medicine, 1750–1820.* Cambridge: Cambridge University Press, 2002.

———. "University Reform in Medical Thought at the End of the Eighteenth Century." *Osiris* 5 (1989): 36–53.

Brooke, John Hedley. "The Fortunes and Functions of Natural Theology." In *Science and Religion: Some Historical Perspectives*, 192–225. Cambridge: Cambridge University Press, 1991.

Brophy, James. "The Public Sphere." In *Germany 1800–1870*, edited by Jonathan Sperber, 185–208. Short Oxford History of Germany. Oxford: Oxford University Press, 2004.

Brose, Eric Dorn. *The Politics of Technological Change in Prussia: Out of the Shadow of Antiquity, 1809–1848.* Princeton, NJ: Princeton University Press, 1993.

Bücking, J. J. H. "Christian Gottfried Beireis." *Zeitgenossen* 2 (1818): 69–122.

Bumann, Waltraud. "Der Wissenschaftsbegriff im deutschen Sprach- und Denkraum." In *Der Wissenschaftsbegriff. Histor. und systemat. Untersuchungen.*, edited by Alwin Diemer, 64–75. Meisenheim am Glan: Hain, 1970.

Cahan, David, ed. *From Natural Philosophy to the Sciences: Writing the History of Nineteenth-Century Science.* Chicago: University of Chicago Press, 2003.

———. "Helmholtz and the Civilizing Power of Science." In *Hermann von Helmholtz and the Foundations of Nineteenth-Century Science*, 559–601. Berkeley: University of California Press, 1994.

———. "Institutions and Communities." In *From Natural Philosophy to the Sciences: Writing the History of Nineteenth-Century Science*, edited by David Cahan, 291–328. Chicago: University of Chicago Press, 2003.

Calinich, Ernst Adolf Eduard. *Philosophische Propädeutik für Gymnasien, Realschulen und höhere Bildungsanstalten.* Dresden: Adler und Dietze, 1847.

Candolle, Augustin Pyramus de. *Regni vegetabilis systema naturale, sive Ordines, genera et species plantarum secundum methodi naturalis normas digestarum et descriptarum.* Paris and London: Treuttel and Würtz, 1818.

Candolle, Augustin Pyramus de, and Kurt Sprengel. *Grundzüge der wissenschaftlichen Pflanzenkunde.* Leipzig: Cnoblauch, 1820.

Caneva, Kenneth L. "From Galvanism to Electrodynamics." *Historical Studies in the Physical Sciences* 9 (1978): 63–160.

———. "Physics and *Naturphilosophie*: A Reconnaissance." *History of Science* 35 (1997): 35–106.

Carus, Carl Gustav. *Neun Briefe über Landschaftsmalerei.* Leipzig: Fleischer, 1831.

———. "Unterschied zwischen descriptiver, geschichtlicher, vergleichender und philosophischer Anatomie." *Litterarischen Annalen der gesammten Heilkunde* 4 (1826): 1–30.

———. "Von den Naturreichen, ihrem Leben und ihrer Verwandtschaft." *Zeitschrift für Natur- und Heilkunde* 1 (1820): 1–72.

———. *Von der Anforderungen an eine künftige Bearbeitung der Naturwissenschaften.* Leipzig: Ernst Fleischer, 1822.

Chartier, Roger. "The Printing Revolution: A Reappraisal." In *Agent of Change: Print Culture Studies after Elizabeth L. Eisenstein*, edited by Sabrina A. Baron, Eric N. Lindquist, and Eleanor F. Shevlin, 397–408. Amherst: University of Massachusetts Press, 2007.

Choulant, Ludwig. *Anleitung zum Studium der Medizin*. Leipzig: Voß, 1829.

Clark, William. *Academic Charisma and the Origins of the Research University*. Chicago: University of Chicago Press, 2006.

———. "The Death of Metaphysics in Enlightened Prussia." In *Sciences in Enlightened Europe*, edited by William Clark, Jan Golinski, and Simon Schaffer, 423–76. Chicago: University of Chicago Press, 1999.

———. "From the Medieval Universitas Scholarium to the German Research University: A Sociogenesis of the German Academic." PhD diss., UCLA, 1986.

Clauss, K. W. *Chronik des Gewerbe-Vereins zu Dresden als Festschrift zur fünfzigjährigen Stiftungsfeier*. Dresden: Wilhelm Hoffmann, 1884.

Cobres. *Büchersammlung zur Naturgeschichte*. 1782.

Coen, Deborah R. *Vienna in the Age of Uncertainty: Science, Liberalism, and Private Life*. Chicago: University of Chicago Press, 2007.

Collini, Stefan. Introduction to *The Two Cultures*, by C. P. Snow, iii–lxxi. Cambridge: Cambridge University Press, 1998.

Confino, Alon. *The Nation as a Local Metaphor: Württemberg, Imperial Germany, and National Memory, 1871–1918*. Chapel Hill: University of North Carolina Press, 1997.

Connolly, John M., and Thomas Keutner, eds. *Hermeneutics versus Science? Three German Views*. Notre Dame, IN: University of Notre Dame Press, 1988.

Conze, Werner, and Jürgen Kocka, eds. *Bildungsbürgertum im 19. Jahrhundert*. 4 vols. Stuttgart: Klett-Cotta, 1985–92.

Cooper, Alix. *Inventing the Indigenous: Local Knowledge and Natural History in Early Modern Europe*. Cambridge: Cambridge University Press, 2007.

Cunningham, Andrew, and Nicholas Jardine. "Introduction: The Age of Reflection." In *Romanticism and the Sciences*, edited by Andrew Cunningham and Nicholas Jardine, 1–9. Cambridge: Cambridge University Press, 1990.

Cunningham, Andrew, and Perry Williams. "De-centering the 'Big Picture': The Origins of Modern Science and the Modern Origins of Science." *British Journal for the History of Science* 26 (1996): 407–32.

Curio, Johann Carl Daniel. "An das Publikum." *Gelehrte Beyträge zu den Braunschweigischen-Anzeigen* 26 (1786): 189.

Dänhardt, Walter. *Festschrift aus Anlaß des hundertjährigen Bestehens der Flora*. Dresden, 1926.

Dann, Otto. "Die Lesegesellschaften des 18. Jahrhunderts und der gesellschaftlichen Aufbruch des deutschen Bürgertums." In *Buch und Leser*, edited by Herbert Georg Göpfert, 160–93. Hamburg: Hauswedell, 1977.

———. *Vereinsbildung und Nationsbildung: Sieben Beiträge*. Cologne: SH-Verlag, 2003.

———, ed. *Vereinswesen und bürgerliche Gesellschaft in Deutschland*. Munich: R. Oldenbourg, 1984.

Daßdorf, Karl Wilhelm. *Beschreibung der vorzüglichsten Merkwürdigkeiten der Churfürstlichen Residenzstadt Dresden und einiger seiner umliegenden Gegenden*. Vol. 2. Dresden: Walther, 1782.

Daston, Lorraine. "The Academies and the Unity of Knowledge: The Disciplining of the Disciplines." In *Die Königlich Preußische Akademie der Wissenschaften zu Berlin im Kaiserreich*, edited by Jürgen Kocka, 61–84. Berlin: Akademie-Verlag, 1999.

———. "Afterword: The Ethos of Enlightenment." In *Sciences in Enlightened Europe*, edited by William Clark, Jan Golinski, and Simon Schaffer, 495–504. Chicago: University of Chicago Press, 1999.

———. "Attention and the Values of Nature in the Enlightenment." In *The Moral Authority of Nature*, edited by Lorraine Daston and Fernando Vidal, 100–126. Chicago: University of Chicago Press, 2004.

———. "Die Quantifizierung der weiblichen Intelligenz." In *Aller Männerkultur zum Trotz: Frauen in Mathematik und Naturwissenschaft*, 69–82. Frankfurt am Main: Campus, 1997.

———. "The Glass Flowers." In *Things That Talk: Object Lessons from Art and Science*, edited by Lorraine Daston, 223–56. New York: Zone Books, 2004.

———. "The Ideal and Reality of the Republic of Letters in the Enlightenment." *Science in Context* 4, no. 2 (1991): 367–86.

Daston, Lorraine, and Peter Galison. *Objectivity*. New York: Zone Books, 2010.

Daston, Lorraine, and Katharine Park. *Wonders and the Order of Nature, 1150–1750*. New York: Zone Books, 1998.

Daston, Lorraine, and H. Otto Sibum. "Scientific Personae and Their Histories." *Science in Context* 16 (2003): 1–8.

Daum, Andreas. "Science, Politics and Religion: Humboldtian Thinking and the Transformations of Civil Society in Germany, 1830–1870." *Osiris* 17 (2002): 107–40.

———. "*Wissenschaft* and Knowledge." In *Germany, 1800–1870*, edited by Jonathan Sperber, 149–53. Oxford: Oxford University Press, 2004.

———. *Wissenschaftspopularisierung im 19. Jahrhundert: Bürgerliche Kultur, naturwissenschaftliche Bildung und die deutsche Öffentlichkeit, 1848–1914*. Munich: R. Oldenbourg, 1998.

Davis, Natalie Zemon. *Women on the Margins: Three Seventeenth-Century Lives*. Cambridge, MA: Harvard University Press, 1995.

Dear, Peter. "*Totius in verba*: Rhetoric and Authority in the Early Royal Society." *Isis* 76 (1985): 145–61.

Dettelbach, Michael. "The Face of Nature: Precise Measurement, Mapping and Sensibility in the Work of Alexander von Humboldt." *Studies in the History and Philosophy of Biology and the Biomedical Sciences* 30, no. 4 (1999): 473–504.

———. "Humboldtian Science." In *Cultures of Natural History*, edited by Nicholas Jardine, James A. Secord, and E. C. Spary, 287–304. Cambridge: Cambridge University Press, 1996.

Diemer, Alwin. "Die Differenzierung der Wissenschaften in die Natur- und die Geisteswissenschaften und die Begrüdung der Geisteswissenschaften als Wissenschaft." In *Beiträge zur Entwicklung der Wissenschaftstheorie im 19. Jahrhundert*. Meisenheim am Glan: Anton Hain, 1968.

Dierbach, Johann Heinrich. *Anleitung zum Studium der Botanik*. Heidelberg: Groos, 1820.

———. *Repertorium botanicum oder Versuch einer systematischen Darstellung der neuesten Leistungen im ganzen Umfange der Pflanzenkunde*. Lemgo: Meyer, 1831.

Dittrich, Heinrich. *Zur Feier der Vermählung des Herrn Dr. Julius Petzholdt mit Fräulein Hermine Reichard: Beiträge zur Erklärung und Kritik des Theokritos*. Dresden: Teubner, 1844.

Doering-Manteuffel, Hans Robert. *Dresden und sein Geistesleben im Vormärz: Ein Beitrag zur Geschichte des kulturellen Lebens in der sächsischen Hauptstadt*. Dresden: Risseverlag, 1935.

Döring, Detlef, and Kurt Nowak, eds. *Gelehrte Gesellschaften im mitteldeutschen Raum, 1650–1820*. Leipzig: Verlag der Sächsischen Akademie der Wissenschaften, 1999.

Dräger, Horst, ed. *Volksbildung in Deutschland im 19. Jahrhundert*. Braunschweig: Westermann, 1979.

Drayton, Richard Harry. *Nature's Government: Science, Imperial Britain, and the "Improvement" of the World*. New Haven, CT: Yale University Press, 2000.

Drechsler, Adolf. "Kurzgefasste Geschichte der naturwissenschaftlichen Gesellschaft Isis zu Dresden." In *Denkschriften der Naturwissenschaftlichen Gesellschaft Isis zu Dresden: Festgabe zur Feier ihres fünfundzwanzigjährigen Bestehens*, edited by Adolf Drechsler, 69–101. Dresden: Verlagsbuchhandlung von Rudolf Kuntze, 1860.

Droysen, Johann Gustav. "Zur Charakteristik der europäischen Krisis." In *Politische Schriften: Im Auftrag der Preussischen Akademie der Wissenschaften*, edited by Felix Gilbert, 302–42. Munich: R. Oldenbourg, 1933.

Du Bois-Reymond, Emil. "Die Humboldt-Denkmäler an der Berliner Universität." [1883]. In *Reden von Emil Du Bois-Reymond*, edited by Estelle Du Bois-Reymond, 249–84. Leipzig: Veit, 1912.

Düding, Dieter. *Organisierter gesellschaftlicher Nationalismus in Deutschland (1808–1847): Bedeutung und Funktion der Turner- und Sängervereine für die deutsche Nationalbewegung*. Studien zur Geschichte des neunzehnten Jahrhunderts, vol. 13. Munich: R. Oldenbourg, 1984.

Eckert, Manfred. *Die schulpolitische Instrumentalisierung des Bildungsbegriffs: Zum Abgrenzungsstreit zwischen Realschule und Gymnasium im 19. Jahrhundert*. Frankfurt: R. G. Fischer, 1984.

BIBLIOGRAPHY

Ehrhart, Friedrich. *Beiträge zur Naturkunde*. Hannover: Osnabrück, 1787.
Eisnerova, Vera. "Botanische Disziplinen." In *Geschichte der Biologie: Theorien, Methoden, Institutionen, Kurzbiographien*, 3rd ed., edited by Ilse Jahn and Erika Krausse, 302–23. Berlin: Spektrum, 2000.
Eley, Geoff, and Keith Nield. *The Future of Class in History: What's Left of the Social?* Ann Arbor: University of Michigan Press, 2007.
Engelhardt, Dietrich von. "Der Bildungsbegriff der Naturwissenschaft des 19. Jahrhunderts." *Bildungsbürgertum im 19. Jahrhundert*, edited by Reinhart Koselleck, 106–16. Stuttgart: Klett Cotta, 1991.
———. *Historisches Bewußtsein in der Naturwissenschaft: Von der Aufklärung bis zum Positivismus*. Freiburg: Alber, 1979.
———. "Romanticism in Germany." In *Romanticism in National Context*, edited by Roy Porter and Mikuláš Teich, 109–33. Cambridge: Cambridge University Press, 1988.
———. "Romantik—im Spannungsfeld von Naturgefühl, Naturwissenschaft, und Naturphilosophie." In *Romantik in Deutschland*, edited by Richard Brinkmann, 64–74. Stuttgart: J. B. Metzler, 1978.
———. "Science, Society and Culture in the Romantic *Naturforschung* around 1800." In *Nature and Society in Historical Context*, edited by Mikuláš Teich, Roy Porter, and Bo Gustafsson, 195–208. Cambridge: Cambridge University Press, 1997.
———, ed. *Zwei Jahrhunderte Wissenschaft und Forschung in Deutschland: Entwicklungen—Perspektiven*. Stuttgart: Wissenschaftliche Verlagsgesellschaft, 1998.
Engelhardt, Ulrich. *"Bildungsbürgertum": Begriffs- und Dogmengeschichte eines Etiketts*. Stuttgart: Klett-Cotta, 1986.
Engeli, Christian. "Stadterweiterungen in Deutschland im 19. Jahrhundert." In *Die Städte Mitteleuropas im 19. Jahrhundert*, edited by Wilhelm Rausch, 47–72. Linz: Rudolf Trauner, 1983.
Engelmann, Wilhelm, and Theodore Christian Friedrich Enslin. *Bibliotheca oeconomica*. Leipzig: W. Engelmann, 1841.
Engels, Friedrich. *Dialektik der Natur*. 2nd ed. Berlin: Dietz, 1955.
Engelsing, Rolf. *Der Bürger als Leser: Lesergeschichte in Deutschland, 1500–1800*. Stuttgart: Metzler, 1974.
Erlin, Matt. *Berlin's Forgotten Future: City, History, and Enlightenment in Eighteenth-Century Germany*. Chapel Hill: University of North Carolina Press, 2004.
Erster Hauptrechenschaftsbericht des Gewerbe-Vereins zu Dresden. N.p., 1836.
Eskildsen, Kasper Risbjerg. "How Germany Left the Republic of Letters." *Journal of the History of Ideas* 65, no. 3 (2004): 421–32.
Festschrift zum 150-jährigen Bestehen der Ökonomischen Sozietät zu Leipzig und der Ökonomischen Gesellschaft im Königreich Sachsen zu Dresden. Leipzig: Alexander Edelmann, 1914.
Fichte, Johann Gottlieb. "Deduzierter Plan einer zu Berlin zu Errichtenden höhern Lehranstalt, die in gehöriger Verbindung mit einer Akademie der

Wissenschaften stehe." In *Die Idee der deutschen Universität: Die fünf Grundschriften aus der Zeit ihrer Neubegründung durch klassischen Idealismus und romantischen Realismus.* Darmstadt: Hermann Gentner, 1956.

Ficinus, Heinrich David August. *Flora der Gegend um Dresden.* 2nd ed. Dresden: Arnold, 1821.

———. *Physik allgemein faßlich dargestellt.* Dresden: P. G. Hilscher, 1828.

Fiedler, Annett. *Die Physikalische Gesellschaft zu Berlin: Vom lokalen naturwissenschaftlichen Verein zu nationalen Deutschen Physikalischen Gesellschaft, 1845–1900.* Berichte aus der Physik. Aachen: Shaker, 1998.

Flechsig, Robert. *Die Struve'schen Mineralwässer in Beziehung zu ihrem durch ärztliche Erfahrung festgestellten Heilwerth gegenüber den natürliche Heilquellen.* Leipzig: F. C. W. Vogel, 1886.

Flint, Robert. *Philosophy as Scientia Scientiarum: A History of Classifications of the Sciences.* New York: Arno, 1975.

Flurl, Mathias. *Rede von dem Einfluße der Wissenschaften insbesondere der Naturkunde auf die Kultur einer Nation.* Munich: Joseph Lindauer, 1799.

Fox, Robert. "The Savant Confronts His Peers: Scientific Societies in France, 1815–1870." In *The Organization of Science and Technology in France, 1808–1914,* edited by Robert Fox, 241–82. Cambridge: Cambridge University Press, 1980.

Fraas, Karl Nikolaus. *Klima und Pflanzenwelt in der Zeit: Ein Beitrag zur Geschichte beider.* Landshut: J. G. Wölfe, 1847.

Freese, E. *Das deutsche Gymnasium, nach den Bedürfnissen der Gegenwart.* Dresden: Arnold, 1845.

Frege, Christian August. *Versuch eines allgemeinen botanischen Handwörterbuchs, lateinisch-deutsch, und deutsch-lateinisch.* Zeitz: W. Webel, 1808.

Frevert, Ute. *Mann und Weib, und Weib und Mann: Geschlechter-Differenzen in der Moderne.* Munich: C. H. Beck, 1995.

Friedman, Michael, and Alfred Nordmann, eds. *The Kantian Legacy in Nineteenth-Century Science.* Cambridge, MA: MIT Press, 2006.

Fritsch, Johann Christian Gottfried, Johann Burkhard Mencke, and Christian Gottlieb Jöcher. *Allgemeines Gelehrten-Lexicon.* Leipzig: in Johann Friedrich Gleditschens Buchhandlung, 1750.

Fryer, Geoffrey. "Jacob Christian Schäffer FRS, a Versatile Eighteenth-Century Naturalist, and His Remarkable Pioneering Researches on Microscopic Creatures." *Notes and Records of the Royal Society* 62, no. 2 (2008): 167–85.

Fuchs, August. "Schriften zur Pädagogik." *Pädagogische Revue* 12 (1846): 22–226, 306–28.

Gagliardo, John G. *From Pariah to Patriot: The Changing Image of the German Peasant, 1770–1840.* Lexington: University Press of Kentucky, 1969.

Galison, Peter. "Introduction: The Context of Disunity." In *The Disunity of Science: Boundaries, Contexts, and Power,* edited by Peter Galison and David J. Stump, 1–33. Stanford, CA: Stanford University Press, 1996.

Galison, Peter, and David J. Stump, eds. *The Disunity of Science: Boundaries, Contexts, and Power.* Stanford, CA: Stanford University Press, 1996.

Gaukroger, Stephen. *The Emergence of a Scientific Culture: Science and the Shaping of Modernity, 1210–1685.* Oxford: Clarendon Press, 2006.

Gesetze der naturwissenschaftlichen Gesellschaft in Dresden. Dresden: Teubner, 1844.

Geyer, H. "Ueber das Verhältniss der Philosophie zu den Naturwissenschaften." In *Populäre Vorlesungen über Naturwissenschaft,* edited by Alexander Petzholdt, 20–40. Leipzig: Carl B. Lorck, 1845.

Geinitz, H. G. "Verzeichnis der Mitglieder des Gewerbe-Vereins in Dresden." In *Fünfte Hauptbericht über das Wirken des Gewerbe-Vereins zu Dresden.* Dresden: Ernst Blochmann & Sohn, [1844].

Geschäftsbericht der naturwissenschaftlichen Gesellschaft in Dresden von den Jahren, 1843–1845. Dresden: Teubner, 1845.

"Geschichte und Verhandlungen des Dresdener Gymnasialvereins." *Vermischte Blätter zur Gymnasialreform* 1 (1847): 29–92.

Gierl, Martin. "Bestandaufnahme im gelehrten Bereich: Zur Entwicklung der 'Historia literaria' im 18. Jahrhundert." In *Denkhorizonte und Handlungsspielräume: Historische Studien für Rudolf Vierhaus zum 70. Geburtstag,* 53–80. Göttingen: Wallstein, 1992.

Gispen, Kees. *New Profession, Old Order: Engineers and German Society, 1815–1914.* Cambridge: Cambridge University Press, 1989.

Gleditsch, Johann. *Theoretisch-praktische Geschichte der Pflanzen.* Berlin: Decker, 1777.

Goldfuß, Georg August. "Die in Vorschlag gebrachte Versammlung." *Isis* (1821): 41–42.

Goldgar, Anne. *Impolite Learning: Conduct and Community in the Republic of Letters, 1680–1750.* New Haven, CT: Yale University Press, 1995.

Golinski, Jan. *Science as Public Culture: Chemistry and Enlightenment in Britain, 1760–1820.* Cambridge: Cambridge University Press, 1992.

Goodman, Dena. *The Republic of Letters: A Cultural History of the French Enlightenment.* Ithaca, NY: Cornell University Press, 1994.

Gordin, Michael. "The Importation of Being Earnest: The Early St. Petersburg Academy of Sciences." *Isis* 91 (2000): 1–31.

Gothein, Marie Luise Schroeter. *Geschichte der Gartenkunst.* [2. Aufl.]. Jena: E. Diederichs, 1926.

Götze, J. A. E. "Nachricht an das Publikum." *Leipziger Magazin zur Naturkunde, Mathematik und Oekonomie* 1 (1781): 420–22.

Götzinger, Wilhelm Lebrecht. *Schandau und seine Umgebungen, oder Beschreibung der sächsischen Schweiz.* 2nd ed. 1812. Reprint, Leipzig: Zentralantiquariat der Deutschen Demokratischen Republik, 1973.

Gradmann, Johann Jacob. *Das gelehrte Schwaben, oder Lexicon der jetzt lebenden schwäbischen Schriftsteller.* Hildesheim: G. Olms, 1979.

Grafton, Anthony. "*Polyhistor* into *Philolog*: Notes on the Transformation of German Classical Scholarship, 1780–1850." *History of Universities* 3 (1983): 159–92.

Gray, Marion W. *Productive Men, Reproductive Women: The Agrarian Household and the Emergence of Separate Spheres during the German Enlightenment*. New York: Berghahn Books, 2000.

Green, Abigail. *Fatherlands: State-Building and Nationhood in Nineteenth-Century Germany*. Cambridge: Cambridge University Press, 2001.

Gregory, Frederick. "Kant's Influence on Natural Scientists in the German Romantic Period." In *New Trends in the History of Science*, edited by R. P. W. Visser, 53–72. Amsterdam: Rudolphi, 1989.

———. *Nature Lost? Natural Science and the German Theological Traditions of the Nineteenth Century*. Cambridge, MA: Harvard University Press, 1992.

———. *Scientific Materialism in Nineteenth-Century Germany*. Dordrecht, Holland: D. Reidel, 1977.

Griep, Wolfgang, and Hans-Wolf Jäger, eds. *Reisen im 18. Jahrhundert: Neue Untersuchungen*. Heidelberg: C. Winter, 1986.

Groß, Reiner. *Die bürgerliche Agrarreform in Sachsen in der ersten Hälfte des 19. Jahrhunderts: Untersuchung zum Problem des Übergangs vom Feudalismus zum Kapitalismus in der Landwirtschaft*. Weimar: Böhlau, 1968.

———. *Geschichte Sachsens*. Dresden and Leipzig: Sonderausgabe der Sächsischen Landeszentrale für politische Bildung, 2007.

Grosse, Johannes. *Hermann Eberhard Richter, der Gründer des deutschen Aerztevereinsbundes*. Leipzig: Otto Wigand, 1896.

Grossmann, Christian. "Separatvotum des Superintendenten Dr. Grossmann aus Leipzig vom 13. Juli 1834." In *Der naturwissenschaftliche Unterricht auf Gymnasien*, edited by Hermann Richter and Ludwig Reichenbach, 173–87. Dresden: Arnold, 1847.

Habermas, Jürgen. *The Structural Transformation of the Public Sphere: An Inquiry into a Category of Bourgeois Society*. Translated by Thomas Burger. Cambridge, MA: MIT Press, 1991.

Haenel, Erich, and Eugen Kalkschmidt, eds. *Das alte Dresden: Bilder und Dokumente aus zwei Jahrhunderten*. 3rd ed. Leipzig: Schmidt & Günther, 1941.

Hagen, William W. *Ordinary Prussians: Brandenburg Junkers and Villagers, 1500–1840*. Cambridge: Cambridge University Press, 2002.

Hagner, Michael, and Manfred Laubichler, eds. *Der Hochsitz des Wissens: Das Allgemeine als wissenschaftlicher Wert*. Zurich and Berlin: Diaphanes, 2006.

Hamm, Ernst P. "Goethes Sammlungen auspacken: Das Öffentliche und das Private im naturgeschichtlichen Sammeln." In *Sammeln als Wissen. Das Sammeln und seine wissenschaftsgeschichtliche Bedeutung*, edited by Anke te Heesen and E. C. Spary, 85–114. Göttingen: Wallstein, 2001.

Hankins, Thomas L. *Science and the Enlightenment*. Cambridge: Cambridge University Press, 1985.

Hansen, James Roger. "Scientific Fellowship in a Swiss Community Enlightenment: A History of Zurich's Physical Society, 1746–1798." PhD diss., Ohio State University, 1981.

Hardtwig, Wolfgang. *Genossenschaft, Sekte, Verein in Deutschland*. Vol. 1, *Vom Spätmittelalter bis zur Französischen Revolution*. Munich: C. H. Beck, 1997.

———. "Strukturmerkmale und Entwicklungstendenzen des Vereinswesen in Deutschland, 1789–1848." In *Vereinswesen und bürgerliche Gesellschaft in Deutschland*, edited by Otto Dann, 11–53. Munich: R. Oldenbourg, 1984.

Harleß, Johann Christian Friedrich. *Die Verdienste der Frauen um Naturwissenschaft, Gesundheits- und Heilkunde, so wie auch um Länder-, Völker-, und Menschenkunde, von der ältesten Zeit bis auf die neueste*. Göttingen: Vandenhoeck-Ruprecht, 1830.

Harwood, Jonathan. *Technology's Dilemma: Agricultural Colleges between Science and Practice in Germany, 1860–1934*. Frankfurt am Main: Peter Lang Publishing, 2005.

Harzer, August. "Kleine Fußwanderungen in den umliegenden Gegenden von Dresden." *Ueber Berg und Tal* 35 (1912): 73–78, 88–90, 101–2, 124–26; and 37 (1914): 207–8.

Harzer, Carl August Friedrich. *Naturgetreue Abbildungen der vorzüglichsten, essbaren, giftigen und verdächtigen Pilze*. Dresden: E. Pietzsch, 1942.

Hasse, Friedrich Christian August. *Dresden und die umliegende Gegend*. 2nd ed. Dresden: In der Arnoldischen Buch- und Kunsthandlung, 1804.

Hassler, Johann Georg. *Vollständiges lateinisch-teutsches und teutsch-lateinisches Pflanzen-Lexicon*. 1812.

Haugen, Kristine Louise. "Academic Charisma and the Old Regime." *History of Universities* 22 (2007): 199–228.

Haupt, H. G., and F. Lenger. "Liberalismus und Handwerk in Frankreich und Deutschland um die Mitte des 19. Jahrhunderts." In *Liberalismus im 19. Jahrhundert: Deutschland im europäischen Vergleich*, edited by Dieter Langewiesche, 305–31. Göttingen: Vandenhoeck & Ruprecht, 1988.

Hausen, Karin. "Die Polarisierung der 'Geschlechtscharaktere.'" In *Sozialgeschichte der Familie in der Neuzeit Europas: Neue Forschungen*, edited by Werner Conze, 363–93. Stuttgart: Klett, 1976.

Haushofer, Hans. *Die deutsche Landwirtschaft im technischen Zeitalter*. Stuttgart: Eugen Ulmer, 1963.

Haymann, Christian Johann Gottfried. *Dresdens theils neuerlich verstorbene theils jetzt lebende Schriftsteller und Künstler*. Dresden: Walther, 1809.

Heerde, Dietrich. *Karl Benjamin Preusker (1786–1871): Ein Heimatforscher und Volksbildungsfreund*. Grossenhain: Rat der Stadt, 1986.

Heidelberger, Michael. "Some Patterns of Change in the Baconian Sciences of Early 19th-Century Germany." In *Epistemological and Social Problems of the Sciences in the Early Nineteenth Century*, edited by H. N. Jahnke and Michael Otte, 3–18. Dordrecht, Holland: D. Reidel, 1981.

Heilbron, J. L. *Electricity in the 17th and 18th Centuries: A Study of Early Modern Physics*. Berkeley: University of California Press, 1979.
———. "Physics and Its History at Göttingen around 1800." In *Göttingen and the Development of the Natural Sciences*, edited by Nicolaas Rupke, 50–71. Göttingen: Wallstein, 2002.
Hellmuth, Eckart, and Wolfgang Piereth. "Germany, 1760–1815." In *Press, Politics and the Public Sphere in Europe and North America, 1760–1820*, edited by Hannah Barker and Simon Burrows, 69–92. Cambridge: Cambridge University Press, 2002.
Helmholtz, Hermann von. "On the Relation of Natural Science to Science in General." In *Science and Culture: Popular and Philosophical Essays*, edited by David Cahan, 76–95. Chicago: University of Chicago Press, 1995.
Hennig, Lothar, ed. *Von der Romantik bis zur Gründerzeit: Bürgerkultur im 19. Jahrhundert in Bamberg*. Bamberg: Fruhauf, 1997.
Hermand, Jost. "Die touristische Erschliessung und Nationalisierung des Harzes im 18. Jahrhundert." In *Reise und soziale Realität am Ende des 18. Jahrhunderts*, edited by Wolfgang Griep and Hans-Wolf Jäger, 169–87. Heidelberg: C. Winter, 1983.
Herminghouse, Patricia. "Women and the Literary Enterprise in Nineteenth-Century Germany." In *German Women in the Eighteenth and Nineteenth Centuries: A Social and Literary History*, edited by Ruth Ellen B. Joeres and Mary Jo Maynes, 78–93. Bloomington: Indiana University Press, 1986.
Hertz, Deborah Sadie. *Jewish High Society in Old Regime Berlin*. Modern Jewish History. New Haven, CT: Yale University Press, 1988.
Heuser, Regina, and Emil Zott, eds. *Die streitbaren Gelehrten: Justus von Liebig und die Preußischen Universitäten*. Berlin: ERS, 1992.
Heusinger, Carl Friedrich. *Grundriss der Encyklopädie und Methdologie der Natur- und Heilkunde*. Eisenach: Christian Friedrich Bärecke, 1839.
Hilpert, Johann Wolfgang. *Zum Andenken an Dr. Jacob Sturm, den Ikonographen der deutschen Flora und Fauna*. Nuremberg: Von der naturhistorischen Gesellschaft zu Nürnberg ihren Mitgliedern gewidmet, 1849.
Hobsbawm, E. J., and T. O. Ranger. *The Invention of Tradition*. Cambridge: Cambridge University Press, 1983.
Hochadel, Oliver. *Öffentliche Wissenschaft: Elektrizität in der deutschen Aufklärung*. Göttingen: Wallstein, 2003.
Hoelder, Christian Gottlieb, and Adolphe Peschier. *Mozin's deutsch-französisch und französisch-deutsches Hand-Wörterbuch*. Stuttgart: J. G. Cotta, 1868.
Hoffmann, Stefan-Ludwig. *Civil Society, 1750–1914*. Basingstoke, UK: Palgrave Macmillan, 2006.
———. *The Politics of Sociability: Freemasonry and German Civil Society, 1840–1918*. Ann Arbor: University of Michigan Press, 2007.
Hohendahl, Peter Uwe, ed. *Geschichte der deutschen Literaturkritik, 1730–1980*. Stuttgart: J. B. Metzler, 1985.

Homburg, Ernst. "Two Factions, One Profession: The Chemical Profession in German Society, 1780–1870." In *The Making of the Chemist: The Social History of Chemistry in Europe, 1789–1914*, edited by David M. Knight and Helge Kragh, 39–76. Cambridge: Cambridge University Press, 1998.

Hörz, Herbert, ed. *Brückenschlag zwischen zwei Kulturen: Helmholtz in der Korrespondenz mit Geisteswissenschaftlern und Künstlern*. Marburg an der Lahn: Basilisken Presse, 1997.

Houghton, Bernard. *Scientific Periodicals: Their Historical Development, Characteristics, and Control*. Hamden, CT: Linnet, 1975.

Hubrig, Hans. *Die patriotischen Gesellschaften des 18. Jahrhunderts*. Weinheim: Beltz, 1957.

Hufbauer, Karl. *The Formation of the German Chemical Community, 1720–1795*. Berkeley: University of California Press, 1982.

———. "Social Support for Chemistry in Germany during the Eighteenth Century: How and Why Did It Change?" *Historical Studies in the Physical Sciences* 3 (1971): 205–31.

Hull, Isabel V. *Sexuality, State, and Civil Society in Germany, 1700–1815*. Ithaca, NY: Cornell University Press, 1997.

Humpert, Magdalene. *Bibliographie der Kameralwissenschaften*. Cologne: B. Pick, 1937.

Hundert Jahre Geschichte der Arnoldischen Buchhandlung zu Dresden von 1790 bis 1890. Dresden: Arnold, 1890.

Hunter, Johann. *Beiträge zur Naturgeschichte der Wallfischarten*. Translated by Johann Gottlob Schneider. Leipzig: Schäffer, 1795.

Hunziker, Scott Brandon. "Associations, Demonstrations, Recreations: Working People and the Revolutionary Experience in Saxony, 1830–1850." PhD diss., University of North Carolina, 2004.

Ijsewijn, Jozef. *Companion to Neo-Latin Studies*. Vol. 1, *History and Diffusion of Neo-Latin Literature*. 2nd ed. Leuven, Belgium: Leuven University Press, 1990.

Ijsewijn, Jozef, and Dirk Sacré. *Companion to Neo-Latin Studies*. Vol. 2, *Literary, Linguistic, Philological, and Editorial Questions*. Leuven, Belgium: Leuven University Press, 1998.

Im Hof, Ulrich. *Das gesellige Jahrhundert: Gesellschaft und Gesellschaften im Zeitalter der Aufklärung*. Munich: Beck, 1982.

Immermann, Karl Leberecht. *Die Epigonen: Familienmemoiren in neun büchern*. Düsseldorf: J. E. Schaub, 1836.

Inkster, Ian, and Jack Morrell. *Metropolis and Province: Science in British Culture, 1780–1850*. London: Hutchinson, 1983.

Jablonsky, Karl Gustav. *Natursystem aller bekannten in- und ausländischen Insekten*. Vol. 1. Berlin: J. Pauli, 1785.

Jäck, Heinrich Joachim. *Leben und Werke der Künstler Bambergs*. Erlangen: Palm & Enke, 1821.

———. *Pantheon der Literaten und Künstler Bambergs*. Bamberg: im Komptoir der Zeitung, 1843.
Jäckel, Gunter. "Die Entdeckung der Stadtlandschaft." *Dresdner Hefte* 58 (1999): 16–26.
———. "Von den 'sächsischen Frauenzimmer' um 1800." *Dresdner Hefte* 62 (2000): 18–28.
Jackson, Myles W. "Can Artisans Be Scientific Authors? The Unique Case of Fraunhofer's Artisanal Optics and the German Republic of Letters." In *Scientific Authorship: Credit and Intellectual Property in Science*, edited by Mario Biagioli and Peter Galison, 113–32. London: Routledge, 2003.
———. *Harmonious Triads: Physicists, Musicians, and Instrument Makers in Nineteenth-Century Germany*. Cambridge, MA: MIT Press, 2006.
———. *Spectrum of Belief: Joseph von Fraunhofer and the Craft of Precision Optics*. Cambridge, MA: MIT Press, 2000.
Jacob, Margaret C., and Larry Stewart. *Practical Matter: Newton's Science in the Service of Industry and Empire, 1687–1851*. Cambridge, MA: Harvard University Press, 2004.
Jacobi, Friedrich Heinrich. *Ueber gelehrte Gesellschaften, ihren Geist und Zweck*. Munich: E. A. Fleischmann, 1807.
Jäger, Hans-Wolf. *Europäisches Reisen im Zeitalter der Aufklärung*. Neue Bremer Beiträge, vol. 7. Heidelberg: Winter, 1992.
———. "Philologisch-historische Fächer." In *Handbuch der deutschen Bildungsgeschichte*, edited by Christa Berg, 192–203. Munich: C. H. Beck, 1987.
Jahn, Ilse. "Die Rolle der Gesellschaft Naturforschender Freunde zu Berlin im interdisziplinären Wissenschaftsaustausch des 19. Jahrhundert." *Sitzungberichte der Gesellschaft Naturforschender Freunde zu Berlin* 31 (1991): 3–13.
Jahn, Ilse, and Erika Krausse, eds. *Geschichte der Biologie: Theorien, Methoden, Institutionen, Kurzbiographien*. 3rd ed. Berlin: Spektrum, 2000.
Jahresbericht über die königlichen Studien-Anstalten zu Bamberg. Bamberg: Georg Romauld Klebadel, 1822.
Jardine, Nicholas. "*Naturphilosophie* and the Kingdoms of Nature." In *Cultures of Natural History*, edited by Nicholas Jardine, James A. Secord, and E. C. Spary, 230–45. Cambridge: Cambridge University Press, 1996.
———. *The Scenes of Inquiry: On the Reality of Questions in the Sciences*. Oxford: Clarendon Press, 1991.
Jaumann, Herbert. "*Respublica litteraria* / Republic of Letters: Concept and Perspectives of Research." In *Die europäische Gelehrtenrepublik im Zeitalter des Konfessionalismus*, edited by Herbert Jaumann, 11–19. Wiesbaden: Harrassowitz, 2001.
Jeismann, Karl-Ernst. "Das höhere Knabenschulwesen." In *Handbuch der deutschen Bildungsgeschichte*. Vol. 3, *1800–1870*, edited by Karl-Ernst Jeismann and Peter Lundgreen, 152–80. München: C. H. Beck, 1987.

Jeismann, Karl-Ernst and Peter Lundgreen, eds. *Handbuch der deutschen Bildungsgeschichte*. Vol. 3, *1800–1870, Von der Neuordnung Deutschlands bis zur Gründung des Deutschen Reiches*. Munich: C. H. Beck, 1987.

Jenisch, Daniel. *Cultur-Charakter und Cultur-Geschichte des achtzehnten Jahrhunderts*. Berlin: Königl. Preuß. Akad. Kunst- u. Buchhandlung, 1800.

Jentzsch, Rudolf. *Der deutsch-lateinische Büchermarkt nach den Leipziger Ostermess-Katalogen von 1740, 1770 und 1800 in seiner Gliederung und Wandlung*. Leipzig: Voigtländer, 1912.

"Johann Heinrich Seidel und seine Pflanzen: Ein Gärtner um Wende des 18. Jahrhunderts." *Sitzungsberichte und Abhandlungen, Flora zu Dresden*, N.F. 7 (1904): 51–69.

Johnson's English Dictionary. Boston: Nathan Hale, 1835.

Johnston, James F. W. "Meeting of the Cultivators of Natural Science and Medicine at Hamburgh [sic], in September 1830." *Edinburgh Journal of Science*, n.s., 4 (1831): 189–244.

Jordanova, Ludmilla. "Sex and Gender." In *Inventing the Human Sciences*, edited by Christopher Fox, Roy Porter, and Robert Wokler, 152–83. Berkeley: University of California Press, 1995.

Jungnickel, Christa. "The Royal Saxon Society of Sciences: A Study of Nineteenth-Century German Science." PhD diss., Johns Hopkins University, 1978.

———. "Teaching and Research in the Physical Sciences and Mathematics in Saxony, 1820–1850." *Historical Studies in the Physical Sciences* 10 (1979): 3–47.

Jurkowitz, Edward. "Helmholtz and the Liberal Unification of Science." *Historical Studies in the Physical and Biological Sciences* 32, no. 2 (2002): 291–317.

Kamphausen, Georg, and Thomas Schnelle. *Die Romantik als naturwissenschaftliche Bewegung*. Bielefeld: B. K. Verlag, 1982.

Kant, Immanuel. *Metaphysical Foundations of Natural Science*. Translated by Michael Friedman. Cambridge: Cambridge University Press, 2004.

———. "Über den Gemeinspruch: Das mag in der Theorie richtig sein, taugt aber nicht für die Praxis [1793]." *Über Theorie und Praxis*, edited by Dieter Henrich, 39–87. Frankfurt am Main: Suhrkamp, 1967.

Kanz, Kai Torsten. "'. . . man wieß nur was man eine' Mann schreiben soll mit dem man einmal persönlich verhandelt hat': Zum Briefwechsel Goethes mit Christian Gottfried Nees von Esenbeck." In *Naturwissenschaften um 1800: Wissenschaftskult in Jena-Weimar*, 203–15. Weimar: Hermann Böhlaus Nachfolger, 2001.

———. *Nationalismus und internationale Zusammenarbeit in den Naturwissenschaften: Die deutsch-französischen Wissenschaftsbeziehungen zwischen Revolution und Restauration, 1789–1832*. Stuttgart: Franz Steiner, 1997.

Kranke, Kurt. *Zur Geschichte der Freimaurerei in Dresden*. Dresden: Verein für Regionale Politik und Geschichte Dresdens, 2000.

Karsten, Wenceslaus Johann Gustav. *Einleitung zur gemeinnützlichen Kenntniß der Natur.* Halle: Renger, 1783.

———. *Kurzer Entwurf der Naturwissenschaft: Vornehmlich ihres chymisch-mineralogischen Theils: Mit Kupfern.* Halle: Renger, 1785.

Kaschuba, Wolfgang. "Die Fußreise—von der Arbeitswanderung zur bürgerlichen Bildungsbewegung." In *Reisekultur: Von der Pilgerfahrt zum modernen Tourismus,* edited by Hermann Bausinger, Klaus Beyrer, and Gottfried Korff, 165–73. Munich: C. H. Beck, 1991.

———. "German *Bürgerlichkeit* after 1800: Culture as Symbolic Practice." In *Bourgeois Society in Nineteenth-Century Europe,* edited by Jürgen Kocka and Allan Mitchell, 393–422. Oxford: Berg, 1993.

Keferstein, Christian. *Erinnerungen aus dem Leben eines alten Geognosten und Ethnographen, mit Nachrichten über die Familie Keferstein.* Halle: E. Anton, 1855.

Kehn, Wolfgang. "'Die Schönheiten der Natur gemeinschaftlich betrachten': Zum Zusammenhang von Freundschaft, ästhetischer Naturerfahrung und 'Gartenrevolution' in der Spätaufklärung." In *Frauenfreundschaft, Männerfreundschaft: Literarische Diskurse im 18. Jahrhundert,* edited by Wolfram Mauser and Barbara Becker-Cantarino, 167–93. Tübingen: M. Niemeyer, 1991.

Kennedy, Ildephon. "Abhandlungen von einigen in Baiern gefundenen Beinen." *Neue philosophische Abhandlungen der Baierischen Akademie der Wissenschaften* 4 (1785): 1–48.

Kiesewetter, Hubert. *Industrialisierung und Landwirtschaft: Sachsens Stellung im regionalen Industrialisierungsprozeß.* Cologne: Böhlau, 1988.

Kim, Mi Gyung. "'Public' Science: Hydrogen Balloons and Lavoisier's Decomposition of Water." *Annals of Science* 63, no. 3 (2006): 291–318.

Kirchner, Joachim. *Das deutsche Zeitschriftenwesen: Seine Geschichte und seine Probleme.* 2nd ed. Wiesbaden: O. Harrassowitz, 1958.

Kleeberg, Bernhard. *Theophysis: Ernst Haeckels Philosophie des Naturganzen.* Cologne: Böhlau, 2005.

Klein, Felix. "Gauss' wissenschaftliches Tagebuch, 1796–1814." In *Festschrift zur Feier des Hundertjährigen Bestehens der Königlichen Gesellschaft der Wissenschaften zu Göttingen,* 1–44. Berlin: Weimann, 1901.

Klein, Lawrence Eliot, and Anthony J. La Vopa, eds. *Enthusiasm and Enlightenment in Europe, 1650–1850.* San Marino, CA: Huntington Library, 1998.

Klein, Ursula. "Apothecary-Chemists in Eighteenth-Century Germany." In *New Narratives in Eighteenth-Century Chemistry,* edited by Lawrence Principe, 97–137. Dordrecht, Holland: Springer, 2007.

Kleine-Natrop, H. E. *Das heilkundige Dresden: Dresdner Chirurgenschulen und medizinischen Lehrstätten in drei Jahrhunderten.* Dresden and Leipzig: Theodor Steinkopff, 1964.

Kleinert, Andreas. "Die Naturforschende Gesellschaft zu Halle." *Acta Historica Leopoldina* 36 (2000): 246–72.

Kleinhans, Bernd. *Der "Philosoph" in der neueren Geschichte der Philosophie: "Eigentlicher Philosoph" und "vollendeter Gelehrter": Konkretionen des praktischen Philosophen bei Kant und Fichte.* Würzburg: Königshausen & Neumann, 1999.

Kleist, Heinrich von. *Heinrich von Kleist: Sämtliche Briefe.* Stuttgart: Reclam, 1999.

Klemm, Volker. *Agrarwissenschaften in Deutschland.* St. Katharinen: Scripta Mercaturae, 1992.

Klopstock, Friedrich Gottlieb. *Die deutsche Gelehrtenrepublik.* Hamburg: J. J. C. Bode, 1774.

Knight, David. *The Age of Science: The Scientific World-View in the Nineteenth Century.* Oxford: Basil Blackwell, 1986.

Knight, David, and Helge Kragh, eds. *Making of the Chemist: The Social History of Chemistry in Europe, 1789–1914.* Cambridge: Cambridge University Press, 1998.

Koch, Johann Friedrich Wilhelm. *Botanisches Handbuch für deutsche Liebhaber der Pflanzenkunde überhaupt und fur Gartenfreunde, Apotheker und Oekonomen insbesondere.* Magdeburg: Heinrichshofen, 1797.

Köchly, Hermann. *Über das Princip des Gymnasialunterrichtes der Gegenwart.* Dresden: Arnold, 1845.

———. *Zur Gymnasialreform.* Dresden: Arnold, 1846.

Kocka, Jürgen. "The European Pattern and the German Case." In *Bourgeois Society in Nineteenth-Century Europe*, edited by Jürgen Kocka and Allan Mitchell, 3–39. Oxford: Berg, 1993.

———. *Industrial Culture and Bourgeois Society: Business, Labor, and Bureaucracy in Modern Germany.* New York: Berghahn, 1999.

Kocka, Jürgen, and Allan Mitchell. *Bourgeois Society in Nineteenth-Century Europe.* Oxford: Berg, 1993.

Koerner, Joseph Leo. *Caspar David Friedrich and the Subject of Landscape.* New Haven, CT: Yale University Press, 1990.

Koerner, Lisbet. "Carl Linnaeus in His Time and Place." In *Cultures of Natural History*, edited by Nicholas Jardine, James A. Secord, and E. C. Spary, 56–81. Cambridge: Cambridge University Press, 1996.

———. "Goethe's Botany: Lessons of a Feminine Science." *Isis* 84 (1993): 470–95.

———. *Linnaeus: Nature and Nation.* Cambridge, MA: Harvard University Press, 1999.

König, G. "Naturwissenschaften." In *Historisches Wörterbuch der Philosophie*, 6:641–50. Basel: Schwabe, 1984.

"Königliche Bestätigung der Gesellschaft und ihrer Statuten." *Schriften der Naturforschenden Gesellschaft zu Leipzig* 1 (1822): 1–6.

Koselleck, Reinhart. "Einleitung." In *Geschichtliche Grundbegriffe.* Vol. 1. Stuttgart: Klett-Cotta, 1979.

———. "On the Anthropological and Semantic Structure of *Bildung*." In *The Practice of Conceptual History: Timing History, Spacing Concepts*. Translated by Todd Samuel Presner. Stanford, CA: Stanford University Press, 2002.

Košenina, Alexander. Afterword to *Ueber Pedanterie und Pedanten, als eine Warnung für die Gelehrten des 18. Jahrhunderts*, 20–28. Hannover: Revonnah, 1996.

Kreutzberg, Karl Josef. "Ueber den Einfluß der Naturwissenschaften auf die Industrie und das Leben." *Gewerbe-Blatt für Sachsen* 1 (1838).

Kreyssig, Friedrich Ludwig. *De l'usage des eaux minérales naturelles et artificielles*. Leipzig: F. A. Brockhaus, 1829.

Kronick, David A. *A History of Scientific and Technical Periodicals: The Origins and Development of the Scientific and Technical Press, 1665–1790*. 2nd ed. Metuchen, NJ: Scarecrow Press, 1976.

Krueger, Rita. *Czech, German, and Noble: Status and National Identity in Habsburg Bohemia*. New York: Oxford University Press, 2009.

Kruger, Lorenz, ed. *Universalgenie Helmholtz: Rückblick nach 100 Jahren*. Berlin: Akademie Verlag, 1994.

Krüger, Marcus Salomon. *Bibliographia botanica*. Berlin: Haude und Spencer, 1841.

Kuhn, Dorothea, ed. "Christian Nees von Esenbeck, XI Präsident der Leopoldina, an Johann Friedrich Cotta, 1816–1818." *Acta Historica Leopoldina* 9 (1975): 69–92.

Kuhn, Oskar. "Das Naturalienkabinett des alten Klosters Banz." *Fränkische Blätter für Geschichtsforschung und Heimatpflege* 4, no. 8 (1952).

Kunz, Georg. *Verortete Geschichte: Regionales Geschichtsbewusstsein in den deutschen Historischen Vereinen des 19. Jahrhunderts*. Göttingen: Vandenhoeck & Ruprecht, 2000.

Lambert, Johann Heinrich. *Deutscher gelehrter Briefwechsel*. Vol. 3, edited by Johann Bernoulli. Berlin: bey dem Herausgeber, 1783.

Lang, C. *Beschreibung des Plauenschen Grundes*. 2nd ed. Dresden: Beger, 1812.

Langewiesche, Dieter. *Liberalismus in Deutschland*. Frankfurt am Main: Suhrkamp, 1988.

Larson, Anne. "Equipment for the Field." In *Cultures of Natural History*, edited by Nicholas Jardine, James A. Secord, and E. C. Spary, 358–77. Cambridge: Cambridge University Press, 1996.

Laurop, C. P. *Handbuch der Forst- und Jagdliteratur: Von den ältesten Zeiten bis Ende des Jahres 1828 systematisch geordnet*. Erfurt and Gotha: Henning, 1830.

La Vopa, Anthony J. *Fichte: The Self and the Calling of Philosophy, 1762–1799*. Cambridge: Cambridge University Press, 2001.

———. *Grace, Talent, and Merit: Poor Students, Clerical Careers, and Professional Ideology in Eighteenth-Century Germany*. Cambridge: Cambridge University Press, 1988.

———. "Herder's Publikum: Language, Print, and Sociability in Eighteenth-Century Germany." *Eighteenth-Century Studies* 29 (1995): 5–24.

———. "Review: Conceiving a Public: Ideas and Society in Eighteenth-Century Europe." *Journal of Modern History* 64, no. 1 (March 1, 1992): 79–116.

———. "Specialists against Specialization: Hellenism as Professional Ideology in German Classical Studies." In *German Professions, 1800–1950*, edited by Geoffrey Cocks and Konrad Hugo Jarausch, 30–45. New York: Oxford University Press, 1990.

Lekan, Thomas M. *Imagining the Nation in Nature: Landscape Preservation and German Identity, 1885–1945*. Cambridge, MA: Harvard University Press, 2004.

Lekan, Thomas M., and Thomas Zeller, eds. *Germany's Nature: Cultural Landscapes and Environmental History*. New Brunswick, NJ: Rutgers University Press, 2005.

Lempa, Heikki. *Beyond the Gymnasium: Educating the Middle-Class Bodies in Classical Germany*. Lanham, MD: Lexington Books, 2007.

Lenoir, Timothy. *Instituting Science: The Cultural Production of Scientific Disciplines*. Stanford, CA: Stanford University Press, 1997.

———. *The Strategy of Life: Teleology and Mechanics in Nineteenth-Century German Biology*. Dordrecht, Holland: D. Reidel, 1982.

Lenz, Max. *Geschichte der Königlichen Friedrich-Wilhelms-Universität zu Berlin*. 4 vols. Halle: Buchhandlung des Waisenhauses, 1910–18.

Lepenies, Wolf. *Das Ende der Naturgeschichte: Wandel kultureller Selbstverständlichkeiten in den Wissenschaften des 18. und 19. Jahrhunderts*. Hanser Anthropologie. Munich: C. Hanser, 1976.

Levinger, Matthew Bernard. *Enlightened Nationalism: The Transformation of Prussian Political Culture, 1806–1848*. Oxford: Oxford University Press, 2002.

Leysser, Friedrich Wilhelm von. *Beyträge zur Beförderung der Naturkunde*. Halle: Trampe, 1774.

Liebig, Justus. *Ueber das Studium der Naturwissenschaften und über den Zustand der Chemie in Preußen*. Braunschweig: Friedrich Vieweg und Sohn, 1840.

Lindau, W. A. *Rundgemälde der Gegend um Dresden und die umliegende Gegend*. Dresden: Arnold, 1920.

Lindemann, Friedrich. *Die wichtigsten Mängel des Gelehrtenschulwesens im Königreichs Sachsen*. Zittau and Leipzig: Birr und Nauwerck, 1834.

Lindemann, Mary. *Health and Healing in Eighteenth-Century Germany*. Baltimore: Johns Hopkins University Press, 1996.

Lindenfeld, David. *The Practical Imagination: The German Sciences of State in the Nineteenth Century*. Chicago: University of Chicago Press, 1997.

Lindner, Konrad. "Die Linnéische Sozietät zu Leipzig: Über einige Wirkungen der mitteldeutschen Gelehrtengesellschaft um 1800." In *Gelehrte Gesellschaften im mitteldeutschen Raum, 1650–1820*, edited by Detlef Döring and Kurt Nowak, 211–29. Leipzig: Verlag der Sächsischen Akademie der Wissenschaften, 1999.

Link, Heinrich Friedrich. *Ideen zu einer philosophischen Naturkunde.* Breslau: J. F. Korn, 1814.

Lips, Alexander. *Die Nürnberg-Fürther Eisenbahn in ihren nächsten Wirkungen und Resultaten.* Riegel: Wiener, 1836.

Lloyd, Hannibal Evans, and Georg Heinrich Noehden. *New Dictionary of the English and German Languages.* Hamburg: A. Campe, 1836.

Lorenz, Martina. "Funktionsbestimmungen von Naturlehre in bayerischen und fränkischen Zeitschriften in der Aufklärung." *Zeitschrift für bayerische Landesgeschichte* 57 (1994): 383–404.

———. "Physik im Hamburgischen Magazin (1747–1767): Publizistische Utopie und Wirklichkeit." *Zeitschrift des Vereins für Hamburgische Geschichte* 80 (1994): 13–46.

Lösch, Stephan. *Prof. Dr. Adam Gengler, 1799–1866: Die Beziehungen des Bamberger Theologen zu J. J. J. Döllinger und J. A. Möhler.* Würzburg: Kommissionsverlag F. Schöningh, 1963.

Löwe, Johann Carl Friedrich. *Handbuch der theoretischen und praktischen Kräuterkunde.* Wroclaw: Gottlieb Löwe, 1787.

Lowood, Henry. *Patriotism, Profit, and the Promotion of Science in the German Enlightenment: The Economic and Scientific Societies, 1760–1815.* New York: Garland, 1991.

Lundgreen, Peter. *Techniker in Preussen während der frühen Industrialisierung: Ausbildung und Berufsfeld einer entstehenden sozialen Gruppe.* Berlin: Colloquium, 1975.

Lustig, A. J. "Cultivating Knowledge in Nineteenth-Century English Gardens." *Science in Context* 13 (2000): 155–81.

Lynn, Michael R. *Popular Science and Public Opinion in Eighteenth-Century France.* Manchester, UK: Manchester University Press, 2006.

MacLean, Michael. "History in a Two-Cultures World: The Case of the German Historians." *Journal of the History of Ideas* 49, no. 3 (1988): 473–94.

———. "Johann Gustav Droysen and the Development of Historical Hermeneutics." *History and Theory* 21 (1982): 347–65.

Mager, Karl. "Schriften zur Pädagogik." *Pädogogische Revue* 12 (1846): 204–23; 13 (1846): 41–69.

Mah, Harold. "Phantasies of the Public Sphere: Rethinking the Habermas of the Historians." *Journal of Modern History* 72, no. 1 (2000): 153–82.

Marchand, Suzanne L. *Down from Olympus: Archaeology and Philhellenism in Germany, 1750–1970.* Princeton, NJ: Princeton University Press, 1996.

———. *German Orientalism in the Age of Empire: Religion, Race, and Scholarship.* Cambridge: Cambridge University Press, 2010.

Martini, Friedrich Heinrich Wilhelm. "Vorbericht." *Beschäftigungen der Berlinischen Gesellschaft Naturforschender Freunde* 1 (1775): iii–xi.

Martino, Alberto, and Marlies Stützel-Prüsener. "Publikumsschichten, Lesegesellschaften und Leihbibliotheken." In *Deutsche Literatur: Eine Sozialgeschichte*, edited by Horst Albert Glaser, Ursula Liebertz-Grün, and Ingrid

Bennewitz. Vol. 5, *Zwischen Revolution und Restauration: Klassik und Romantik*, 45–57. Reinbek bei Hamburg: Rowohlt, 1980.

Martius, Carl Friedrich Philipp von. "Johann Salomo Christoph Schweigger." In *Akademische Denkreden*. Leipzig: Friedrich Fleischer, 1866.

Maurer, Michael. "Die pädagogische Reise: Auch eine Tendenz der Reiseliteratur in der Spätaufklärung." In *Europäisches Reisen im Zeitalter der Aufklärung*, edited by Hans-Wolf Jäger, 54–69. Heidelberg: Winter, 1992.

McClellan, James E., III. *Science Reorganized: Scientific Societies in the Eighteenth Century*. New York: Columbia University Press, 1985.

———. "Scientific Institutions and the Organization of Science." In *Eighteenth-Century Science*, 4:87–128. Cambridge History of Science. Cambridge: Cambridge University Press, 2003.

McClelland, Charles E. *State, Society, and University in Germany, 1700–1914*. Cambridge: Cambridge University Press, 1980.

McNeely, Ian F. *The Emancipation of Writing: German Civil Society in the Making, 1790s–1820s*. Berkeley: University of California Press, 2003.

Meinel, Christoph. "Die wissenschaftliche Fachzeitschrift: Struktur- und Funktionswandel eines Kommunikationsmediums." In *Fachschrifttum, Bibliothek und Naturwissenschaft im 19. und 20. Jahrhundert*, edited by Christoph Meinel, 137–55. Wiesbaden: Harrassowitz, 1997.

Mencke, Johann Burkhard. *Compendiöses Gelehrten-Lexicon*. Leipzig: J. F. Gleditsch und Sohn, 1715.

Mendelsohn, Everett. "The Emergence of Science as a Profession in Nineteenth-Century Europe." In *The Management of Scientists*, edited by Karl Hill, 3–48. Boston: Beacon Press, 1964.

Merkwürdigkeiten Dresdens und der Umgegend. Dresden and Leipzig, 1829.

Merton, Robert K. "De-gendering the 'Man of Science': The Genesis and Epicene Character of the Word 'Scientist.'" In *Sociological Visions*, edited by Kai Erikson, 225–53. Lanham, MD: Rowman and Littlefield, 1997.

Meyer-Krentler, Eckhardt. "Freundschaft im 18. Jahrhundert: Zur Eröffnung in die Forschungsdiscussion." In *Frauenfreundschaft, Männerfreundschaft: Literarische Diskurse im 18. Jahrhundert*, edited by Wolfram Mauser and Barbara Becker-Cantarino, 1–22. Tübingen: M. Niemeyer, 1991.

Miltitz, Friedrich von. *Handbuch der botanischen Literatur: Für Botaniker, Bibliothekare, Buchhändler und Auctionatoren*. Berlin: A. Rücker, 1829.

Mitchell, Timothy. *Art and Science in German Landscape Painting, 1770–1840*. Oxford: Clarendon Press, 1993.

"Mitglieder des Königlichen Sächsischen Alterthums-Vereins." *Mittheilungen des Königlichen Sächsischen Vereins für Erforschung und Erhaltung der vaterländischen Alterthümer* 1 (1835): 76–79.

Mohl, Hugo von. "Rede gehalten bei der Eröffnung der naturwissenschaftlichen Facultät der Universität Tübingen." [1863]. In *Quellen zur Gründungsgeschichte der Naturwissenschaftlichen Fakultät in Tübingen*, edited by Wolf Freiherr von Engelhardt, 187–208. Tübingen: J. C. B. Mohr, 1963.

Möller, Horst. *Vernunft und Kritik: Deutsche Aufklärung im 17. und 18. Jahrhundert*. 1st ed. Neue historische Bibliothek. Frankfurt am Main: Suhrkamp, 1986.

Mommertz, Monika. "The Invisible Economy of Science: A New Approach to the History of Gender and Astronomy at the Eighteenth-Century Berlin Academy of Sciences." In *Men, Women and the Birthing of Modern Science*, edited by Judith P. Zinsser, 159–78. DeKalb: Northern Illinois University Press, 2005.

Monjau, Heide. "'Auffahrt und Rettung': Der Ballonfahrerin Wilhelmine Reichard." *Dresdner Hefte* 55 (1998): 4–10.

Morrell, Jack, and Arnold Thackray. *Gentlemen of Science: Early Years of the British Association for the Advancement of Science*. Oxford: Clarendon Press, 1981.

Morus, Iwan Rhys, Simon Schaffer, and James A. Secord. "Scientific London." In *London: World City, 1800–1840*, edited by Celina Fox, 129–42. New Haven, CT: Yale University Press, 1992.

Mössler, Johann Christoph. *Gemeinnütziges Handbuch der Gewächskunde*. Altona: Bey J. F. Hammerich, 1815.

Muhs, Rudolf. "Zwischen Staatsreform und politischem Protest: Liberalismus in Sachsen zur Zeit des Hambacher Festes." In *Liberalismus in der Gesellschaft des deutschen Vormärz*, Wolfgang Schieder, 195–238. Göttingen: Vandenhoeck & Ruprecht, 1983.

Müller, Johann Traugott. *Einleitung in die Oekonomische und Physikalische Bücherkunde*. Leipzig: Schwickert, 1780.

Müller, Johannes. *Die wissenschaftlichen Vereine und Gesellschaften Deutschlands im neunzehnten Jahrhundert*. Berlin: A. Asher & Co., 1883.

Müller-Sievers, Helmut. "Skullduggery: Goethe and Oken, Natural Philosophy and Freedom of the Press." *Modern Language Quarterly* 59, no. 2 (1998): 231–59.

Müller-Wille, Staffan. *Botanik und weltweiter Handel: Zur Begründung eines natürlichen Systems der Pflanzen durch Carl von Linné, 1707–78*. Berlin: VWB, 1999.

———. "Nature as a Marketplace: The Political Economy of Linnaean Botany." *History of Political Economy* 35 (2003): 154–72.

———. "Walnuts in Hudson Bay, Coral Reefs in Gotland: The Colonialism of Linnaean Botany." In *Colonial Botany: Science, Commerce and Politics in the Early Modern World*, 35–48. Philadelphia: University of Pennsylvania Press, 2005.

Nachricht von dem Leben und Wirken Dr. Burkhard Wilhelm Seiler. Dresden: Teubner, 1844.

Nägeli, Carl. "Ueber die gegenwärtige Aufgabe der Naturgeschichte, insbesondere der Botanik." *Zeitschrift für wissenschaftliche Botanik* 1 (1844).

"Naturwissenschaft." *Allgemeine Deutsche Real-Encyclopädie*. 6th ed. Vol. 6. Leipzig: F. A. Brockhaus, 1824.

Nau, Bernhard Sebastian von, and Georg Zinner. *Tabellarischer Entwurf der Naturgeschichte*. Mainz: Hof- und Universitäts-Buchdruckerey, 1784.

Naumann, Arno. *Dresdens Gartenbau bis zur Gründungszeit der "Flora," Gesellschaft für Botanik und Gartenbau in Dresden: Eine Festschrift zur Siebzigsten Stiftungsfeier der Genossenschaft "Flora."* Dresden: Arthur Schönfeld, 1898.

Nees von Esenbeck, C. G. "Von der Metamorphose der Botanik." *Isis* 2 (1818): 991–1008.

"Nekrolog [H. G. Ludwig Reichenbach]." *Sitzungsberichte der Naturwissenschaftlichen Gesellschaft Isis* 19 (1879): 1–8.

Neue Statuten der Königlich Preußischen Akademie gemeinnütziger Wissenschaft zu Erfurt. Erfurt, 1819.

Nickol, Thomas, ed. *Johann Wolfgang von Goethe und Christian Gottfried Daniel Nees von Esenbeck*. Halle: Deutsche Akademie der Naturforscher Leopoldina, 1997.

Nipperdey, Thomas. *Deutsche Geschichte, 1800–1866: Bürgerwelt und starker Staat*. Munich: C. H. Beck, 1983.

———. "Verein als soziale Struktur in Deutschland im späten 18. und 19. Jahrhundert: Eine Fallstudie zur Modernisierung." In *Gesellschaft, Kultur, Theorie: Gesammelte Aufsätze zur neueren Geschichte*, 174–205. Göttingen: Vandenhoeck & Ruprecht, 1976.

Noble, David F. *A World without Women: The Christian Clerical Culture of Western Science*. New York: Knopf, 1992.

Nummedal, Tara E. *Alchemy and Authority in the Holy Roman Empire*. Chicago: University of Chicago Press, 2007.

Nyhart, Lynn K. *Biology Takes Form: Animal Morphology and the German Universities, 1800–1900*. Chicago: University of Chicago Press, 1995.

———. "Civic and Economic Zoology in Nineteenth-Century Germany: The 'Living Communities' of Karl Möbius." *Isis* 89 (1998): 605–30.

———. *Modern Nature: The Rise of the Biological Perspective in Germany*. Chicago: University of Chicago Press, 2009.

———. "Natural History and the New Biology." In *Cultures of Natural History*, edited by Nicholas Jardine, James A. Secord, and E. C. Spary, 426–43. Cambridge: Cambridge University Press, 1996.

———. "Writing Zoologically: The *Zeitschrift für wissenschaftliche Zoologie*." In *Literary Structure of Scientific Argument: Historical Studies*, 43–71. Philadelphia: University of Pennsylvania Press, 1991.

Nyhart, Lynn K., and Thomas H. Broman, eds. *Science and Civil Society*. Osiris 17, 2002.

O'Boyle, Leonore. "Klassische Bildung und soziale Struktur in Deutschland zwischen 1800 und 1848." *Historische Zeitschrift* 207, no. 3 (1968): 584–608.

———. "The Problem of an Excess of Educated Men in Western Europe, 1800–1850." *Journal of Modern History* 42, no. 4 (1970): 471–95.

Ochsenheimer, Ferdinand. *Die Schmetterlinge Sachsens, mit Rücksichten auf alle bekannte europäische Arten*. Leipzig: Schwickert, 1806.
Oken, Lorenz. "Für die Aufnahme der Naturwissenschaften in den allgemeinen Unterricht." *Isis* (1829): 1225–37.
———. "Link mit Sprengel und Schrader." *Isis* 1 (1817): 2029–33.
———. *Ueber den Werth der Naturgeschichte, besonders für die Bildung der Deutschen*. Jena: Friedrich Frommann, 1809.
Olesko, Kathryn Mary. "Civic Culture and Calling in the Königsberg Period." In *Universalgenie Helmholtz: Rückblick nach 100 Jahren*, edited by Lorenz Krüger, 22–42. Berlin: Akademie, 1994.
———. *Physics as a Calling: Discipline and Practice in the Königsberg Seminar for Physics*. Ithaca, NY: Cornell University Press, 1991.
———, ed. *Science in Germany: The Intersection of Institutional and Intellectual Issues*. Osiris 5, 1989.
Paletschek, Sylvia. "Adelige und bürgerliche Frauen, 1770–1870." In *Adel und Bürgertum in Deutschland, 1770–1848*, edited by Elisabeth Fehrenbach, 159–85. Munich: R. Oldenbourg, 1994.
———. "Verbreitete sich ein Humboldt'sches Modell an den deutschen Universitäten im 19. Jahrhundert?" In *Humboldt international: Der Export des deutschen Universitätsmodells im 19. und 20. Jahrhundert*, edited by Rainer Christoph Schwinges, 75–104. Basel: Schwabe, 2001.
Parthier, Benno. *Die Leopoldina: Bestand und Wandel der ältesten deutschen Akademie*. Halle: Die Akademie, 1994.
Penny, H. Glenn. "Fashioning Local Identities in an Age of Nation-Building: Museums, Cosmopolitan Visions, and Intra-German Competition." *German History* 17 (1999): 489–505.
———. *Objects of Culture: Ethnology and Ethnographic Museums in Imperial Germany*. Chapel Hill: University of North Carolina Press, 2002.
———. "Wissenschaft in einer polyzentrischen Nation: Der Fall der deutschen Ethnologie." In *Wissenschaft und Nation in der europäischen Geschichte*, edited by Ralph Jessen and Jakob Vogel, 80–96. Frankfurt am Main: Campus, 2002.
Personalbestand und Büchersammlung der Gesellschaft für Natur- und Heilkunde in Dresden. Dresden: B. G. Teubner, 1841.
Peschier, Adolphe. *Wörterbuch der französischen und deutschen Sprache*. Stuttgart: J. G. Cotta, 1862.
Petzholdt, Alexander. *Die galvanische Vergoldung, Versilberung, Verkupferung usw.* Dresden and Leipzig: Arnold, 1842.
———, ed. *Populäre Vorlesungen über Naturwissenschaft*. Leipzig: Carl B. Lorck, 1845.
Petzholdt, Julius. *Der Plauensche Grund*. Dresden: Ernst Blockmann, 1842.
———. Johanna Wilhelmine Siegmundine Reichard. *Neuer Nekrolog der Deutschen* 26 (1848): 195–98.

———. *Nachricht von der Bibliothek des Gewerbvereines zu Dresden*. Dresden: Ernst Blochmann, 1843.
Pfetsch, Frank R. *Zur Entwicklung der Wissenschaftspolitik in Deutschland, 1750–1914*. Berlin: Duncker und Humblot, 1974.
Phillips, Denise. "Friends of Nature: Urban Sociability and Regional Natural History in Dresden, 1800–1850." *Osiris* 18 (2003): 43–59.
———. "Science, Myth and Eastern Souls: J. S. C. Schweigger and the Society for the Spread of Natural Knowledge and Higher Truth." *East Asian Science, Technology and Medicine* 26 (2007): 40–67.
"Physik." *Physkalisches Wörterbuch*. Vol. 7. Leipzig: E. B. Schwickert, 1833.
Pickstone, John V. "Working Knowledges before and after circa 1800." *Isis* 98, no. 3 (2007): 489–516.
Pinkard, Terry P. *German Philosophy, 1760–1860: The Legacy of Idealism*. Cambridge: Cambridge University Press, 2002.
"Plan und Gesetze der Naturforschenden Gesellschaft in Halle." *Abhandlungen der Hallischen Naturforschenden Gesellschaft* 1 (1783): xxi–xl.
Planert, Ute. "From Collaboration to Resistance: Politics, Experience, and Memory of the Revolutionary and Napoleonic Wars in Southern Germany." *Central European History* 39 (2006): 676–705.
Plan und Gesetze nebst dem Verzeichniss der jetztlebenden Mitglieder der Gesellschaft Naturforschender Freunde. Berlin, n.d.
Poggi, Stefano, and Maurizio Bossi, eds. *Romanticism in Science, Science in Europe, 1790–1840*. Dordrecht, Holland: Kluwer, 1994.
Pörksen, Uwe. "Der Übergang vom Gelehrtenlatein zur deutschen Wissenschaftssprache: Zur frühen deutschen Fachliteratur und Fachsprache in den naturwissenschaftlichen und mathematischen Fächern (ca. 1500–1800)." *Zeitschrift für Literaturwissenschaft und Linguistik* 51–52 (1983): 227–58.
Porter, Roy. *The Creation of the Modern World: The Untold Story of the British Enlightenment*. New York: Norton, 2000.
———. "Science, Provincial Culture and Public Opinion in Enlightenment England." *British Journal for Eighteenth-Century Studies* 3 (1980): 20–46.
Porter, Roy, and Mikuláš Teich, eds. *The Enlightenment in National Context*. Cambridge: Cambridge University Press, 1981.
Preusker, Karl Benjamin. *Andeutungen über Sonntags- und Gewerbeschulen, Vereine, Bibliotheken, und andere Förderungsmittel des vaterländischen Gewerbefleißes und der Volksbildung im Allgemeinen*. Leipzig: C. F. Hartman, 1834.
———. *Bürgerhalle: Anstalten und Einrichtungen zur gewerblichen, so wie zur allgemeinen Fortbildung des Bürgerstandes*. Meißen: C. C. Klinkicht & Sohn, 1847.
———. "Der Gewerbegeist im hermetisch-verschlossenen Glase." *Gewerbe-Blatt für Sachsen* 4 (1839).
———. *Der Sophien-Ducaten, oder des Tischlers Gustav Walthers Lehrjahre*. Leipzig: J. C. Hinrich, 1845.

———. *Gutenberg und Franklin: Eine Festgabe zum vierten Jubiläum der Erfindung der Buchdruckerkunst*. Leipzig: Heinrich Weinedel, 1840.
Prizelius, Johann Gottfried. *Vollständige Pferdewissenschaft*. Leipzig: Bey Weidmanns Erben und Reich, 1777.
Raschig, Franz. "Sendschreiben des Herrn Rektor M. Raschig in Zwickau an Dr. H. Köchly." *Vermischte Blätter zur Gymnasialreform* 1 (1847): 1–10.
Reddy, William. "The Structure of a Cultural Crisis: Thinking about the Commodification of Cloth in France before and after the Revolution." In *The Social Life of Things: Commodities in Cultural Perspective*, edited by Arjun Appadurai, 261–84. New York: Cambridge University Press, 1986.
Reichel, Friedrich. *Dresdens Umgebung nebst einem Wegweiser durch die Gegenden der sächsischen Schweiz*. Dresden: F. D. Reichel, 1818.
———. *Standorte der seltneren und ausgezeichneten Pflanzen in der Umgegend von Dresden*. Dresden and Leipzig: Arnold, 1837.
Reichel, Friedrich, and Ewald Dietrich. *Darstellungen der Heilquellen und Cur- und Badeorte des Königreichs Sachsen*. Dresden: Walther, 1824.
Reichenbach, H. G. Ludwig. *Blicke in das Leben der Thierwelt, verglichen mit dem Leben des Menschen*. Dresden: Arnold, 1843.
———. *Botanik für Damen, Künstler und Freunde der Naturwissenschaft überhaupt*. Leipzig: Carl Cnobloch, 1828.
———. *Catechismus der Botanik*. Leipzig: Baumgarten, 1825.
———. "Das Gymnasium und die Naturkunde." In *Der naturwissenschaftliche Unterricht auf Gymnasien*, edited by Hermann Richter and Ludwig Reichenbach, 67–88. Dresden: Arnold, 1847.
———. *Das naturhistorische Museum in Dresden*. Leipzig: Wagner, 1836.
———. "Denkschrift der Gesellschaft für Natur- und Heilkunde in Dresden." In *Der naturwissenschaftliche Unterricht auf Gymnasien*, edited by Hermann Richter and Ludwig Reichenbach, 1–16. Dresden and Leipzig: Arnold, 1847.
———. *Flora saxonica: Die Flora von Sachsen, ein botanisches Excursionsbuch*. Dresden: Arnold, 1842.
———. *Iconographia botanica exotica*. Leipzig: F. Hofmeister, 1827.
———. *Iconographia botanica, seu, Plantae criticae*. Leipzig: F. Hofmeister, 1823.
Reichenbach, H. G. Ludwig, and Johann Christoph Mössler. *Johann Christian Mösslers Handbuch der Gewächskunde: Enthaltend eine Flora von Deutschland, mit Hinzufügung der wichtigsten ausländischen Cultur-Pflanzen*. 2nd ed. Altona: Bei J. F. Hammerich, 1827.
Reid, Alexander. *Dictionary of the English Language*. 17th ed. Edinburgh: Oliver & Boyd, 1863.
Reill, Peter Hanns. *Vitalizing Nature in the Enlightenment*. Berkeley: University of California Press, 2005.
Retallack, James N., ed. *Saxony in German History: Culture, Society, and Politics, 1830–1933*. Social History, Popular Culture, and Politics in Germany. Ann Arbor: University of Michigan Press, 2000.

———. "Society and Politics in Saxony in the Nineteenth and Twentieth Century." *Archiv für Sozialgeschichte* 38 (1998): 396–457.
Reuß, Christian Friedrich. "Abhandlungen wie die Naturkunde der Grund zu einer wohleingerichteten Oekonomie, und wie groß der Einfluß derselben in dieser Wissenschaft ist." *Beschäftigung der Berlinischen Gesellschaft naturforschender Freunde* 3 (1777): 3–28.
———. *Dictionarium botanicumn*. Leipzig: Bey Christian Gottlob Hilscher, 1781.
———. *Dictionarium botanicum*. 2nd ed. Leipzig: Bey Christian Gottlob Hilscher, 1786.
Richards, Robert J. "Nature Is the Poetry of the Mind; or, How Schelling Solved Goethe's Kantian Problems." In *The Kantian Legacy in Nineteenth-Century Science*, edited by Michael Friedman and Alfred Nordmann, 27–50. Cambridge, MA: MIT Press, 2006.
———. *The Romantic Conception of Life: Science and Philosophy in the Age of Goethe*. Chicago: University of Chicago Press, 2002.
———. *The Tragic Sense of Life: Ernst Haeckel and the Struggle over Evolutionary Thought*. Chicago: University of Chicago Press, 2008.
Richter, G. M. "Hermann Eberhard Richter: Leben und Werk eines grossen Dresdner Arztes." PhD diss., Medizinische Akademie Carl Gustav Carus, 1964.
Richter, Hermann. "Denkschrift der Gesellschaft Isis." In *Der naturwissenschaftliche Unterricht auf Gymnasien*, edited by Hermann Richter and Ludwig Reichenbach, 17–67. Dresden and Leipzig: Arnold, 1847.
Richter, Hermann, and Ludwig Reichenbach, eds. *Der naturwissenschaftliche Unterricht auf Gymnasien*. Dresden and Leipzig: Arnold, 1847.
Richter, Otto, ed. *Dresdens Umgebung in Landschaftsbildern aus dem Anfange des neunzehnten Jahrhunderts*. Dresden: Römmler & Jonas, 1902.
Richter, Paul Emil. *Verzeichniss selbständiger Werke in der Kgl. öffentlichen Bibliothek zu Dresden welche sich auf die ganze sächsische Schweiz oder einzelne Theile derselben beziehen*. Dresden: C. C. Meinhold & Söhne, 1880.
Ringer, Fritz K. *The Decline of the German Mandarins: The German Academic Community, 1890–1933*. Cambridge, MA: Harvard University Press, 1969.
———. *Education and Society in Modern Europe*. Bloomington: Indiana University Press, 1979.
Riskin, Jessica. *Science in the Age of Sensibility: The Sentimental Empiricists of the French Enlightenment*. Chicago: University of Chicago Press, 2002.
Ritvo, Harriet. *The Platypus and the Mermaid, and Other Figments of the Classifying Imagination*. Cambridge, MA: Harvard University Press, 1997.
Roberts, Lissa. "Situating Science in the Dutch Enlightenment." In *The Sciences in Enlightened Europe*, edited by William Clark, Jan Golinski, and Simon Schaffer, 350–88. Chicago: University of Chicago Press, 1999.
Roche, Daniel. "Natural History in the Academies." In *Cultures of Natural History*, edited by Nicholas Jardine, James A. Secord, and E. C. Spary, 127–44. Cambridge: Cambridge University Press, 1996.

Röhling, Johann Christoph. *Deutschlands Flora: Zum bequemen Gebrauche beim Botanisiren, nebst einer erklärenden Einleitung in die botanische Kunstsprache zum Besten der Anfänger.* Bremen: F. Wilmans, 1796.

Römer, Johann Jacob. *Versuch eines möglichst vollständigen Wörterbuchs der botanischen Terminologie.* Zurich: Orell, Füßli und Comp., 1816.

Ross, Sydney. "Scientist: The Story of a Word." *Annals of Science* 18, no. 2 (1962): 65–85.

Rosseaux, Ulrich. *Freiräume: Unterhaltung, Vergnügen und Erholung in Dresden, 1694–1830.* Norm und Struktur, vol. 27. Cologne: Böhlau, 2007.

Roth, Albrecht Wilhelm. *Abhandlung über die Art und Nothwendigkeit die Natur-Geschichte in Schulen zu behandeln.* Nuremberg: George Peter Monath, 1779.

Rousseau, Jean-Jacques. *Abhandlungen, welche bey der Akademie zu Dijon im Jahr 1750 den Preis über folgende von der Akademie vorgelegte Frage davon getragen hat, Ob die Wiederherstellung der Wissenschaften und Künste etwas zur Läuterung der Sitten beygetragen hat?* [1752]. Translated by Johann Daniel Tietz. St. Ingbert: Röhring, 1997.

Rowland, Herbert, and Karl J. Fink, eds. *The Eighteenth-Century German Book Review.* Heidelberg: C. Winter, 1995.

Rudolphi, Caroline Christiane Louise. *Gemälde weiblicher Erziehung.* Heidelberg: Mohr und Winter, 1807.

Rudwick, Martin. "Minerals, Strata and Fossils." In *Cultures of Natural History*, edited by Nicholas Jardine, James A. Secord, and E. C. Spary, 287–304. Cambridge: Cambridge University Press, 1996.

Rüger, Alexander. "Populäre Naturwissenschaft in Nürnberg am Ende des 18. Jahrhunderts." *Berichte zur Wissenschaftsgeschichte* 5 (1982): 171–91.

Rupke, Nicolaas A. *Alexander von Humboldt: A Metabiography.* Frankfurt am Main: Peter Lang, 2005.

Ruppius, Heinrich Bernhard. *Flora Jenensis.* Frankfurt am Main: Ernestum Claud Bailliar, 1718.

Sachse, Carl Traugott. "Bericht der Isis." *Allgemeine deutsche naturhistorische Zeitung* 2 (1847): 183–274.

———. "Bericht über die Wirksamkeit der 'Isis.'" *Allgemeine deutsche naturhistorische Zeitung* 1 (1846): 181–201.

Sakurai, Ayako. "Science, Identity, and Urban Reinvention in a Mercantile City-State: The Associational Culture of Nineteenth-Century Frankfurt am Main." PhD diss., Cambridge University, 2006.

Schäffer, Jacob Christian. *Erläuterte Vorschläge zur Ausbesserung und Förderung der Naturwissenschaft.* Regensburg: Johann Leopold Montag, 1764.

Schaffer, Simon. "Genius in Romantic Natural Philosophy." In *Romanticism and the Sciences*, edited by Andrew Cunningham and Nicholas Jardine, 82–98. Cambridge: Cambridge University Press, 1990.

———. "Natural Philosophy and Public Spectacle in the Eighteenth Century." *History of Science* 21 (1983): 1–43.

Schama, Simon. *Landscape and Memory.* New York: Knopf, 1995.

Scheuchzer, J. J. *Physica, oder Natur-Wissenschaft*. Zurich, 1701.
Scheuner, Ulrich. "Staatsbild und politische Form in der romantischen Anschauung in Deutschland." In *Romantik in Deutschland*, edited by Richard Brinkmann, 70–89. Stuttgart: J. B. Metzler, 1978.
Schiebinger, Londa L. "Gender and Natural History." In *Cultures of Natural History*, edited by Nicholas Jardine, James A. Secord, and E. C. Spary, 163–77. Cambridge: Cambridge University Press, 1996.
———. *The Mind Has No Sex? Women in the Origins of Modern Science*. Cambridge, MA: Harvard University Press, 1989.
Schindling, Anton. *Bildung und Wissenschaft in der frühen Neuzeit, 1650–1800*. Munich: R. Oldenbourg, 1994.
Schlegel, Friedrich. *Kritische Schriften und Fragmente*. Vol. 1, *1794–1797*, edited by Ernst Behler and Hans Eichner. Paderborn: Ferdinand Schöningh, 1958.
Schleiden, M. J. *Grundzüge der wissenschaftlichen Botanik*. Leipzig: Wilhelm Engelmann, 1842.
Schleiden, M. J., and E. E. Schmid. *Encyclopädie der gesammten theoretischen Naturwissenschaften in ihrer Anwendung auf die Landwirthschaft*. Braunschweig: Friedrich Vieweg und Sohn, 1850.
Schleiermacher, Friedrich. "Gelegentliche Gedanken über Universitäten im deutschen Sinn." [1808]. In *Die Idee der deutschen Universität: Die fünf Grundschriften aus der Zeit ihrer Neubegründung durch klassischen Idealismus und romantischen Realismus*, edited by Ernst Anrich, 219–308. Darmstadt: Hermann Gentner, 1956.
Schlosser, Johann Georg. *Ueber Pedanterie und Pedanten, als eine Warnung für die Gelehrten des 18. Jahrhunderts*. Basel: Carl August Serini, 1787.
Schmaus, Warren. "Understanding and Explanation in France." In *Historical Perspectives on* Erklären *and* Verstehen, edited by Uljana Feest, 101–20. Dordrecht, Holland: Springer, 2010.
Schmid, Carl. *Allgemeine Encyklopädie und Methodologie der Wissenschaften*. Jena: Akademische Buchhandlung, 1810.
Schmid, Christian Heinrich. *Abriß der Gelehrsamkeit für Enzyklopädie Vorlesungen*. Berlin, 1783.
Schmiel, Martin. "Landwirthschaftliches Bildungswesen." In *Handbuch der deutschen Bildungsgeschichte*. Vol. 3, *Von der Neuordnung Deutschlands bis zur Gründung des Deutschen Reiches, 1800–1870*, edited by Karl-Ernst Jeismann and Peter Lundgreen, 306–10. Munich: C. H. Beck, 1987.
Schnädelbach, Herbert. *Philosophy in Germany, 1831–1933*. Cambridge: Cambridge University Press, 1984.
Schneider, Ute. *Friedrich Nicolais Allgemeine Deutsche Bibliothek als Integrationsmedium der Gelehrtenrepublik*. Wiesbaden: Harrassowitz, 1995.
Schneiders, Werner. "Der Philosophiebegriff des philosophischen Zeitalters: Wandlungen im Selbstverständnis der Philosophie von Leibniz bis Kant."

In *Wissenschaften im Zeitalter der Aufklärung*, edited by Rudolf Vierhaus. Göttingen: Vandenhoek und Ruprecht, 1985.

Schnieber, Hans-Rudolf. "Entwicklung des Zierpflanzenbaues von 1800–1939 am Beispiel Dresden." Dr. rer. hort., Technische Hochschule Hannover, 1958.

Schöler, Walter. *Geschichte des naturwissenschaftlichen Unterrichts im 17. bis 19. Jahrhundert*. Berlin: Walter de Gruyter, 1970.

Schön, Erich. "Weibliches Lesen: Romanleserinnen im späten 18. Jahrhundert." In *Untersuchungen zum Roman von Frauen um 1800*, edited by Elke Klainau and Claudia Opitz, 20–41. Tübingen: M. Niemeyer, 1990.

Schrank, Franz von Paula. *Allgemeine Anleitung, die Naturgeschichte zu studieren*. Munich: J. B. Strohl, 1783.

Schröder, Johann Samuel. "Einige Bermerkungen für die Sammler der Papilionen." *Mannigfaltigkeiten* 4 (1773): 490–96.

Schubert, J. A. "Ueber Beharrungsvermögen oder Trägheit der Masse." In *Populäre Vorlesungen über Naturwissenschaft*, edited by Alexander Petzholdt, 41–59. Dresden: Carl B. Lorck, 1845.

Schubring, Gert, ed. *"Einsamkeit und Freiheit" neu besichtigt: Universitätsreformen und Disziplinenbildung in Preussen als Modell für Wissenschaftspolitik im Europa des 19. Jahrhunderts*. Stuttgart: F. Steiner, 1991.

———. "Mathematisch-naturwissenschaftliche Fächer." In *Handbuch der deutschen Bildungsgeschichte*. Vol. 3, *1800–1870*, edited by Karl-Ernst Jeismann and Peter Lundgreen, 508–51. Munich: C. H. Beck, 1987.

———. "The Rise and Decline of the Bonn Natural Science Seminar." *Osiris* 5 (1989): 57–93.

Schultes, J. A., and J. J. Römer. *Systema vegetabilium*. Stuttgart: J. G. Cotta, 1817.

"Schul- und Unterrichtswesen." *Leipziger Repertorium der deutschen und ausländischen Literatur* 4 (1846): 60.

Schuster, John A., and Richard R. Yeo, eds. *Politics and Rhetoric of Scientific Method*. Dordrecht: D. Reidel, 1986.

Schütz, Johann Carl. *Kurze Beschreibung des Zinnstockwerks zu Altenberg*. Leipzig: Sommer, 1789.

Schweigger, J. S. C. *Einleitung in die Mythologie auf dem Standpunkte der Naturwissenschaft*. Halle: Bei Eduard Anton, 1836.

———. "J. S. C. Schweiggers 'Vorschläge zum Besten der Leopoldinisch-Carolinischen Akademie.'" Edited by Rudolph Zaunick. *Nova Acta Leopoldina* 29 (1964): 7–36.

Secord, Anne. "Botany on a Plate: Pleasure and the Power of Pictures in Promoting Early Nineteenth-Century Scientific Knowledge." *Isis* 93 (2002): 28–57.

Seibert, Peter. *Der literarische Salon: Literatur und Geselligkeit zwischen Aufklärung und Vormärz*. Stuttgart: Metzler, 1993.

Seit, Stefan. "Michael von Deinlein, 1800–1875." In *Die Bamberger Erzbischöfe: Lebensbilder*, edited by Josef Urban, 149–53. Bamberg: Archiv des Erzbistums Bamberg, 1997.

Selwyn, Pamela Eve. *Everyday Life in the German Book Trade: Friedrich Nicolai as Bookseller and Publisher in the Age of Enlightenment, 1750–1810*. University Park: Pennsylvania State University Press, 2000.

Shaffer, Elinor. "Romantic Philosophy and the Organization of Disciplines." In *Romanticism and the Sciences*, edited by Andrew Cunningham and Nicholas Jardine, 38–54. Cambridge: Cambridge University Press, 1990.

Shapin, Steven. "The Image of the Man of Science." In *Eighteenth-Century Science*, 4:159–83. Cambridge History of Science. Cambridge: Cambridge University Press, 2003.

———. *Never Pure: Historical Studies of Science as If It Was Produced by People with Bodies, Situated in Time, Space, Culture, and Society, and Struggling for Credibility and Authority*. Baltimore: Johns Hopkins University Press, 2010.

———. *A Social History of Truth: Civility and Science in Seventeenth-Century England*. Science and Its Conceptual Foundations. Chicago: University of Chicago Press, 1994.

Shapin, Steven, and Simon Schaffer. *Leviathan and the Air-Pump: Hobbes, Boyle, and the Experimental Life*. Princeton, NJ: Princeton University Press, 1985.

Sheehan, James J. *German History, 1770–1866*. Oxford: Oxford University Press, 1993.

———. *German Liberalism in the Nineteenth Century*. Chicago: University of Chicago Press, 1978.

———. "The German States and the European Revolution." In *Revolution and the Meaning of Freedom in the Nineteenth Century*, 246–79. Stanford, CA: Stanford University Press, 1996.

Shteir, Ann B. *Cultivating Women, Cultivating Science: Flora's Daughters and Botany in England, 1760–1860*. Baltimore: Johns Hopkins University Press, 1996.

Sibum, H. Otto. "Experimentalists in the Republic of Letters." *Science in Context* 16 (2003): 89–120.

Siemens, Werner. "Das naturwissenschaftliche Zeitalter." In *Von der Naturforschung zur Naturwissenschaft: Vorträge, gehalten auf Versammlungen der Gesellschaft Deutscher Naturforscher und Ärzte*, edited by Hans-Jochem Autrum. Berlin: Springer, 1987.

Skinner, Quentin. "Language and Social Change." In *Meaning and Context: Quentin Skinner and His Critics*, edited by James Tully, 97–134. Princeton, NJ: Princeton University Press, 1988.

Sloan, Philip R. "The Buffon-Linnaeus Controversy." *Isis* 67 (1976): 356–75.

Smellie, William, and Anton August Heinrich Lichtenstein. *Philosophie der Naturgeschichte*. Translated by Eberhard August Wilhelm von Zimmermann. Berlin: Voss, 1791.

Smith, Helmut Walser. *The Continuities of German History: Nation, Religion, and Race across the Long Nineteenth Century.* Cambridge: Cambridge University Press, 2008.

Smith, Pamela H. *The Body of the Artisan: Art and Experience in the Scientific Revolution.* Chicago: University of Chicago Press, 2004.

———. *The Business of Alchemy: Science and Culture in the Holy Roman Empire.* Princeton, NJ: Princeton University Press, 1994.

Smith, Pamela H., and Paula Findlen, eds. *Merchants and Marvels: Commerce, Science, and Art in Early Modern Europe.* New York: Routledge, 2002.

Smith, Roger. "British Thought on the Relation between the Natural Sciences and the Humanities, 1870–1910." In *Historical Perspectives on* Erklären *and* Verstehen, edited by Uljana Feest. Dordrecht, Holland: Springer, 2010.

Snelders, H. A. M. "Oersted's Discovery of Electromagnetism." In *Romanticism and the Sciences*, edited by Andrew Cunningham and Nicholas Jardine, 228–40. Cambridge: Cambridge University Press, 1990.

Snell Karl. *Skizze einer philosophischen Begründung des Gymnasial-Unterrichts.* Dresden: Karl Wagner, 1833.

Snyder, Laura J. *Reforming Philosophy: A Victorian Debate on Science and Society.* Chicago: University of Chicago Press, 2006.

Sondermann, Ernst Friedrich. *Karl August Böttiger, literarischer Journalist der Goethezeit in Weimar.* Mitteilungen zur Theatergeschichte der Goethezeit, vol. 7. Bonn: Bouvier, 1983.

Spary, E. C. "Identity Problems: On the History of Societies." Essay review. *Studies in the History and Philosophy of Biology and the Biomedical Sciences* 22, no. 3 (1991): 533–38.

———. "Political, Natural and Bodily Economies." In *Cultures of Natural History*, edited by Nicholas Jardine, James A. Secord, and E. C. Spary, 178–96. Cambridge: Cambridge University Press, 1996.

Spary, Emma C. "Forging Nature in the Republican Museum." In *Faces of Nature in the Enlightenment*, edited by Lorraine Daston, and Gianna Pomata, 163–80. Berlin: BWV, 2003.

Sperber, Jonathan. "*Bürger, Bürgerlichkeit, Bürgerliche Gesellschaft*: Studies of the German (Upper) Middle Class and Its Sociocultural World." *Journal of Modern History* 69 (1997): 271–97.

Speyer, O. *Bericht über die Thätigkeit des Vereins für Naturkunde zu Cassel vom 18. April 1847 bis 18. April 1860.* Cassel: Trömer & Dietrich, 1861.

Stadtlexikon Dresden. Dresden: Verlag der Kunst, 1998.

Stafleu, Frans Antonie. *Linnaeus and the Linnaeans: The Spreading of Their Ideas in Systematic Botany, 1735–1789.* Utrecht: International Association for Plant Taxonomy, 1971.

Starke, Holger. "Von der Residenzstadt zum Industriezentrum: Die Wandlung der Dresdner Wirtschaftskultur im 19. und 20. Jahrhundert." *Dresdner Hefte* 18, no. 61 (2000): 3–15.

Statuten der Isis. Dresden: Carl Ramming, 1844.

Statuten der naturforschenden Gesellschaft zu Bamberg. Bamberg: Reindl, 1834.
Statuten des Gewerbevereins in Dresden. Dresden, 1834.
Steffens, Henrik. *Grundzüge der philosophischen Naturwissenschaft.* Berlin: Im Verlage der Realschulbuchhandlung, 1806.
———. "Ueber das Verhältniß unserer Gesellschaft zum Staate." In *Schriften, alt und neu.* Breslau: Josef Max, 1821.
———. "Ueber die Bedeutung eines freien Vereins für Wissenschaft und Kunst." In *Schriften, alt und neu.* Breslau: Josef Max, 1821.
Stegmann, Hans. *Die Harmonie zu Dresden, 1786–1936.* Dresden, 1936.
Steinbach, Christoph Ernst. *Vollständiges Deutsches Wörterbuch.* Breslau: Johann Jacob Korn, 1734.
Stevens, Peter F. *The Development of Biological Systematics: Antoine-Laurent de Jussieu, Nature, and the Natural System.* New York: Columbia University Press, 1994.
Stewart, Jill. "The Culture of the Water Cure in Nineteenth-Century Austria, 1800–1914." In *Water, Leisure and Culture: European Historical Perspectives,* edited by Susan C. Anderson and Bruce H. Tabb, 23–36. Oxford: Berg, 2002.
Stewart, Larry. "Feedback Loop: A Review Essay on the Public Sphere, Pop Culture, and the Early-Modern Sciences." *Canadian Journal of History* 42 (2007): 463–83.
———. "A Meaning for Machines: Modernity, Utility, and the Eighteenth-Century British Public." *Journal of Modern History* 70 (1998): 259–94.
Stichweh, Rudolf. "Universität und Öffentlichkeit: Zur Semantik des Öffentlichen in der frühneuzeitlichen Universitätsgeschichte." In *Öffentlichkeit im 18. Jarhundert,* edited by Hans-Wolf Jäger, 103–16. Göttingen: Wallstein, 1997.
———. *Zur Entstehung des modernen Systems wissenschaftlicher Disziplinen: Physik in Deutschland, 1740–1890.* Frankfurt am Main: Suhrkamp, 1984.
Strickland, Stuart. "Galvanic Disciplines: The Boundaries, Objects and Identities of Experimental Science in the Era of Romanticism." *History of Science* 33 (1995): 449–68.
Struve, Friedrich August. *Ueber die Nachbildung der natürlichen Heilquellen.* Dresden: Arnold, 1824.
Suchow, Gustav. *Systematische Encyklopädie und Methodologie der theoretischen Naturwissenschaft.* Halle: C. A. Schwertschke und Sohn, 1839.
Suckow, Georg Adolph. *Anfangsgründe der theoretischen und angewandten Botanik.* Leipzig: Weidmanns Erban und Reich, 1797.
Sutton, Geoffrey V. *Science for a Polite Society: Gender, Culture, and the Demonstration of Enlightenment.* Boulder, CO: Westview Press, 1995.
Tebben, Karin, ed. *Beruf—Schriftstellerin: Schreibende Frauen im 18. und 19. Jahrhundert.* Göttingen: Vandenhoeck & Ruprecht, 1998.
Te Heesen, Anke. "Vom naturgeschichtlichen Investor zum Staatsdiener: Sammler und Sammlungen der Gesellschaft naturforschender Freunde zu

Berlin um 1800." In *Sammeln als Wissen: Das Sammeln und seine wissenschaftsgeschichtliche Bedeutung*, edited by Anke te Heesen and E. C. Spary, 62–84. Göttingen: Wallstein, 2001.

Tenfelde, Klaus. "Lesegesellschaften und Arbeiterbildungsvereine: Ein Ausblick." In *Lesegesellschaften und bürgerliche Emanzipation*, edited by Otto Dann, 253–74. Munich: C. H. Beck, 1981.

Terrall, Mary. "The Culture of Science in Frederick the Great's Berlin." *History of Science* 28 (1990): 333–64.

———. "The Uses of Anonymity in the Age of Reason." In *Scientific Authorship: Credit and Intellectual Property in Science*, edited by Mario Biagioli and Peter Galison, 91–112. New York: Routledge, 2003.

Thackray, Arnold. "Natural Knowledge in Cultural Context: The Manchester Model." *American Historical Review* 79, no. 3 (1974): 672–709.

Thackray, Arnold, and Jack Morrell, eds. *Gentlemen of Science: Early Correspondence of the British Association for the Advancement of Science*. London: Royal Historical Society, 1984.

Thamer, H.-U. "Emanzipation und Tradition: Zur Ideen- und Sozialgeschichte von Liberalismus und Handwerk, 1800–1850." In *Liberalismus in der Gesellschaft des deutschen Vormärz*, edited by Wolfgang Schieder, 55–73. Göttingen: Vandenhoeck & Ruprecht, 1983.

Thiersch, Friedrich. *Ueber den angeblichen Jesuitismus und Obscurantismus des Bayerischen Schulplanes*. Suttgart: J. C. Cotta, 1830.

Titze, Hartmut. "Die zyklische Überproduktion von Akademikern im 19. und 20. Jahrhundert." *Geschichte und Gesellschaft* 10, no. 1 (1984): 92–121.

Toellner, R. "Entstehung und Programm der Göttinger Gelehrten Gesellschaft unter besonderer Berücksichtigung des Hallerschen Wissenschaftsbegriffes." In *Der Akademiegedanke im 17. und 18. Jahrhundert*, edited by Fritz Hartmann and Rudolf Vierhaus, 97–115. Bremen and Wolfenbüttel: Jacobi, 1977.

Tolkemitt, Brigitte. "Knotenpunkte im Beziehungsnetz der Gebildeten." In *Ordnung, Politik und Geselligkeit der Geschlechter im 18. Jahrhundert*, edited by Ulrike Weckel, Brigitte Tolkemitt, Claudia Opitz, and Olivia Hochstrasser, 167–202. Göttingen: Wallstein, 1998.

Trentmann, Frank, ed. *Paradoxes of Civil Society: New Perspectives on Modern German and British History*. New York: Berghahn Books, 2000.

Trumpler, Maria. "Verification and Variation: Patterns of Experimentation in Investigations of Galvinism in Germany, 1790–1800." *PSA: Proceedings of the Biennial Meeting of the Philosophy of Science Association* 2 (1996): 75–84.

Tuchman, Arleen. *Science, Medicine, and the State in Germany: The Case of Baden, 1815–1871*. Oxford: Oxford University Press, 1993.

Tuchman, Arleen Marcia. "Institutions and Disciplines: Recent Work in the History of German Science." *Journal of Modern History* 69, no. 2 (1997): 298–319.

Turner, R. Steven. "The *Bildungsbürgertum* and the Learned Professions in Prussia, 1770–1830." *Histoire Social—Social History* 13 (1980): 105–35.
———. "German Science, German Universities: Historiographical Perspectives from the 1980s." In *"Einsamkeit und Freiheit" neu besichtigt: Universitätsreformen und Disziplinenbildung in Preussen als Modell für Wissenschaftspolitik im Europa des 19. Jahrhunderts*, edited by Gert Schubring, 24–36. Stuttgart: F. Steiner, 1991.
———. "The Great Transition and the Social Patterns of German Science." *Minerva* 25, nos. 1–2 (1987): 56–76.
———. "The Growth of Professorial Research in Prussia, 1818 to 1848." *Historical Studies in the Physical Sciences* 3 (1971): 137–82.
———. "Historicism, *Kritik*, and the Prussian Professoriate." In *Philologie und Hermeneutik im 19. Jahrhundert*, edited by Mayotte Bollack and Heinz Wismann, 450–89. Vol. 2. Göttingen: Vandenhoeck & Ruprecht, 1983.
———. "The Prussian Universities and the Research Imperative, 1806–1848." PhD diss., Princeton University, 1973.
———. "University Reformers and Professorial Scholarship in Germany, 1760–1806." In *The University in Society*. Vol. 2. Princeton, NJ: Princeton University Press, 1974.
"Ueber den Don Karlos." *Journal aller Journale* 9–10 (1787): 162–88.
"Ueber Gymnasialreform." *Blätter für literarische Unterhaltung* 2 (1845): 1239–40.
Umbach, Maiken. "Culture and *Bürgerlichkeit* in Eighteenth-Century Germany." In *Cultures of Power in Europe during the Long Eighteenth Century*, 180–99. Cambridge: Cambridge University Press, 2007.
Van Dülmen, Andrea. *Das irdische Paradies: Bürgerliche Gartenkultur der Goethezeit*. Cologne: Böhlau, 1999.
Van Dülmen, Richard. *Society of the Enlightenment: The Rise of the Middle Class and Enlightenment Culture in Germany*. Cambridge: Polity, 1992.
Van Horn Melton, James. *The Rise of the Public in Enlightenment Europe*. Cambridge: Cambridge University Press, 2001.
Varrentrapp and Wenner. Book advertisement in *Bibliothek der gesammten Naturkunde* 1 (1790): 170–73.
Verzeichniß der auf der vereinigten Hallischen und Wittenbergischen Friedrichs-Universität . . . zu haltenden Vorlesungen. Halle, 1829–40.
Verzeichniß der Mitglieder der Flora, Gesellschaft für Botanik und Gartenbau. Dresden, 1828.
"Verzeichniss der Mitglieder der Naturforschenden Gesellschaft." *Schriften der Naturforschenden Gesellschaft zu Leipzig* 1 (1822): 6–11.
Vick, Brian E. *Defining Germany: The 1848 Frankfurt Parliamentarians and National Identity*. Cambridge, MA: Harvard University Press, 2002.
Vickery, Amanda. "Golden Age to Separate Spheres?" *Historical Journal* 36, no. 2 (1993): 383–414.

Viereck, Regina. *"Zwar sind es weibliche Hände": Die Botanikerin und Pädagogin Catharina Helena Dörrien, 1717–1795.* Frankfurt am Main: Campus, 2000.
Vierhaus, Rudolf. "Bildung." In *Geschichtliche Grundbegriffe,* edited by Otto Brunner, Werner Conze, and Reinhart Koselleck, 1:508–51. Stuttgart: Ernst Klett, 1972.
———. "Die Organisation wissenschaftlicher Arbeit: Gelehrte Sozietäten und Akademien im 18. Jahrhundert." In *Die Königlich Preußische Akademie der Wissenschaften zu Berlin im Kaisserreich,* edited by Jürgen Kocka, 3–21. Berlin: Akademie, 1999.
———. "'Patriotismus'—Begriff und Realität einer moralisch-politischen Haltung." In *Deutsche patriotische und gemeinnützige Gesellschaften,* edited by Rudolf Vierhaus, 9–29. Wolfenbütteler Forschungen, vol. 8. Munich: Kraus International Publishing, 1980.
Virchow, Rudolf. "Die naturwissenschaftliche Methode und die Standpunkte in der Therapie." *Archiv für pathologische Anatomie und Physiologie und für klinische Medicin* 2 (1849): 3–7.
———. *Ueber die nationale Entwicklung und Bedeutung der Naturwissenschaften.* Berlin: August Hirschwald, 1865.
Volkens, G. "Die Geschichte des Botanischen Vereins der Provinz Brandenburg, 1859–1909." *Verhandlungen des Botanischen Vereins der Provinz Brandenburg* 51 (1909): 1–86.
Volkmann, Johann Jacob. *Italienische Bibliothek, oder Sammlung der merkwürdigsten kleinen Abhandlungen zur Naturgeschichte, Oekonomie und dem Fabrikwesen, aus den neuesten italienischen Monatsschriften.* Leipzig: Caspar Fritsch, 1778.
Voss, Jürgen. "Die Akademien als Organisationsträger der Wissenschaften im 18. Jahrhundert." *Historische Zeitschrift* 231 (1980): 43–74.
Wachter, Friedrich. *General-Personal-Schematismus der Erzdiözese Bamberg, 1007–1907.* Bamberg: Johann Nagengast, 1908.
Wagner, Rudolph. *Grundriß der Encyklopädie und Methodologie der medizinischen Wissenschaften.* Erlangen: J. J. Palm und Ernst Enke, 1838.
Wakefield, Andre. "The Apostles of Good Police: Science, Cameralism, and the Culture of Administration in Central Europe, 1656–1800." PhD diss., University of Chicago, 1999.
———. *The Disordered Police State: German Cameralism as Science and Practice.* Chicago: University of Chicago Press, 2009.
———. "Police Chemistry." *Science in Context* 12 (1999): 231–67.
Walker, Mack. *German Home Towns: Community, State, and General Estate, 1648–1871.* Ithaca, NY: Cornell University Press, 1998.
Waquet, Françoise. *Latin; or, The Empire of the Sign: From the Sixteenth to the Twentieth Century.* London: Verso, 2001.
Watanabe-O'Kelly, Helen. *Court Culture in Dresden: From Renaissance to Baroque.* Houndmills, Basingstoke, UK: Palgrave, 2002.

Weckel, Ulrike. "Der Fiberfrost des Freiherrn: Zur Polemik gegen weibliche Gelehrsamkeit und ihren Folgen für die Geselligkeit der Geschlechter." In *Geschichte der Mädchen- und Frauenbildung*, edited by Elke Kleinau and Claudia Opitz, 360–72. Vol. 1. Frankfurt am Main: Campus, 1996.

———. "'Der mächtige Geist der Assoziation': Ein- und Ausgrenzungen bei der Geselligkeit der Geschlechter im späten 18. und frühen 19. Jahrhundert." *Archiv für Sozialgeschichte* 38 (1998): 57–77.

———. "A Lost Paradise of a Female Culture? Some Critical Questions Regarding the Scholarship on Late Eighteenth- and Early Nineteenth-Century German Salons." *German History* 18, no. 3 (2000): 310–36.

———. *Zwischen Häuslichkeit und Öffentlichkeit: Die erste deutschen Frauenzeitschriften im späten 18. Jahrhundert und ihr Publikum*. Tübingen: Max Niemeyer, 1998.

Wehler, Hans Ulrich. *Deutsche Gesellschaftsgeschichte*. Vol. 2, *Von der Reformära bis zur industriellen und politischen "Deutschen Doppelrevolution," 1815–1845/9*. Munich: C. H. Beck, 1987.

Weis, Eberhard. *Die Säkularisation der bayerischen Klöster, 1802/03: Neue Forschungen zu Vorgeschichte und Ergebnissen*. Munich: Verlag der Bayerischen Akademie der Wissenschaften, 1983.

Wengenroth, Ulrich. "Science, Technology, and Industry." In *From Natural Philosophy to the Sciences: Writing the History of Nineteenth-Century Science*, edited by David Cahan, 221–51. Chicago: University of Chicago Press, 2003.

Whaley, Joachim. "The Transformation of *Aufklärung*: From the Idea of Power to the Power of Ideas." In *Cultures of Power in Europe during the Long Eighteenth Century*, edited by Hamish Scott and Brendan Simms, 158–80. Cambridge: Cambridge University Press, 2007.

Whitman, James. "The Last Generation of Roman Lawyers in Germany." In *The Uses of Greek and Latin: Historical Essays*, edited by A. C. Dionisotti, Anthony Grafton, and Jill Kraye, 212–25. London: Warburg Institute, 1988.

Winter, Emma L. "Between Louis and Ludwig: From the Culture of French Power to the Power of German Culture, 1789–1848." In *Cultures of Power in Europe during the Long Eighteenth Century*, 348–68. Cambridge: Cambridge University Press, 2007.

Wise, M. Norton. "On the Relation of Physical Science to History in Late Nineteenth-Century Germany." In *Functions and Uses of Disciplinary Histories*, edited by Loren Graham, Peter Weingart, and Wolf Lepenies, 3–34. Dordrecht, Holland: D. Reidel, 1983.

———, ed. *The Values of Precision*. Princeton, NJ: Princeton University Press, 1997.

Wittmann, Reinhard. *Geschichte des deutschen Buchhandels*. 2nd ed. Beck'sche Reihe 1304. Munich: C. H. Beck, 1999.

Wood, Paul. "Science, the Universities, and the Public Sphere in Eighteenth-Century Scotland." *History of Universities* 13 (1994): 99–135.

Woodmansee, Martha. *The Author, Art, and the Market: Rereading the History of Aesthetics.* New York: Columbia University Press, 1994.

Wucherer, Gustav Friedrich. *Über das Verhältniss des Studiums der Naturlehre zur übrigen wissenschaftlichen Ausbildung.* Freiburg: Herder, 1813.

Yearbook of Scientific and Learned Societies of Great Britain and Ireland. London: Charles Griffin and Co., 1884.

Yeo, Richard R. "Classifying the Sciences." In *Eighteenth-Century Science.* Vol. 4. Cambridge History of Science. Cambridge: Cambridge University Press, 2003.

———. *Encyclopaedic Visions: Scientific Dictionaries and Enlightenment Culture.* Cambridge: Cambridge University Press, 2001.

———. *Science in the Public Sphere: Natural Knowledge in British Culture, 1800–1860.* Aldershot, UK: Ashgate, 2001.

Zaunick, Rudolph. "Gründung und Gründer der Naturwissenschaftlichen Gesellschaft Isis in Dresden vor hundert Jahren." *Sitzungsberichte und Abhandlungen der Naturwissenschaftlichen Gesellschaft Isis zu Dresden* (1934): 20–49.

Zaunstöck, Holger. *Sozietätslandschaft und Mitgliederstrukturen: Die mitteldeutschen Aufklärungsgesellschaften im 18. Jahrhundert.* Hallesche Beiträge zur europäischen Aufklärung 9. Tübingen: Niemeyer, 1999.

———. "Zur Einleitung: Neue Wege in der Sozietätgeschichte." In *Sozietäten, Netzwerke, Kommunikation: Neue Forschungen zur Vergesellschaftung im Jahrhundert der Aufklärung,* 1–10. Halle: Niemeyer, 2003.

Zaunstöck, Holger, and Markus Meumann. *Sozietäten, Netzwerke, Kommunikation: Neue Forschungen zur Vergesellschaftung im Jahrhundert der Aufklärung.* Tübingen: Niemeyer, 2003.

Zeise, Roland. "Die bürgerliche Umwälzung: Zentrum der proletarischen Parteibildung." In *Geschichte Sachsens,* edited by Karl Czok, 332–80. Weimar: Hermann Böhlaus Nachfolger, 1989.

Ziche, Paul. "Von der Naturgeschichte zur Naturwissenschaft: Die Naturwissenschaften als eigenes Fachgebiet an der Universität Jena." *Berichte zur Wissenschaftsgeschichte* 21 (1998): 251–63.

Ziche, Paul, and Peter Bornschlegell. "Wissenschaftskultur in Briefen: F. A. C. Grens antiphlogistische Bekehrung, galvanische Experimentalprogramme und internationale Wissenschaftsbeziehungen in Briefen an die Jenaer 'Naturforschende Gesellschaft.'" *NTM* 8 (2000): 149–69.

Zimmerman, Andrew. *Anthropology and Antihumanism in Imperial Germany.* Chicago: University of Chicago Press, 2001.

Ziolkowski, Theodore. *German Romanticism and Its Institutions.* Princeton, NJ: Princeton University Press, 1990.

Zunck, Hermann Leopold. *Die natürlichen Pflanzensysteme geschichtlich entwickelt.* Leipzig: Hinrich, 1840.

Index

Academy of the Arts (Dresden), 79, 81, 183
Academy of Sciences: Bavaria, 1, 103; Göttingen, 122; Paris, 5, 44; Prussia, 12, 53, 178
academy professors, 79, 96, 103, 183, 207–12
aesthetic taste, science as a source of: in the Enlightenment, 33–34, 78–79; in liberal writings, 214–17, 255; during the Romantic Era, 151–52, 154, 158–61, 163, 165–70, 172–76
agricultural education, 19–20, 92, 94, 207, 208, 225
agricultural improvement, 20, 65, 89–96, 92, 168, 207, 208, 225. *See also* economic societies
Allgemeine Deutsche Bibliothek, 50, 129, 267n7
Allgemeine Literatur-Zeitung, 51, 129, 151
amateur science, 12–13, 14, 84–85, 119–20, 157, 165–66, 177–80
Ammon, Friedrich August von, 109
anatomy, 4, 5, 7, 143, 216, 238
anonymity, 17, 50–51, 130–31, 138–39
Anschauung, 248–49
apothecaries: in early nineteenth-century learned networks, 102, 106, 153, 169; as members of regional natural history societies, 184–85, 186, 190, 191, 192, 195, 199; as *Naturforscher* in the eighteenth century: 61–62, 65–66
art and *Wissenschaft*, 151, 160–61, 163, 167–70, 172–76. *See also* aesthetic taste, science as a source of
Association of German Agriculturalists, 96
Arenswald, C. F. von, 55
artisans: as artists, 215–17; hostility to science, 226; as members of economic or manufacturing societies, 92, 94, 208, 210; as *Naturforscher*, 61, 64–65, 69–70, 72, 92, 103–6, 186, 191, 204, 206, 221–24; utility of science to, 70, 209–11, 219–22, 223–24
associational life, 20–21, 145–46. *See also* civil society and natural science; learned world; public sphere
authorship, 124–28, 130–31, 138–39, 144–45, 154–56
automata, 76

Baader, Klement Alois, 125
Bacon, Francis, 56, 233, 243
ballooning, 76, 255
Bamberg, Nature-Researching Society of, 195–98
Bartsch, Moritz, 183
Bastian, Adolph, 231, 249
Bavaria, 194–200
Bavarian school plan of 1829, 236, 250
Becker, Wilhelm, 80

349

Beckmann, Johann, 35, 95
Beger, August, 207, 217, 218
Beireis, Gottfried Christoph, 128
Berlin, Physical Society of, 191
Berlin, Society of Nature-Researching Friends in, 12, 34–35, 36, 37, 38, 44, 47, 73, 118, 119, 139, 140, 179; correspondence networks of, 45, 58, 126, 135–36; during the Prussian Reform Era, 90, 100–101; publication patterns of, 140–41, 142; status differences among members, 67–69; ties to Freemasonry, 67
Bernoulli, Johann, 32, 51
Bescherer, Julius, 185
Bessel, Friedrich Wilhelm, 186, 191
Beuth, Christian Peter Wilhelm, 216, 225
Bildung, 15–16, 63; natural science as a form of, 82–83, 96–97, 105, 109, 117, 136, 149–51, 164, 181, 200, 297–98n13; women's influence on, 171–72, 233
Bildungsbürgertum: relation to *Besitzbürgertum*, 254–55; use as a social label, 15, 21–22, 63, 179–80, 183, 219, 252
Bischof, Wilhelm, 189
Blankenburg, Natural Scientific Society of, 7, 181
Blumenbach, Johann Friedrich, 55, 56–57
Bonn, Natural History Association of, 184
botany, 7, 37, 73–74, 83, 102; Latinity and, 75, 111; philological works, 112; Romantic versions of, 151, 152–53, 155, 158–73; theory and practice in, 106–7; *wissenschaftliche Botanik*, 176, 187
bourgeois identity, 12, 21, 42, 53, 55, 146, 150, 264n53, 266n73, 297–98n13
Breslau, Silesian Society for the Culture of the Fatherland, 87
Brewster, David, 5
British Association for the Advancement of Science, 5, 8, 262n23
British science, comparisons with Germany, 4–8, 23–24, 58–59, 125, 146–47, 172, 228, 229–30, 269n37–38
Brockhaus lexicon, 28, 115, 149
Bücking, J. J. H., 128

Calinich, Eduard, 229
cameralism, 35, 182, 269n38
Carus, Carl Gustav, 87–88, 99, 108, 116, 149, 168, 238
Cassel, natural history society in, 185

Catholic priests, as members of scientific societies, 185, 194–98
chemistry, 2, 4, 64, 66; natural philosophy and, 31–32, 33, 34, 36; popularity of, 70, 73, 75, 84, 148, 161, 190; theory and practice in, 91, 106, 108–9; the unity of natural science and, 116; utility of, 204, 206, 207, 208, 210, 222, 236, 255
civility, 133–38
civil society and natural science, 20–22, 35–36, 52–53, 70, 97, 264n53. *See also* associational life; learned world; public sphere
classical philology: compared with natural science, 206, 245–46; ideals of *Wissenschaft* and, 281n45; *Naturforscher* as practitioners of, 112–13. *See also* Latinity; neohumanism
classification of the sciences in the early modern period, 2, 22, 27–28, 30–32
collecting, 36, 65, 66–67, 76, 78, 80, 83–85, 117, 127, 128, 143, 257; divergences between low and high science, 187–89; regional natural history and, 177, 199–201, 210; Romantic guidelines for, 164–65
commodification, 17–18, 47–49, 51
Congrès scientifiques de France, 8
consumer culture, 17–18, 52, 61–62, 117, 179
correspondence networks: in eighteenth-century learned culture, 16, 29, 46–47, 51, 57, 58, 68, 83–84; nineteenth-century changes in, 105–6, 126–27, 128, 177–78
cosmopolitan ideals, 22, 23–24; in the regional natural history milieu, 182, 186, 193, 194, 197–98, 200; in the Romantic era, 120–24, 285n33
Cotta, Johann, 156–62, 175
court culture, 53, 55, 60–62, 64–66, 82, 150, 153
Crell, Lorenz, 34, 66
criticism, 50–51, 129–31, 132–39

Danzig, Nature-Researching Society of, 190
Dierbach, Johann Heinrich, 92, 107, 141
Dilthey, Wilhelm, 240
Dittrich, Heinrich, 254
Döllinger, Ignaz, 158

Dörrien, Catharina Helene, 69
Dresden: court culture in, 60–62; Flora Society, 167–72; Gymnasialverein, 242–47; Isis Society, 177–78, 183–85, 188, 193; Manufacturing Association, 220, 203, 207–10, 212, 224, 255, 256; Mineralogical Society, 137; Natural Scientific Society, 210–13, 255, 256; relationship to countryside, 77–78; Society for Natural and Medical Knowledge, 101–3, 105, 107–9, 117, 126, 136–40, 151, 167, 185, 188; Society for Saxon Antiquities, 181; status as a scientific center, 24–25, 120; Technical Academy, 212; urban culture of, 24–25, 62–63
Droysen, Johann, 30, 240
Du Bois-Reymond, Emil, 6–7, 120, 249, 250
dynastic loyalty, 24, 181–82, 194

economic societies: divergence from learned societies, 89–96; eighteenth-century similarities to learned societies, 35–37. *See also* manufacturing associations
educated estates, as a social category, 19–20, 21–22, 58, 63, 149–51
educated middle classes. See *Bildungsbürgertum*
educational reform: 19–20, 257; effect on the concept of "natural science," 229–31, 239–47; in Prussian, 90. *See also* agricultural education; normal schools; technical education
Ehrhart, Friedrich, 34, 55, 72, 74
Einseidel, Detlev von, 93, 181
Einseitigkeit, 116–17, 167, 174, 233
electricity, 3, 33, 57, 76, 111, 153, 202, 205, 249
emotion, natural science and, 33–34, 258. *See also* aesthetic taste, science as a source of
engineers, 179, 184–85. *See also* technical education
Enlightenment science, 2–3, 30–39, 50–53, 57–59, 64, 68, 70–71, 74, 76, 83; nineteenth-century legacy of, 197, 219–22; Romantic critiques of, 86–89, 91–92, 98–100, 105, 113–14, 129, 131–32, 135, 138, 166
erudition, 15–16, 41, 109–13. *See also* Latinity

experimental philosophy, 44–45, 61, 75, 137–38
experimentation, 2, 66, 76, 95, 137–38, 210, 213, 242, 281n55
expertise: learnedness and, 52–53, 91–96, 105–6, 268n13; technical, 62, 64–65, 179, 184–85, 203–4, 210, 255–57. *See also* under *Wissenschaft*

factory owners, 103–4, 107–9, 210, 256
farmers, 21, 38, 61, 67, 70, 71, 72, 94, 96, 110, 204, 209
Fichte, Johann Gottlieb, 132–33, 214
Ficinus, Andreas, 60–61, 62, 63, 66, 83, 177
Ficinus, Heinrich, 83–84, 109
floras, 11, 74, 83–84, 102, 189
Flurl, Mathias: 1–2, 37–38, 46, 231
forestry, 20, 35, 38, 70, 91, 92, 107, 167, 184, 185, 199, 242
Fraas, Karl, 179, 180
Frankfurt, Senckenberg Society of, 104, 122, 193
Fraunhofer, Joseph, 20, 103–4, 144
Frederick II, King of Prussia, 53
Freemasonry, 67–68, 88, 145–46
French science, comparisons with Germany, 4–8, 23–24, 53–54, 55, 125, 146–47, 228, 269n38
Friedrich August II, King of Saxony, 181, 244
friendship: scientific communication and, 45–47, 134–35; shared love of nature and, 34, 76–77, 82

gardeners, 65, 69, 70, 102, 153, 169, 173
gardening, 37, 76–77, 78–79, 107, 167, 169, 220, 226, 227
Gauss, Carl Friedrich, 111
gebildete Publikum, das. *See* educated estates, as a social category
Gehler, J. S. T., 27–29
Gelehrte. *See* learned identity
gender: as a boundary marker in the learned world, 53–55, 153–56, 261–62n14; cameralism and, 290n20. *See also* women
Geschichtliche Grundbegriffe (Koselleck), 3
Gesellschaft Deutscher Naturforscher und Ärzte, 5, 8, 12, 87, 88, 118, 120, 123, 133–34, 202
gift exchange, 17, 49

351

INDEX

Gilbert, Ludwig Wilhelm, 142, 144
Gleim, Johann Wilhelm Ludwig, 45
Goethe, Johann Wolfgang von, 11, 57, 102; morphology and, 88, 115–16; as a scientific model, 151, 157, 159, 160–61, 168, 176, 216–17. *See also* Weimar classicism
Goldfuß, Georg August, 131–32, 133, 158, 161
Görlitz: Nature-Researching Society of, 120; Ornithological Society of, 119
Göttingischen Gelehrten Anzeigen, 50
Gössel, Johann Heinrich, 183, 190, 191
Götzinger, Wilhelm Lebrecht, 80, 81
Gradmann, Johann Jacob, 125–26
Gross- und Neuschönau, Saxonia Society of, 184
gymnastic societies, 21, 145, 240, 252

Habermas, Jürgen, 17, 42, 267–68n12
Halle, Natural Scientific Association for Saxony and Thüringen, 191–93, 227
Halle, Nature-Researching Society of, 34–35, 54, 101, 117, 191–93; publication patterns of, 141, 179, 227
Haller, Albrecht von, 83
Hamburg, 193
handbooks, introductory, 73–74, 162
Harleß, Johann Christian, 154–56
Harzer, Carl August Friedrich, 81–83, 177–78, 180, 183–84, 186, 189, 190, 191
Haymann, Gottfried, 63–64
Hegel, G. F. W., 214
Hegelianism, 10, 13, 239–40
Heimat, 182, 194, 252
Helmholtz, Hermann von, 6–7, 11, 178, 249
Herder, Johann Gottfried, 55
Hilpert, Johann Wolfgang, 11, 13, 14, 24
history, 12, 23, 181–82, 239–41
Holy Roman Empire, 88, 90, 110
Huber, Therese, 153, 158–62, 176
Humboldt, Alexander von, 11, 116, 130, 151, 175, 245, 249

ideadlism, mid-century decline of, 214–15, 247–48, 283n93. See also *Naturphilosophie*; Romantic science
Immermann, Karl, 124, 129
industrialization, 20; natural science and, 202–4, 206–13; *Wissenschaft* and, 202–3, 213–25

information overload, 126
institutionalization of science, 9, 11, 40–41
Isis, 129–31, 236

Jablonsky, Karl Gustav, 49
Jacobi, Friedrich Heinrich, 99, 100, 105, 123, 135
Jena, Nature-Researching Society, 45, 100
Jews, as members of scientific societies, 68
Jöcher, Christian Gottlieb, 124, 126
Johnston, James, 5, 121
Jussieu, Antoine Laurent de, 74

Kant, Immanuel, 31–33, 38, 40, 115, 160, 214
Karsten, Wenceslaus Johann Gustav, 30–33, 38, 46
Keferstein, Christian, 192
Klopstock, Friedrich Gottlieb, 32, 42, 47, 51
Köchly, Hermann, 235, 236, 239–43
Königsberg, Physical-Economic Society in, 36
Koselleck, Reinhard, 3
Kreutzberg, Karl Josef, 202, 203, 205, 206
Kreyssig, Friedrich Ludwig, 109

Lambert, Johann Heinrich, 32, 51
Latin, use of, 53, 55–57, 109–12, 254, 276–77n64
Latinity: as a boundary marker of the learned world, 53, 55–57, 109–13; criticism of, 71–72. *See also* classical philology; neohumanism
learned estate, 1, 15–16, 23, 41–42, 61, 100, 150, 270–71n67
learned identity, 14–21, 152–53; practical knowledge and, 38–39, 98–100, 186; threats to, 39–40, 98–99; markers of, 84–85, 125–28, 125–28, 136–37, 143–46, 200–201
learned practices, spread of, 61–62, 72–85, 89–90, 125–28, 178–81, 200–201, 205–6, 221–24, 227–28, 255
learned world, 13–22, 29–30, 251–53; social composition of, 61–63, 150, 257–58; tensions in, 84–85, 177–81, 226–27. *See also under* public sphere
lecturing, 75–76, 210–13, 255–56
Leipzig, 24; book fair in, 74; Economic Society of, 36, 92–96 (*see also under*

352

Saxony); Linnaean Society, 36, 100; Nature-Researching Society of, 103, 104, 119, 144
leisure, science as a form of, 61–63, 72–83, 120, 179, 183–84, 190, 223, 267n7
Leopoldina, 90, 110–11
Leuckart, Friedrich Sigismund, 123
Leysser, Friedrich Wilhelm von, 34, 70, 71, 73
liberalism, 23, 97, 182, 199, 204–5, 213, 215–25, 243–44
Lichtenberg, Georg, 76
Liebig, Justus von, 206, 207, 208, 210, 212, 213, 236, 250
Link, Heinrich, 116
Linnaean Society (London), 8. *See also* under Leipzig
Linnaeus, Carl, 48, 49, 51, 52, 57, 74, 163
Lippe, Natural Scientific Association of, 180
Literary and Philosophical Societies (British), 8, 58–59, 147, 269
low scientific culture, 178–79

Mach, Ernst, 249
Mager, Karl, 236
manufacturing associations, 219–22
materialism, 10, 263n33
Marburg, Society for the Pursuit of All the Natural Sciences, 120
Martini, Friedrich, 34, 44, 45, 58, 69
mathematics, 10, 36, 102, 208, 234; applied, 2, 27, 28, 46, 148; definitions of natural science and, 6–7, 30–33
mathematization of science, 6, 9, 268n16
Maupertius, Pierre Louis Moreau de, 53
mechanics, 6, 208
Mendel, Gregor, 184
Merian, Maria Sibylla, 19
Metaphysical Foundations of Natural Science (Kant), 31–33
Methodologie, 237–38
mining, 20, 35, 38, 62, 64, 70, 80, 86, 102, 103, 167, 225–26
Mohl, Hugo von, 250
Morgenblatt für gebildete Stände, 153, 156–62
morphology, 7, 88, 115, 187
Müller, Johannes (bibliographer), 29
Müller, Johannes (physiologist), 243
Muncke, Georg Wilhelm, 28

Nagel, Carl, 189
Nägeli, Carl, 187
Napoleonic Wars, 91, 100, 123, 124
national identity, 21, 22, 23–24, 96, 120–24; as an inspiration for the unity of natural science, 23–24, 121; regional natural history as a vehicle for, 200
natural history, 2, 7, 27, 31–32, 34, 37–38, 74, 86; "general" forms of, 91–92; regional, 36, 177–201
natural philosophy, 2, 4, 8, 19, 20, 27–29, 31–32, 35, 36, 37, 38, 46, 52, 64, 70, 72–74, 76, 85, 86, 95, 98, 110, 115–16, 153–54, 204, 208, 230
natural science: definitions of, 2–3, 9, 27–29, 30–33, 115–17, 149–50, 180–81, 202–5, 245; unity of, 6, 9, 31–32, 86–90, 115–17, 135–36, 139–40, 147–48, 238–39, 256–58
natural scientific associations: as a German peculiarity, 7–9; publication patterns of, 140–43; as vehicles for the production of credit, 145, 186
natural scientific method, 37–38, 229–31, 243–45, 249, 301n8
natural theology, 34. *See also* religion
nature as aesthetic and emotional object, 76–79, 254–55, 258–59. *See also* aesthetic taste, science as a source of
Naturforscher, characteristics, 4–5, 25, 29, 33–34, 165–66
Naturphilosophie, 87–88, 116, 154, 159–61, 175–76, 214–15
Naturwissenschaft/Geisteswissenschaft distinction, 228–31, 239–47, 249–51
"*naturwissenschaftlich*" as an adjective, 3, 231, 252–53, 301n8
Nees von Esenbeck, Christian, 87–88, 110–11, 151–53, 156–62, 174–75, 176
neohumanism, 229, 231–36
Neumann, Franz, 249
Newton, Isaac, 186
Nicolai, Friedrich, 50–51, 52
nobility: as an audience for science, 21, 80, 151; as improving landlords, 20, 65; as members of learned associations, 68–69, 94–95, 173; as *Naturforscher*, 54, 60, 61, 62, 64, 66–67, 69; as scientific patrons, 50, 62, 69, 154
normal schools, 19–20, 179, 185, 257
Nuremberg, Natural History Society in, 12

objectivity, 34, 237
Ochsenheimer, Ferdinand, 55, 84–85, 117
Oekonomie, 35, 103
Oken, Lorenz, 14, 87–88, 91, 96–98, 116, 120–21, 129–32, 133–34, 135, 236, 286n54
Ohm, Georg Simon, 249
oral communication in science, 132–33, 265n62
Ørsted, Hans Christian, 111, 144
Ostwald, Wilhelm, 249

Palatinate, the, Pollichia Society, 188, 198–200
patriotic-economic societies. *See* economic societies
patriotism, as an enlightened ideal, 37
Paula Schrank, Franz von, 49
periodicals, 24, 124–25, 142–43
Petzholdt, Alexander, 209, 255–57
Petzholdt, Julius, 254–58
philosophy, 2, 115–16, 212. *See also* natural philosophy; *Naturphilosophie*
physica. *See* natural philosophy
physical science, nineteenth-century definitions of 4, 7
physics, 3, 4, 39, 75, 106, 116, 144, 148, 190, 208, 222, 245, 249. *See also* mathematics; mechanics; natural philosophy
physiology, 4, 5, 148, 213
physicians, university-educated, 61, 62, 64, 92, 94, 101, 109, 212, 281n50
Pleininger, Theodor, 188
political centralization, effects on science from, 122, 147, 181–82, 193–200
political identity, 193–200. *See also* dynastic loyalty; liberalism; national identity; regional identity
polycentrism, as a feature of German cultural life, 23–24, 146–47
popularization, 12–13, 43, 152–53, 211–12
positive knowledge, 233–34, 247–48
Pötzsch, Christian, 72–73
practical knowledge, 35–38, 89–96, 152, 225–26. *See also under* learned identity; *Wissenschaft*
precision, 2
Preusker, Karl, 204, 209–10, 216, 218, 221–22, 223–24
print culture: expansion of, 19, 39–52, 84, 124–27, 132–33, 218–19, 271–72n84;

face-to-face sociability and, 24, 135–36
productive estate (*Nahrungsstand*) and science, 36, 100, 185, 219, 235
professionalization of science, 12–13, 43, 165–66, 218, 268n13
public, learned or scientific v. educated, 18, 29–30, 39–59, 151–53, 157, 161–62, 174
public sphere, 1, 3; the learned world and, 16–22; 29–30; 39–53, 152–53, 267–68n12

railroads, 202, 203
realism, 214–15; as an educational philosophy, 229
Réaumur, René Antoine Ferchault de, 57
Regensburg, Botanical Society of, 119
regional identity, 120–24, 145. *See also* dynastic loyalty
Reichard, Gottfried, 75–76, 255–56
Reichard, Hermine, 254–58
Reichard, Wilhelmine, 75–76, 255–56
Reichel, Friedrich, 188
Reichenbach, Ludwig, 13, 151–53, 162–73, 174–75, 188, 189, 190, 242–47
Reil, Johann Christian, 142
religion: Protestant-Catholic tensions, 199; science and, 10, 33–34, 154, 164, 263n33
republic of letters. *See* learned world
research revolution, 41
Reuß, Johann, 73
review journals, 50, 125, 129, 251–52
Richter, Hermann, 185, 242–47
Romantic science, 154, 283n93, 284n6, 285n33; general science of nature and, 87–88, 118; in liberal writings, 217; literary public and, 169–70, 174–76; low scientific culture and, 214–15; persistence in the 1830s and 1840s, 236, 243; practical knowledge and, 97; relationship of science to aesthetic experience, 79, 151. See also *Naturphilosophie*; Weimar classicism
Roßmäßler, Emil, 242–47
Rudolphi, Caroline, 153–54
Ruge, Arnold, 239

Sachse, Carl, 186–87
salons, 14, 52–53, 170–71, 176
Sattelzeit, 3, 20

Saxony, Economic Society in the Kingdom of, 93–96, 102–3, 105
Saxony, political and economic conditions in, 92–96, 206
Schäffer, Jacob Christian, 57, 70, 71
Schelling, Friedrich Wilhelm Joseph, 115
Schiller, Friedrich, 158, 176, 217
Schlegel, Friedrich, 173
Schleiden, Matthias Jacob, 140, 175–76, 208
Schleiermacher, Fredrich, 99
Schlosser, Johann Georg, 39–41, 44, 54
schoolteachers, 179, 183–85, 190
Schröter, Johann Samuel, 84
Schubert, J. A., 207, 209
Schwägerichen, Christian Friedrich, 144
Schweigger, August Friedrich, 112–13, 143
Schweigger, J. S. C, 10, 111, 137–38, 139
"science," nineteenth-century translations, 4
scientific community, the, 10, 267n7
Scientific Revolution, 61, 137, 270n61
"scientist," nineteenth-century usages, 4–5, 25, 262n18
secularization, 10
Seidel, Johann Heinrich, 102
Seiler, B. W., 139, 142
Seven Years' War, 60, 63
Siemens, Werner von, 1–2, 16, 30, 215
Skinner, Quentin, 30
Snell, Karl, 245
sociability, mixed-gender, 54, 170–71, 176
Society of Arts, Manufactures and Commerce (London), 8; Society of German Natural Researchers and Doctors. See *Gesellschaft deutscher Naturforscher und Ärzte*
Sonderweg, 25
specialization, 118–20, 133, 139, 189–90
state bureaucracies: economic improvement and, 94–95, 167–69, 172; education for, 234; expansion of, 20, 184; *Naturforscher* as state officials, 59, 64; ties to manufacturing associations, 219; ties to regional natural history associations, 193–95, 198–99
Steffens, Henrik, 87–88, 99, 116, 123, 134, 135, 136
Stichweh, Rudolf, 28
Strauss, David Friedrich, 10
Struve, Friedrich, 107–9

Struve, Friedrich Georg, 123
Sturm, Jacob, 11–14, 17, 19, 24, 178
Stuttgart, *Verein für vaterländische Naturkunde*, 188
Switzerland, 36
systematics, 3, 37, 111, 163, 167, 174, 187, 189, 213, 221

technical education, 19–20, 35–36, 99, 184–86, 207–12
Technologie, 203, 219–22
temporalization of science, 268n16
Thaer, Albrecht, 92, 94, 95
Thiersch, Friedrich, 236, 250
tourism, 77, 81–82

utility: Romantic criticisms of, 86, 98–100, 106–7; as a scientific value, 33–39, 58, 214, 269n37
universities, 11–12, 23, 61; ideals of *Wissenschaft* and, 97–98; masculine culture of, 54; practical sciences and, 35, 89; relationship to learned societies, 98–99, 191–93
University of Berlin, 97, 98, 100–101, 132
University of Bonn, 103, 133, 152, 157
University of Breslau, 11, 14
University of Halle, 191–93
University of Jena, 32
University of Königsberg, 112
University of Leipzig, 101
University of Tübingen, 189, 250
university professors: as members of learned societies, 100–101, 118, 131–33; relationship to low science, 191–93; role in the German scientific elite, 11–12, 19

Virchow, Rudolf, 14, 120, 178
Voigt, C. G., 184
Voigtland, the, Society for Natural and Medical Knowledge, 138–39

Weber, Ernst Heinrich Weber, 143
Weimar classicism, 151–52, 158, 174, 216–17, 289n7
Westphalian Society of Nature-Researching Friends, 71–72
Whewell, William, 5
Wissenschaft: definition, 3–6, 149, 220; economic reform and, 224–25; learned

Wissenschaft (continued)
    identity and, 16, 96–100, 109–10, 132, 166; nineteenth-century translations, 4, 6; popularization of, 215–23; practical knowledge and, 99–100, 103–7, 113–14, 152, 168–69, 208–10; unity of, 86–90, 135, 252;
Wolff, Christian, 2
women: as an audience for science, 39, 53–55, 153–56, 162, 163–72, 174, 211; as authors, 155–56; as members of learned societies, 54, 68–69, 172–73;
    as *Naturforscher*, 54, 70, 156. *See also* gender; sociability, mixed-gender
Wucherer, Gustav, 99, 100, 116
Württemberg, *Gesellschaft für vaterländische Naturkunde*, 187–88

Young Hegelians, 239–40

Zedler, J. H., 27
zoology, 7, 123, 148, 222, 245; *wissenschaftliche Zoologie*, 187
Zurich, Physical Society, 64, 100